INTRODUCTION TO

ARTIFICIAL LIFE

In Memory
of
Michael Adami
(1959–1984)

CHRISTOPH ADAMI

INTRODUCTION TO ARTIFICIAL LIFE

ONLINE FILES

Springer TELOS THE ELECTRONIC LIBRARY OF SCIENCE

Christoph Adami
California Institute of Technology
W.K. Kellogg Radiation Laboratory
Pasadena, CA 91125
USA

Publisher: Allan M. Wylde
Publishing Associate: Keisha Sherbecoe
Product Manager: Ken Quinn
Production Supervisor: Steven Pisano
Manufacturing Supervisor: Joe Quatela
Cover Illustration: Jim Barry

Library of Congress Cataloging-in-Publication Data
Adami, Christoph.
Introduction to artificial life / Christoph Adami.
p. cm.
Includes bibliographical references (p.) and index.
ISBN 0-387-94646-2 (hardcover : alk. paper)
1. Artificial intelligence. 2. Artificial life. 3. Neural networks (Computer science) I. Title
QA35.A3 1998
570′.13—dc21 97-37605

Printed on acid-free paper.

Typeset from the author's LaTeX files by Integre Technical Publishing Co., Inc., Albuquerque, NM.
Printed and bound by Hamilton Printing Co., Rensselaer, NY.
Printed in the United States of America.

9 8 7 6 5 4 3 2 1

ISBN 0-387-94646-2 Springer-Verlag New York Berlin Heidelberg SPIN 10525997

THE ELECTRONIC LIBRARY OF SCIENCE

TELOS, The Electronic Library of Science, is an imprint of Springer-Verlag New York with publishing facilities in Santa Clara, California. Its publishing program encompasses the natural and physical sciences, computer science, mathematics, economics, and engineering. All TELOS publications have a computational orientation to them, as TELOS' primary publishing strategy is to wed the traditional print medium with the emerging new electronic media in order to provide the reader with a truly interactive multimedia information environment. To achieve this, every TELOS publication delivered on paper has an associated electronic component. This can take the form of book/diskette combinations, book/CD-ROM packages, books delivered via networks, electronic journals, newsletters, plus a multitude of other exciting possibilities. Since TELOS is not committed to any one technology, any delivery medium can be considered. We also do not foresee the imminent demise of the paper book, or journal, as we know them. Instead we believe paper and electronic media can coexist side-by-side, since both offer valuable means by which to convey information to consumers.

The range of TELOS publications extends from research level reference works to textbook materials for the higher education audience, practical handbooks for working professionals, and broadly accessible science, computer science, and high technology general interest publications. Many TELOS publications are interdisciplinary in nature, and most are targeted for the individual buyer, which dictates that TELOS publications be affordably priced.

Of the numerous definitions of the Greek word "telos," the one most representative of our publishing philosophy is "to turn," or "turning point." We perceive the establishment of the TELOS publishing program to be a significant step forward towards attaining a new plateau of high quality information packaging and dissemination in the interactive learning environment of the future. TELOS welcomes you to join us in the exploration and development of this exciting frontier as a reader and user, an author, editor, consultant, strategic partner, or in whatever other capacity one might imagine.

TELOS, The Electronic Library of Science
Springer-Verlag Publishers
3600 Pruneridge Avenue, Suite 200
Santa Clara, CA 95051

THE ELECTRONIC LIBRARY OF SCIENCE

TELOS Diskettes

Unless otherwise designated, computer diskettes packaged with TELOS publications are 3.5″ high-density DOS-formatted diskettes. They may be read by any IBM-compatible computer running DOS or Windows. They may also be read by computers running NEXTSTEP, by most UNIX machines, and by Macintosh computers using a file exchange utility.

In those cases where the diskettes require the availability of specific software programs in order to run them, or to take full advantage of their capabilities, then the specific requirements regarding these software packages will be indicated.

TELOS CD-ROM Discs

For buyers of TELOS publications containing CD-ROM discs, or in those cases where the product is a stand-alone CD-ROM, it is always indicated on which specific platform, or platforms, the disc is designed to run. For example, Macintosh only; Windows only; cross-platform, and so forth.

TELOSpub.com (Online)

Interact with TELOS online via the Internet by setting your World-Wide-Web browser to the URL: *http://www.telospub.com*.

The TELOS Web site features new product information and updates, an online catalog and ordering, samples from our publications, information about TELOS, data-files related to and enhancements of our products, and a broad selection of other unique features. Presented in hypertext format with rich graphics, it's your best way to discover what's new at TELOS.

TELOS also maintains these additional Internet resources:

gopher://gopher.telospub.com
ftp://ftp.telospub.com

For up-to-date information regarding TELOS online services, send the one-line e-mail message:

send info

to: *info@TELOSpub.com*.

Preface

What makes living systems alive? This is a question that has been asked for as long we have been contemplating the world around us, and despite breathtaking advances in physics, chemistry, and genetics in this century, it is still a question that eschews a definite answer. Historically, life and the physical world have been studied largely independently, with little overlap. As this millenium comes to a close, it is apparent that our knowledge of the physical inanimate world dwarfs what we have learned about the living state.

Life is so diverse and complex that it seems impossible to extract general principles that might govern each and any living system. The physical world, on the other hand—while also displaying diversity and complexity in its phenomena—yields to analysis, because we can *deconstruct* complex physical systems and study aspects of them in isolation. Such an approach appears to be hopeless as far as living systems are concerned. In almost all cases, a deconstructed living system is no longer alive. When taking apart life, it disappears in its constituents. To make matters worse, the *simplest* living system, namely that which has been a precursor to all living systems here, was replaced by much more complicated ones over three billion years ago. Attempts to reconstruct it are as yet unsuccessful and will probably remain so for some time. If we want to learn more about the general principles, we need an instance of life that is not of this earth to compare and extract the similarities and differences. While the exploration of the planets and asteroids of our solar system may ultimately yield such all-important evidence, at present we are faced with learning about life from only one instance: the terrestrian variety.

Fortunately, it appears that the unrelenting growth of the power of modern computers we are witnessing has opened up an entirely unexpected avenue: the construction of an *artificial* living system. This has created the possibility of designing and conducting dedicated experiments with such systems that could otherwise only be performed with much hardship, if at all. At the same time, such artificial living systems have rekindled interest in the idea of formulating a set of "general principles of the living state" that are quite *independent* of a particular implementation. Such a theory of simple living systems should equally well predict the outcome of experiments performed on the protean living system that gave rise to life on earth (the speculative RNA world), putative ancient life on Mars, or those worlds in which information is coded in binary strings compiled to programs that have the ability to self-replicate. The latter is an instance of Artificial Life. In pursuing such an endeavor, we need to extricate those aspects of livings systems that are *independent* of the particular substrate (carbon-based vs. computational chemistry) from those that clearly are *not*. This approach assumes that there is a *universality* in the processes that give rise to life, and that given this universality, life can be implemented in any medium that can give rise to such processes. The idea of universality has been very fruitful in the understanding of the principles of chaos, for example. The hope is that the simplest artificial living systems treated in this book can do for our understanding of the basic principles governing biological life what the one-dimensional iterated map has done for chaos theory.

There is no question that this new field of Artificial Life will necessarily be an interdisciplinary one, straddling the classical fields of biology, chemistry, and physics, as well as the more recent field of computer science. The study of artificial living systems requires tools that are not in the toolbox of the average biologist—nor, for that matter, of any other scientist who studies only one particular discipline. In this book, I discuss such tools and concepts and apply them when possible. The subjects treated range from information theory and statistical mechanics, included as primers with applications to living systems; biochemistry of in vitro evolution; computational complexity in cellular automata and complexity measures for symbolic sequences; theory and phenomenology of self-organized criticality, percolation theory, fitness landscapes; and then finally dedicated computational experiments carried out with the avida Artificial Life software included on this book's CD-ROM. From this vantage point, this book is truly an *Introduction*. However, the field of Artificial Life is much broader than the subject matter broached in this book. Only a glimpse of the panoply of subjects that constitute the "braid of ALife" is offered in the introductory chapters. For a more complete reference, the reader should consult the recent overview [Langton, 1995].

Notwithstanding the range of scientific disciplines, a single unifying concept binds all those disparate views of life together: that the phenomena we observe, and which seem magical at times, are rooted in physical theory and can be understood using such tools. At the same time, we need to understand that we are only at the beginning of an endeavor that will shape the science of the twenty-first century: understanding biology using the tools of the science of the twentieth century, namely physics and computers. It is hoped that the interplay between theory and experiment, the two pillars of scientific endeavor that have supported the triumph of physics, will lead us further along the road to discover the *general* principles of the living state, but also to uncover aspects of *particular* complex systems that have remained hidden, obscured by the difficulty to perform dedicated experiments.

This book has grown out of lectures given to advanced undergraduate- and graduate-level students in Computation and Neural Systems and in Physics at Caltech since 1995. Most of the chapters should be accessible to such a target audience. Chapters 6–8 as well as 10–11 contain material that will also be of interest to the more specialized reader. Each chapter closes with an overview of the subject matter treated and the main conclusions reached. These overviews can also be read *before* embarking on each chapter. Generally, the material can be mastered armed only with a background in fundamental methods of statistical physics, knowledge of basic computer architecture and programming skills, as well as a rudimentary knowledge of biology. An extensive bibliography points to sources that go into more depth on the topics treated in the more advanced chapters.

Acknowledgements

This book is the result of my fascination with the power of Darwin's principles of evolution, first transplanted into the computer for everybody to see by Tom Ray in his tierra software. I have been, as have the majority of students that have taken my class, awed by the simplicity of the principles and the complexity they entail. Observing populations of self-replicating computer programs adapt to an artificial world has infallibly elicited the same sense of wonder in the observer: "Evolution works!" Besides Tom Ray, there are a number of people who deserve special thanks for leading me on to and along this path. Steve Koonin slipped Tom Ray's paper into a stack of literature that he thought would arouse my curiosity and wile away the time on a flight from Los Angeles to New York. Even though he may not have foreseen then where this would lead, he has steadfastly supported and encouraged my explorations in this arena (even though the focus at the Kellogg Radiation Laboratory is generally nuclear physics).

The first version of the avida software was designed in collaboration with C. Titus Brown and Charles Ofria, and written by Titus. This first attempt taught us a great deal about the problems associated with "living software" and paved the way for Charles to write the definitive version included in this book. To both Titus and Charles I owe an immense debt of gratitude, on the one hand because I do not have the time and perseverance to write and test such a complex piece of software, on the other because they have been loyal companions on this trek, and a critical conscience at the same time. The appeal of this book draws largely from the avida software that comes with it. All major portions of it have been written and maintained by Charles, a task that is sometimes unrewarding but for which he deserves my respect and my thanks. Portions of the code have been contributed by Travis Collier, who also reliably helped in preparing figures, and in the TAing of the course.

The porting of the code to the Windows environment was facilitated by Dennis Adler of Microsoft Research in collaboration with Charles. I would like to thank Dennis not only for his support of this effort, but also for his inquisitiveness. I am grateful to colleagues, friends, and students who have offered suggestions for improving the book or have contributed to it by collaboration, especially Liubo Borissov, Titus Brown, Nicolas Cerf, Johan Chu, Amy Forth, Mike Haggerty, Chris Langton, Jörg Lemm,

Charles Ofria, Heinz G. Schuster, Chuck Taylor, and all the students of CNS/Phy 175. Any remaining misconceptions (and Germanisms), however, remain entirely my own. Thanks are also due to my publisher Allan Wylde at TELOS, for his dedication to the project and the stoicism maintained throughout, and to Keisha Sherbecoe for an always cheerful attitude. Finally, I would like to thank Taylor Kelsaw for enduring the slings and arrows of this adventure and accepting the toll it has taken on time that was meant to be shared, as well as for his unflagging support and encouragement.

Christoph Adami

Contents

Contents of the CD-ROM

The CD-ROM accompanying *Introduction to Artificial Life* contains files and programs that are executable on a variety of platforms.

All files ending in .txt are readable across all platforms, as are the Java applets and the HTML file al.html. It is recommended to begin using the CD-ROM by first reading the legal.txt and readme.txt files, and then pointing your browser toward al.html.

The avida software is provided for Windows95/NT and UNIX platforms, but not for Macintosh. Several programs included on the CD-ROM are written in C or C++ and can be compiled on all platforms with a little work. However, only source code is included for those, not executables. A complete directory structure for the files on the CD can be found on the CD-ROM itself.

Contents

- legal.txt — Copyright, permissions, and disclaimers
- readme.txt — Explanations about the CD-ROM
- al.html — HTML file containing links, comments, and applets for every chapter
- avida — Directory containing all files to compile and execute avida, as well as the manual in PostScript

- cover.jpg JPEG file of cover art
- MPEG Directory containing movie files
- Percolation Directory containing files and programs
- Sandpile Directory containing files and programs

The avida Software

Avida is the name of the Artificial Life "platform," or experimental testbed, which is used predominantly for experiments by the ALife group at Caltech, and which is included on the CD-ROM accompanying this book. The avida system was inspired by tierra (which was developed by Tom Ray in 1991). Ray's construction is taken one step further in avida, as the latter constitutes a bona fide experimental platform that is used to test ideas and predictions relating to simple living systems. Moreover, such simple populations can be analyzed using methods of statistical physics, which allows predictions about the global behavior of populations of self-replicating strings from first principles.

Avida is the result of an attempt to construct the most simple system that displays the basic properties of life. Simply put, avida is a population of self-replicating strings of computer-code subject to random mutations, adapting to a complex information-rich landscape.

Other Software

- Sandpile consists of C and C++ programs that implement the 2D sandpile of Bak, Tang, and Wiesenfeld.
- Percolation contains Java applets that implement 2D percolation on Euclidean lattices.
- Movies contain MPEG movie-clips of waves of propagation in avida, Karl Sims's virtual creatures, and the self-replicating loops of Langton. The clips about avida are also provided in QuickTime.

CHAPTER ONE

Flavors of Artificial Life

Life is a tale told by an idiot, full of sound and fury, signifying nothing.
W. Shakespeare

Progress in the study of living systems has historically been hampered by the diversity and complexity of life itself, as well as the difficulty of setting up controlled conditions in dedicated experiments. Ultimately, most experiments with living systems that were geared towards understanding the living state have involved the deconstruction, or *analysis* of said state. Scientists have not achieved, or dared to achieve, the construction, or *synthesis*, of life from nonliving materials. Historically, the thought itself has been heretic, and it was not until the early 1920s that the idea of protean living systems, and thus the origin of life from nonliving materials, was proposed [Oparin, 1938; Haldane, 1929]. From then on, efforts were mostly concentrated on showing how the main ingredients of the living state, i.e., amino and hydroxy acids, can be formed under circumstances thought to be prevalent on earth shortly after its formation. In now-famous experiments [Miller, 1953], a mixture of methane, ammonia, water vapor, and hydrogen was pumped through a water solution and subjected to electrical discharges (to simulate lightning flashes on earth), or ultraviolet light or heat. In all such experiments the amino acids form readily and in abundance. However, the subsequent steps from amino acids to polypetides, on to enzyme-driven metabolisms and the replication of the genetic code, are still uncertain.

A combination of frustration with the current state of affairs coupled with curiosity sparked by advances in technology has opened an entirely new approach: the emulation, simulation, and construction of living systems, collectively known as *Artificial Life*. While the modern era of Artificial Life is usually traced back to the first conference by the same name held at Los Alamos in 1988 [Langton, 1989], there is a longer history that consists almost entirely of theoretical work. The advent of fast and affordable computers with the capacity for storing vast amounts of data undoubtedly led to the reemergence of the field, and to its diversification as researchers have attacked more and more problems.

We begin our journey into the study of Artificial Life with a philosophical inquiry, and ask what is the place and what would be the use of a *theory* of the living state. Subsequently, we formulate such a principle and discuss its purpose: to be proven wrong by new evidence, either from extraterrestrial biology or from within. We continue in Section 1.2 by visiting the main branches of Artificial Life that will not be studied here. The intent is not to be exhaustive, but rather to be broad and to put into perspective the subject matter treated here. We visit examples that reflect some of the flavors of Artificial Life that are out there, while keeping in mind that the selection is neither complete nor even representative. Finally, in Section 1.4, we jump back in time and review the theoretical foundations of Artificial Life, which lie in Automata Theory and the work of Turing and von Neumann.

1.1 Whither a Theory of the Living State?

The essence of life has most likely been sought for as long as we have been consciously thinking. Considering that our progress in understanding nearly all aspects of the inanimate world around us has been so spectacular, it might appear as an oversight that there is no consensus as to what separates the living from the nonliving state. Clearly, as scientists we are interested in a nonmetaphysical description of the essence of life and to our astonishment, the present state of this description is still vague and debated.

In the classical scientific fields, progress is achieved by moving from a simple enumeration of observations to classification. For example, a point of light in the dark night sky may first be classified as a *star* (as opposed to other sources of light in the sky) and then later put in the same class as the sun, having understood that the sun and the stars are objects belonging to the same class, but which are at different distances. Yet, creating a class does not necessarily ensure that we have a good understanding of the *characteristics* of this class. Again as an example, in general, we do not have any trouble distinguishing the living from the nonliving, and thus to ascribe all things known to us to any of the two classes. Our problem is to find a minimum number of characteristics that all those systems classified as living have in common, and moreover to have the confidence that this classification does not depend too much on the particular choice of members of the class. In other words, we want the class thus defined (by the minimum number of characteristics) to be *universal*. Clearly, the category "all things twinkling in the night sky" will not do, whereas our current astrophysical definition of star is general enough that few objects exist that we cannot classify. In general, as we shall see using an example from physics, this is only possible if we have a certain *understanding* about the class, which allows *predictions*. Such an understanding is often called a theory, or sometimes, if the realm of application of that theory is clearly limited, a model.

Let us follow the development of the understanding of the concept of *mass* through the history of physics. With the concept of mass, again, it is not difficult to decide between massive or nonmassive; rather, it is the choice of universal characteristics that is all-important. It was Galilei and Newton who, when physics became an exact science, proposed that mass was "that which causes inertia" rather than "that which is heavy," thus gaining predictive power and making the class of all things that are massive much more precise. In fact, such a bold step was never taken in the quest to understand living systems, and it is important to understand why. What characterizes the progress in understanding the concept of mass is a steady removal of layers, with each new concept more precise but still describing the entire class. Progress is achieved from contemplation of experiments geared specifically at addressing theories that arose from the previous set of experiments. For instance, consider what once was considered to be a characteristic of mass, namely that it could only be divided into its indivisible parts (the atoms), and that the mass of the sum of the parts would equal the initial mass. This has been disproved by taking apart massive objects. While we learned that mass can also be converted to energy, we also learned that each subdivision of a massive object was still massive. Moreover, it was commonly assumed (rightfully in this case) that by subdividing the massive object, we progress closer and closer to the most universal massive thing. Arriving at the indivisible building block (the elementary particle) and measuring its characteristics (and inferring why they are there rather than absent), we gain an understanding of the concept of mass.

For physicists, the layers have yielded steadily in a progression from the atom to the nucleus to the nucleon to quarks and leptons. For the living system, step one already fails: we can take apart a living system, but most of the time the parts themselves do not carry the property we are interested in anymore, so an analytic approach is impossible. Thus, there appears to be no elementary living thing: life seems to be a property of a *collection* of components but not a property of the components themselves. Also, putting back together the lifeless parts usually does not reconstruct the initial system. The path that has proven so successful in physics is barred.

Ideally then, we would like to have a glimpse at the system that preceded the prokaryotic metabolism, because if we cannot have an elementary living object, maybe there is such a thing as the *simplest* living system. We would like to see, therefore, whether there was an "RNA world" (a putative period and environment preceding prokaryotic cells dominated by RNA polynucleotides that catalyze their own replication) and how it arose. Reconstructing this system would allow us to perform experiments on one of the (ostensibly) simplest living systems. Furthermore, such a system may inspire a *theory* of what constitutes *essential* ingredients of a living system. In addition, we may insist that such a theory be *universal*, not in the sense that it describes the elementary living thing, but in that it describes the simplest living thing *without* making reference to the *materials* that constitute it, only to the *principles*. Thus, such a theory ought not to be specific as far as the realization of the simplest living system is concerned. For this reason alone it is worthwhile to *construct* the most simple system with what we consider to be important aspects of living systems, such as self-replication, information storage, low entropy, and selection. This system will then be used to *test* predictions of theories of the living state. Both theory and experiment can then,

The *physiological* definition centers on certain functions performed by organisms, such as breathing, moving, etc. Such a definition has been overtaken by time by the discovery of organisms that *do not* partake in such functions.

The *metabolic* definition centers on the exchange of materials between the organism and its surroundings, a definition that also has proved itself to be too narrow to encompass the diversity of life.

The *biochemical* approach defines living systems by their capability to store hereditary information in nucleic acid molecules. Clearly, we recognize an important departure from the previous definitions in a focus on information storage in a particular medium rather than a behavioral characterization. Still, the focus on nucleic acids as a substrate may again prove to be too constricting.

The *genetic* definition focuses on the process of *evolution* as the central defining characteristic of living systems. Such a view does not specify *how* the information is coded, but rather emphasizes the *processes* that this information is subject to, such as mutation, replication, and selection. Such a definition appears to be more satisfying than the purely biochemical one, as it does not prescribe what living systems have to be made up of.

An even more general approach is the *thermodynamic* one, which attempts to define living systems in terms of their ability to maintain low levels of *entropy*, or disorder, only.

BOX 1.1 Definitions of Life

in the time-honored tradition of the sciences, complement each other through trial-and-error. Let us start by considering the common ways by which scientists have attempted to characterize life. We shall then juxtapose them with a principle for living systems that has emerged from considering experiments with artificial living system.

Through the course of time, a number of different ways developed to define or characterize life. In general, this progression of definitions does reflect a *generalization* of the concepts like those we witness in the development of physics concepts. Unlike in physics, however, we are clearly still far from a satisfactory state (see Box 1.1).

The second law of thermodynamics, which we shall study in Chapter 4 because of its relevance to the characterization of living systems, promises that all matter will, as time progresses, reach a state of maximal disorder. Quite obviously, living systems appear to be resisting this approach to chaos rather admirably by producing ordered forms in the midst of disorder. Without a doubt this capability is crucial to living systems. How this aspect can be understood within the physics of entropy and information shall be one of the subjects we will treat as we progress. We will find in Chapter 5 that life has the capability to operate as a *natural Maxwell demon*, thereby sidestepping the second law and evolving towards higher and higher complexity.

Maybe a definition of life is possible focusing on thermodynamics only (see Box 1.1). In our search for a theory, however, it is not enough that we find a characterization that encompasses all living systems, but we would also like to use such a theory to make *predictions* about other *possible* forms of life, and at the same time *exclude* all that is nonlife. To a certain extent, this requires making the thermodynamic definition more precise. In the light of the experiments with Artificial Life that we shall encounter in this book, we might formulate a *principle* of living systems such as:

> Life is a property of an ensemble of units that share information coded in a physical substrate and which, in the presence of noise, manages to keep its entropy significantly lower than the maximal entropy of the ensemble, on timescales exceeding the "natural" timescale of *decay* of the (information-bearing) substrate by many orders of magnitude.

Note first that this principle abandons the idea of life being a property of any single object. Even though this may appear unnecessary (we believe that each one of us can be considered alive), this is significant from a statistical point of view. Such a principle insists that life is an *emergent*, rather than atomistic, phenomenon. Similarly, this principle invokes the natural rate of approach to disorder of the substrate in question, and at the same time insists that there can be no life without this substrate representing information. The natural rate of approach to disorder (the "relaxation time" that we encounter in Chapter 4) for biomolecules, for example, is of the order of hours to days (but sometimes much longer) in a hot and noisy environment, yet the information contained in tRNA

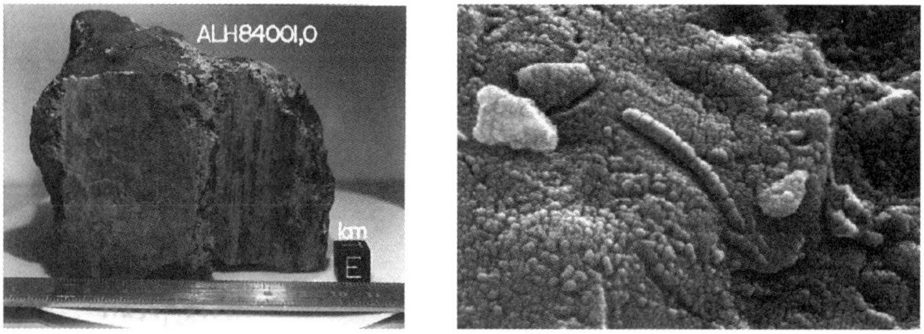

FIGURE 1.1 The meteorite of martian origin ALH84001 (left) which contains fossil-like traces (right) and chemicals not incompatible with a biogenic interpretation (courtesy of NASA).

molecules, for example, has survived with little change for over 3.5 billion years. We know that this information preservation (which leads to the low entropy) is achieved in terrestrial living systems via self-replication; the above principle does not address the *means* of information maintenance, however. Still, it may turn out that replication of information is the *only* process to achieve this feat efficiently; only time will tell.

Far from being a theory, the "principle" is a suggestion whose validity has to withstand the onslaught of experiments and discoveries past and future. For one, we can expect research in Artificial Life to pound at its foundations. Suprisingly enough, this discussion is bound to benefit also from the exciting possibility that ancient life may have existed on Mars [McKay et al., 1996]. This eventuality appears as a harbinger of things to come: evidence of life either independent of or only very distantly related to life on earth, which might challenge or corroborate any principles that we might posit. For example, the ovoids that some believe to be bacterial microfossils (Fig. 1.1, right panel) in the Martian meteorite ALH84001 (Fig. 1.1, left panel), are of a diameter smaller than 0.1 microns, which raises the question about the *minimum number of atoms* needed to maintain a bacterial metabolism while storing and protecting information. The smallest known *terrestrial* microorganism has a diameter of 0.34 microns [Koch, 1996], while theoretical estimates have yielded a minimal diameter of spherical cells of about 0.1 microns [Morowitz, 1992]. The interesting question from an exobiological—but also Artificial Life—point of view is whether such estimates hold for *nonterrestrial* chemistries and metabolisms [Maniloff, 1997], and ultimately if the kind of life that we may find on distant planets, asteroids, or in interstellar material violates or corroborates any of the principles we may formulate.

The use of a principle such as the one above is primarily to serve as a target for thought and experiment: Can it be demolished easily or does it withstand old and new evidence? To a certain extent, the Artificial Life investigations presented in the following chapters seem to uphold such a view of life, but there is little doubt that what we shall learn in the future will replace such a principle by a more educated one, just as the scientific process dictates.

It is quite possible that a theory of the living state that emerged from experiments involving "foreign" living systems will never be quite acceptable, even in the event that it predicts correctly the behavior of the simplest carbon-based living systems. It might, however, give rise to a physical rather than metaphysical definition of life that clearly encompasses more than just life on earth. Still, such a theory may fall short of explaining the exact *origin* of life on earth. The science of biology would have to share such a shortcoming with physics: we know everything about mass, yet its origin is still unclear.

1.2 Emulation and Simulation

This section presents a brief overview of the main branches of Artificial Life that we do not cover here. They usually involve the emulation or simulation of living systems, in whole or by part, without attempting to actually *construct* life. It is important to keep in mind the scope of the entire effort in the field to obtain a feeling on where, in the scheme of things, the computational approach taken in this book is situated. For a more thorough introduction to the diverse areas of inquiry associated with Artificial Life, see Langton (1995), which contains many of the papers that are referenced in this section.

Artificial Life is often described as an attempt to understand high-level behavior from low-level rules. As such, it is a strictly reductionist science: the low-level rules are the laws of physics. However, reduction may stop at a level higher than physical laws, and more importantly, Artificial Life is not strictly about artificial living systems. Rather, the simulation (or emulation) of a living system or parts of a living system with the intent to understand its behavior forms a large part of Artificial Life. On the other end of the spectrum, the precise details of a member of a population may be neglected in order to study the *emergent* properties of living populations, usually in the simulation of many agents.

Simulation of Units

In the emulation of single living agents, the functioning of (or processes affecting) just one or a few units, or just a part of one, is the focus of interest. Such a simulation may be performed by means of computer modeling or engineering. In each case, the analytical or deconstructionist path is or was unsuccessful, and insight into the function of the unit is gained by reconstruction in an artificial medium. A well-known example is Sims's (1994) artificial evolution of swimming motion and morphology in simple simulated animals constructed out of blocks, and the evolution of morphology and strategies for competition.

The object of Sims's study is not the understanding of an emergent phenomenon, but rather the exploration of the consequences of putting together a number of simple but important concepts: simple algorithms that control morphology (as a function of the simulated environment and the required task), evolvable neural networks and sensors, and selection of the fittest. Most of all, the object is not to study a fixed fitness function and let a population adapt to it (the traditional Genetic Algorithm approach) , but rather to let the population be part of the environment, such that the members of the population are *coevolving*, and such that the physical structure of a creature can adapt to its control system *and* vice versa, as well as to each other. Selection of the fittest in Sims' work is implemented by *competition*, i.e., candidates are pitted against each other. In a particular example, the creatures attempt to gain control of a single cube that is placed in the center of the (simulated) world. The simulator also controls the interaction of the creatures with each other and the physics of the world they are in, whereas their morphology and their control system are *evolved*. The morphology and the "brain" of a creature are determined from a *genotype*, shown in Fig. 1.2. There, the bold lines characterize the morphology (a central block connected to two ancillary blocks), while the inner graph describes its neural circuitry. The resulting *phenotype* (the creature's actual morphology and brain) is shown

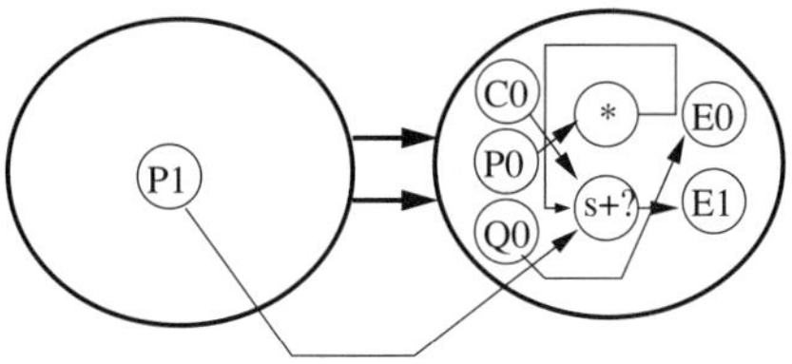

FIGURE 1.2 Example of evolved "nested-graph" genotype. The outer graph in bold describes a creature's morphology, while the inner graph describes its neural circuitry (after Sims, 1994).

in Fig. 1.3. The evolution of virtual creatures starts by creating an initial population of candidates by randomly generating genotypes and checking initially for general viability (like whether their block-arrangements are physically sound and not interpenetrating, for example). Then the remaining creatures are paired off for competition, and their fitness is determined by their success at claiming control over the cube.

The competition is biased in that creatures that have a height advantage must start further away from the block so as to discourage the inelegant solution of just falling over the block. For each generation, the most successful creatures are replicated and take the spot of those that were removed due to low or zero fitness. Then offspring are generated by crossing-over the directed graphs (such as Fig. 1.2) that define their morphology and brain, and mutating them in a probabilistic manner. Fig. 1.4 shows examples of simple evolved creatures in a typical contest situation. The difference in approaches (i.e., morphologies) is astounding, and mirrors the complexity of the fitness landscape that is created by letting the population *itself* shape (to a large extent) the landscape. The results are impressive from a scientific and a visual point of view, and none of the insights obtained could have been gained with standard approaches. Indeed, the study of the evolution of morphology is, for the most part, confined to the study of paleontological records.

An equally impressive example is the emulation of fish in a (simulated) hydrodynamic setting [Terzopoulos et al., 1994]. In this example, the object is not to evolve arbitrary creatures and to study which factors are responsible for adaptability, but rather to model and understand *existing* fish, and understand their behavior as individuals as well as groups. As an alternative to performing experiments with real fish, the animals here are simulated holistically as autonomous agents situated in a simulated

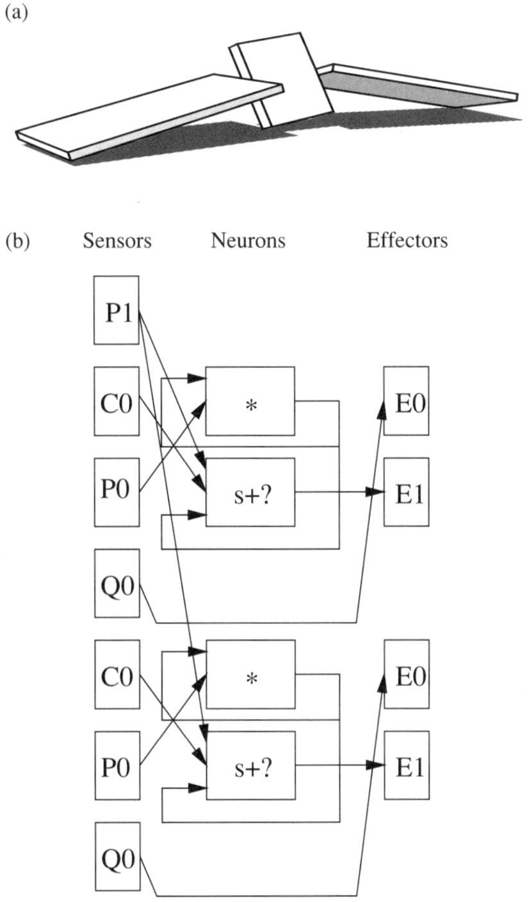

FIGURE 1.3 (a) The phenotype morphology generated from the evolved genotype in Fig. 1.2. (b) The phenotype "brain" generated from the evolved genotype. The effector outputs of this control system cause the morphology shown in (a) to roll forward in tumbling motions (from Sims, 1994).

physical world. The agent has a three-dimensional body with internal muscle actuators and functional fins; sensors, including eyes that can image the environment; and a brain with motor, perception, behavior, and learning centers. In order to match the visual appearance of the fish, realistic textures of fish obtained from photographs are mapped onto the

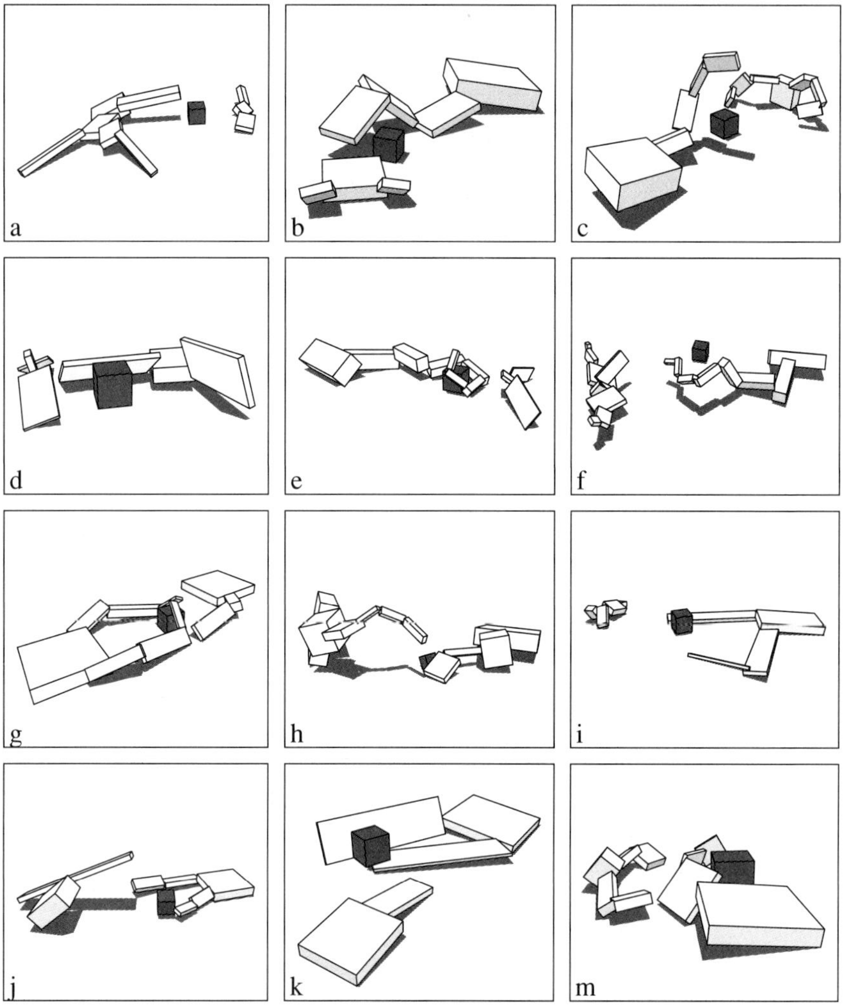

FIGURE 1.4 Evolved competing creatures (from Sims, 1994, with permission).

geometric fish model. The most important aspects of fish control and information flow are then programmed into these creatures (see Fig. 1.5), which subsequently adapt to the environment, finding locomotive and obstacle-avoidance strategies, as well as displaying schooling behavior to escape (emulated) predators.

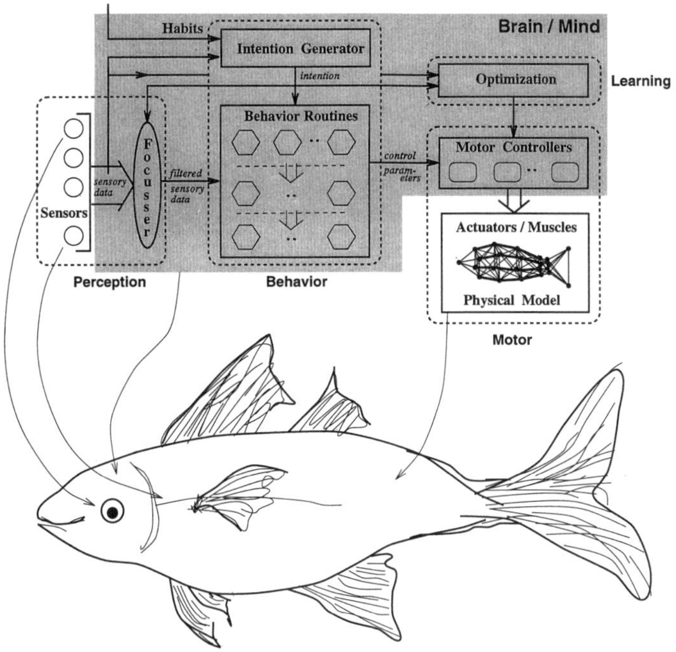

FIGURE 1.5 Control and information flow in artificial fish (from Terzopoulos et al., 1994, with permission).

This use of Artificial Life is of tremendous importance when experiments with real living systems are either difficult or out of the question (for practical or ethical reasons). At the same time, much can be learned about the algorithms that are at work in *real* fish by comparison with the artificial ones.

Parallel to the kind of computational development just described, an impressive engineering effort exists, geared toward the construction of adaptive, autonomous robots. This work differs from the classical robotics approach in that the robotic agent interacts with the environment and learns from it, navigating without detailed course of action. In this sense, the behavior of the robot can be called emergent (see, e.g., Maes, 1994). A breakthrough piece of research in this area was recently accomplished by Barbara Webb (1996) from the University of Edinburgh. Attempting to shed light on the algorithm used by crickets to find their mate in response to their song, the Edinburgh team built a robotic cricket complete with cricket ears (located on its forelegs!); neurons that process the signals; and actuators connecting the brain's output with its locomotory device, the right and left wheels. Asymmetric signals to the wheels make the robot turn, symmetric signals make it go straight. The ears are built to mimic the ingenious cricket design that compares the *phase* of the sound waves (arriving from the forelegs) in the brain in order to judge the distance and direction of the sound. The algorithm that Webb implemented tested an idea that was plausible as a strategy for *phonolocation*: the art of locating a sound and moving towards it. An (artificial) neuron for each ear integrates the chirping received and, if the sum reaches a certain threshold, stops the associated wheel briefly to effect a direction change. If the sound continues, no further response can occur (as the sum is already above threshold), and thus no further directional change can be effected. However, if the sound stops, the signal sum falls back to zero and a new directional change can occur, depending on which of the two neurons reaches the threshold first. Thus, such an algorithm explains why crickets chirp in the first place: they need the intermediary silence to reload their neurons! With this algorithm, Webb and her team were able to test a number of cricket behaviors, and could even emulate the way female crickets move towards one specific male and distinguish between many different cricket songs. Even though the robot itself was built from everyday materials (here, Lego, see Fig. 1.6),

FIGURE 1.6 Robotic cricket (courtesy of Dept. of AI and the Mobile Robots Group, University of Edinburgh).

the analysis of the robot's emergent behavior has yielded important insights on the algorithms Nature uses, and how it is that a cricket finds its mate.

Simulation of Populations

Emergent, or collective, properties of a system are often not apparent in the microscopic rules governing the interaction between the elements of a population or ensemble. What distinguishes living systems from collections of objects of inanimate matter from a purely computational point of view is that ensembles of living agents are for the most part not amenable to a statistical description in terms of macroscopic observables. Even though the microscopic rules are usually simple, they are far more complex than, for example, those governing the collision of molecules in an ideal gas (for which a statistical description is eminently successful). The only recourse, then, is the simulation in parallel of many of the agents, to recognize and analyze the emergent behavior of the population or ensemble. This is not always a straightforward task. The self-organization present in many living systems dictates that most elements of a simulation affect each other in such a manner that they cannot be updated independently (see, e.g., Rasmussen and Barrett, 1995, for a theoretical analysis of this problem). Furthermore, the analysis of the simulation can be complicated by the very fact that the emergent structures can affect the units they are made of in nonlinear fashions. The problems faced here are somewhat comparable to those involved in simulating global weather phenomena: while the main characteristics of the dynamics can probably be described by only a few degrees of freedom, it is often unclear what these are. In the absence of any theory predicting those, they have to be obtained by simulation.

Some examples here are the study of trail-following in foraging ant populations [Goss et al., 1993], and the emergence of swarm intelligence in the building of large structures (such as wasp nests) by members of a population, each acting on local rules only [Theraulaz and Bonabeau, 1995]. The latter provides an intriguing insight into how Nature may accomplish the construction of complicated structures without "knowing" what it is building. In "distributed building," local agents are moving randomly on a three-dimensional cube and, following local rules, can build

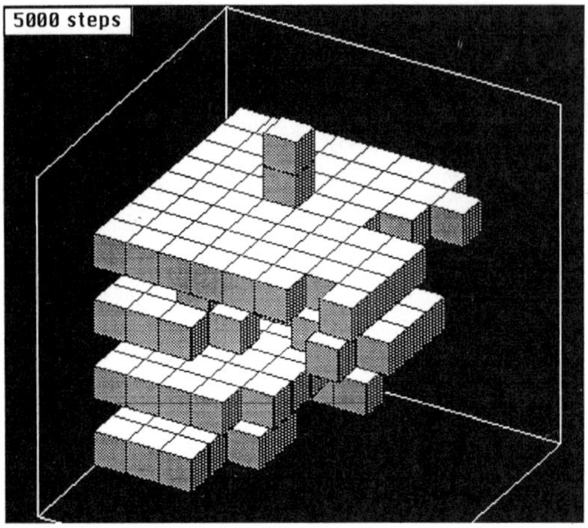

FIGURE 1.7 Artificial nest generated by a swarm of ten artificial wasps depositing bricks (from Bonabeau and Theraulaz, 1994, with permission).

coherent structures. The building activity of termites, for example, does not seem to depend on the interaction between the termites, but rather is dictated by the shape of the building itself. To a certain extent, the building signals to the termites what to build next, i.e., individual behaviors are controlled by previous work. Such an algorithm is reminiscent of the forces that are in play in the biology of development: the concentrations of certain proteins and enzymes give rise to developmental structures, which in turn influence the DNA to produce different proteins and enzymes which further development. Such a local strategy seems to be superior to a centralized one in many circumstances.

To test the ideas of distributed building in the construction of wasp nests, Theraulaz and Bonabeau developed a simple model of artificial agents randomly moving on a three-dimensional lattice. Each agent only possesses the capability to deposit *elementary* bricks, depending on the local configuration of matter. In such a manner, a swarm of agents builds artificial nests using just local rules, as evidenced in Fig. 1.7. In a number of experiments with different rules about where to deposit bricks as a response to which local configuration and for which agent, Theraulaz and Bonabeau were able to show that the organization of individual activities can, in principle, be *totally* directed by the material patterns that the agents encounter (in this case, the nests). Even more, they were able to

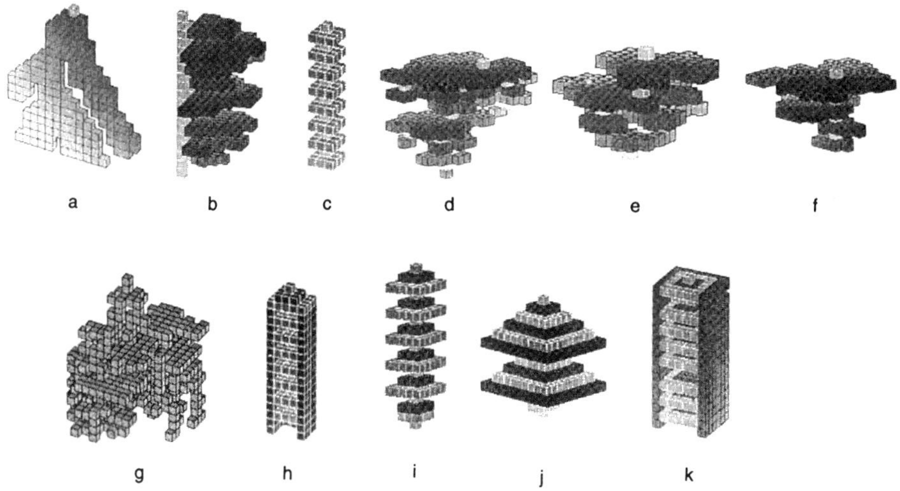

FIGURE 1.8 Wasp nest architectures obtained from simulations of collective building (from Theraulaz and Bonabeau, 1995, with permission).

design local rules that give rise to structures that are strikingly similar to architectures occurring in real wasp species. To a certain extent, such experiments reveal the local rules that the agents in each of the species uses. From a more abstract point of view emphasizing the importance of emergence in the simulation of populations of adaptive agents, it is the coherent structures that are the emerging properties of the local rules. Clearly, the prediction of the structures (see Fig. 1.8) would be difficult, armed just with the local rules and without a simulation of the interaction of the agents.

1.3 Carbon-Based Artificial Life

The construction of living systems out of nonliving parts is clearly the most ambitious of all the areas of Artificial Life. At present, this subfield is split into two largely independent endeavors: the creation of life using the classical building blocks of nature (carbon-based life) and the creation of life using the same principles but a different medium for implementation:

the computer. This section describes two of the "wet" approaches and leaves the computer approach to later sections, as it is our main concern.

Rapid progress in microbiology has brought with it the first serious attempts at recreating a protean living system. Mostly, the effort is centered on recreating the mythical RNA-world, the putative precursor to cellular life, with genetic information stored in single-stranded RNA strings that are the catalysts of their own reproduction (for a review, see, e.g., Hager et al., 1996). Such a molecule seems to have almost magical capabilities. On the one hand, it needs to fold into an RNA polymerase that uses RNA as a *template* and thus copies other RNA molecules; whereas on the other hand, it needs to unfold to act as a template for other replicase molecules. Such molecules are said to "catalyze the template-directed polymerization of RNA"; in other words, they self-replicate. However, no such string or family of strings with the capability of autocatalysis is known. Still, the vast repertoire of possible sequences and their structural complexity effects the hope that such an ancestral string can be found or constructed. Such an evolutionary strategy was followed by Szostak and collaborators [Bartel and Szostak, 1993; Ekland et al., 1995]. This method was considerably improved by Joyce and collaborators [Wright and Joyce, 1997] by allowing the *continuous* evolution of macromolecules in vitro. Because of the importance of this approach and its relevance when comparing the chemistry constructed by these groups to the computational pseudo-chemistry that is the object of most of the following chapters, we will explore in vitro evolution in more detail.

The catalytic activity of the ribozymes (the general term for RNA molecules with catalytic activity) evolved by Bartel and Szostak involves the *self-ligation* of a substrate oligonucleotide to the RNA ligase. In less technical terms, the evolved RNA molecule is able to bind itself to a target molecule at a rate far exceeding unselected ribozymes, and even naturally occurring ones. The target molecule and the self-ligation are chosen to be similar to the polymerization reaction of polymerase proteins, in the hope that such experiments will lead to self-ligating and self-polymerizing molecules. The target molecule has another function, however. After it is successfuly attached to the RNA molecule by the self-ligation process, it serves as a *tag* to identify successful molecules. They can, by virtue of this identifying tag, be selected from the pool of all molecules and amplified by the polymerase chain reaction (PCR) after reverse transcription (see below). The resulting purified pool can then be subjected to mutations to

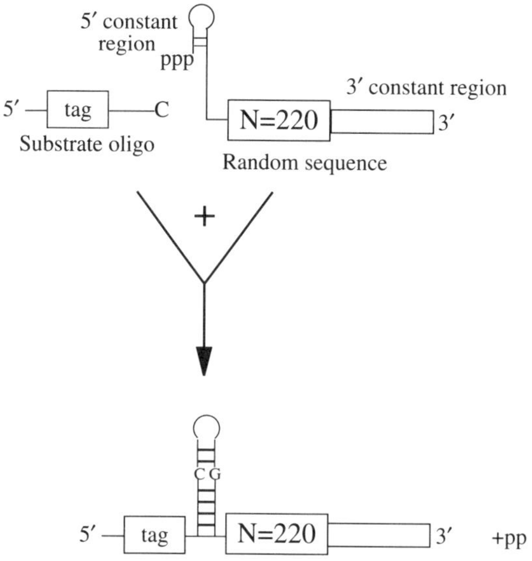

FIGURE 1.9 Ligation of the target substrate to the RNA molecule containing 220 random base pairs. The RNA molecule has a constant (nonrandom) piece at its 5′ end, which is recognized by the substrate.

prepare a generation of fitter molecules to go through another round of selection.

The initial pool of molecules is prepared by attaching a random sequence of length 220 base pairs to a template region that will later bind to the target molecule. Thus, the RNA molecules are only random in the region that is supposed to be catalytically active, not in the 5′ end that is supposed to bind to the 3′ end of the target molecule (see Fig. 1.9). The reaction that takes place in the catalysis is a 3′-5′ phospho-diester linkage that subsequently *releases* the triphosphate (ppp) at the 5′ end. The fact that catalytically active sequences were found starting with a pool of about 10^{15} random sequences is quite astonishing considering that this pool represents a negligible fraction of the 4^{220} possible sequences, and the probability of catalytic motifs (patterns of nucleic acids that the RNA strings can bind to) was estimated to be small, $\sim 4 \times 10^{-19}$. This suggests that a great many RNA structures can catalyze the self-ligation process, and that these structures are evenly distributed in sequence space. Currently, several groups are working on this subject, and are attempting to breed ribozymes that accomplish not only self-binding, but also the cutting and splicing of other ribozymes, in the quest to find a ribozyme that catalyzes its own replication. The existence of such a molecule, if it is discovered, has importance far beyond the study of the origin of life and the physical dynamics of living systems. Such a ribozyme could be crucial in drug-design, because a self-replicator could be used in the adaptive breeding of drugs much more efficiently than the current method where replication and selection has to be carried out painfully by hand. A major step in this direction is due to Wright and Joyce [Wright and Joyce, 1997]. Building on the work of Bartel and Szostak, Joyce's goal was to create a chemistry in which evolution would proceed *without* the intervention of the experimenter who has to select, reverse-transcribe, and amplify the sequences that are successful in solving the task at hand. Let us take a look at this remarkable chemistry.

Joyce started with one of the very active RNA ligases that Bartel and Szostak had bred, but replaced the 5′ end of that ligase with a sequence that can bind to the promoter element of a bacteriophage RNA polymerase: the substrate. In other words, if the RNA ligase manages to bind to this substrate, it carries with it a tag that can be recognized by an RNA polymerase, which could then proceed to *copy* the RNA. This is of course intended to replace the manual PCR amplification of the selected strings. RNAs that lack the promoter element will not be replicated, while RNAs that have it will, in a solution that contains the replicase. Before this can happen, however, another important step must take place. Indeed, RNA polymerase acts on double-stranded *DNA* rather than RNA, so the RNA string that has the attached promoter element must first be converted to DNA by *reverse transcription*.

Reverse transcription is the operation that translates RNA back into DNA, a procedure employed by many viruses, especially the so-called retroviruses. Reverse transcription is automatic if the RNA molecules are in solution with *reverse transcriptase* (a molecule that can be extracted from viruses) but *only* if the RNA is recognized by the reverse transcriptase, so that it can bind to it. Remember that the goal is to create a chemistry where all the steps that were carried out by the experimenter now happen automatically, without interference. For this to happen, the RNA ligase has to bind to a DNA primer, which hybridizes to the other

(3′) end of the RNA. Hybridization is any reaction where DNA and RNA bind together. Note that this happens also in the ligation taking place at the 5′ end, as the promoter element (the substrate) is DNA. After the DNA primer is attached to the RNA, the reverse transcriptase produces a double-stranded DNA molecule that carries the (now double-stranded) promoter element. This can now be recognized by the RNA replicase, which proceeds to make on the average of ten RNA copies from the DNA. Those RNA copies, however, are *lacking* the promoter element (the substrate), which therefore is returned to the population and can partake, along with the newly formed RNAs, in a further round of evolution.

To initiate this breathtaking dance, the RNAs have to have a certain amount of replicative ability at the outset. The changes made in the RNA ligases by Wright and Joyce (in order for it to catalyze the ligation of the promoter), however, reduced the catalytic capabilities tremendously, so a number of rounds of manual stepwise evolution were taken before the continuous evolution was attempted. After those preliminary rounds, the new RNA ligases were added into a reaction mixture that contained substrate, reverse transcriptase from a leukemia virus, DNA primer, and RNA polymerase from the bacteriophage T7. After 60 minutes, a small portion of the mixture was transferred to a fresh reaction vessel, and thus diluted. Indeed, during the 60 minutes the population of RNAs had doubled four times! After transferring the reaction mixture 100 times (this is necessary because the production of RNA uses up materials and increases in concentration exponentially), the overall amplification of the RNA was 3×10^{298}! At the end of the 100 transfers, the doubling time of the RNAs had decreased to 2 minutes by the production (from random mutations) of ever better catalysts. Meanwhile, the RNA were breeding "true" (making exact copies) without selection, reverse transcription, and PCR amplification by the experimenter. In a sense, Wright and Joyce created the first artificial cultures of ribozymes in vitro. As mentioned earlier, the kind of ligase reaction that is catalyzed by the RNA is similar to a polymerization reaction. The hope is therefore that in the future, the ligation could occur between the RNA ligases and NTP (mononucleotide triphosphate) as a substrate, which would open up the possibility that the RNA ligase *itself*, rather than foreign T7 polymerase, could act as a RNA polymerase. Within a decade, this activity may well lead to the resuscitation of an RNA replicase, a molecule that presumably has been extinct for over three billion years.

There is some doubt, however, that such an RNA world can exist in the aqueous environment of an early earth, as water is known to dissolve (lyse) polynucleotides. The belief that self-replicating RNA may have been encased in simple cell membranes made of lipid bilayers has spawned research in the possibility of simple self-replicating compounds, the "core-and-shell" self-reproduction approach [Luisi et al., 1994]. The idea is to join two systems that can be made to replicate. On the one hand, certain bilayered oleate bubbles (i.e., bubbles made out of the same material that most cell membranes are made out of) are known to multiply in the right conditions (i.e., if the right chemicals are present), even though the exact mechanism for their replication is not known. On the other hand, as we have seen, it is possible to replicate RNA polynucleotides. Here, this happens in a solution of $Q\beta$ replicase (a bacterial enzyme) but without the selection imposed in the in vitro evolution experiments (the $Q\beta$ replicase makes copies indiscriminately). If the bilayered bubbles are filled with RNA and the replicase, the core-and-shell systems start to grow in numbers for several generations, with each bubble having roughly the same content of RNA. Yet, such constructions are still far away from any satisfactory protean cell, as the self-reproduction process in those mixtures can only continue as long as excess substrate is present. Also, the mechanisms for core-and-shell reproduction are chemically independent, contrary to what we expect in realistic cells.

1.4 Turing and von Neumann Automata

The foundations of any attempt to create Artificial Life inside of a computer lie in Automata Theory as described by Turing [Turing, 1936] and von Neumann [von Neumann, 1951]. Before we describe von Neumann's theory of self-reproducing automata, we need to detail Turing's construction of a universal automaton, which inspired von Neumann to think about self-reproduction.

A Turing machine is an abstract automaton that can be in one of a finite number of states $(1, \ldots, n)$ and capable (in principle) of reading and writing on a tape of instructions (customarily ones and zeros). Each Turing machine is characterized by the rules by which it changes its state, as a function of both the bit currently being read on an arbitrarily long tape, as well as its own state. The actions (or states) of the Turing machine consist of reading information, moving the read-write head, and writing information onto the tape. Such an abstract, logical automaton can be specified by determining the mapping between a finite number of numbers. Suppose, for example, that we specify the initial state (one out of n) of the automaton by i, and the final state by j. Let us also denote the digit that is underneath the read-write head by e, and the number of digits that the tape will move (or equivalently that the head will move) by p. For the simplest automaton, p can only take on the values $p = -1, 0, +1$, i.e., either the head is stationary, or it moves one digit to the left or right. Finally, let f denote the digit that the Turing machine *writes*, i.e., $f = 0$ or 1. Then, the quintuplet

$$(j, p, f, i, e) \tag{1.1}$$

determines a *rule* for the Turing automaton, and a specification of all the rules (i.e., j, p, f as a function of i, e) determines the automaton uniquely. The importance of Turing's construction lies in the fact that with this specification of automata, he was able to prove that not only could such a machine compute any number, but also that there were *only countably many* such machines, i.e., that they can be listed. This being so, Turing concluded that the description of *any* Turing machine could be provided to another machine (the *universal* machine) as information on the tape, such that the universal machine can emulate any other machine. Thus, *any* computing process can be mapped to the operation of a universal Turing machine, just described above. Apart from its ramifications in the theory of logic systems and the decidability problem (see Box 1.2), the proof that any possible computer could be simulated by this very simple universal Turing machine gave rise to the suggestion (also known as the *Church-Turing thesis*) that all sufficiently complex computers (or computational models) are in essence equivalent—in particular that they are capable of calculating *any* "partial recursive," (or in less mathematical terms: "reasonable") function. One of the first applications of Turing's work was the construction of a formal neuron that could be shown to have the capacity of universal computation and would thus constitute, at least in principle, an adequate building block of a brain.

John von Neumann recognized that the idea of a universal computer could be extended by increasing the number of operations that a Turing

Turing constructed his universal automaton mainly to investigate a famous problem in mathematics, the *Entscheidungsproblem*, or decidability problem. The main question there is whether a machine could be constructed that automatically proves all mathematical theorems. In order to address this problem, Turing mapped the *Entscheidungsproblem* to a more fundamental problem: the halting problem. In essence, he asked whether one could build a machine that would predict when *another* machine is going to halt, which in the language of automata theory means "finishes a calculation and comes up with an answer." He showed that this was in principle impossible, using a version of an argument due to the mathematician Kurt Gödel. The latter had shown that, within any mathematical (or logical) framework that is complex enough (which, for our purposes we can translate with "interesting enough"), there are statements that are patently true, but which *cannot* be proven within the confines of the mathematical or logical framework. Such a theorem can be arrived at by constructing sentences that are semantically self-referential, such as "This statement cannot be proven within the logical framework of the *Principia Mathematica*," and subsequently translating this statement into a mathematical *formula*, which one then attempts to prove. In order to show that no automaton could exist that predicts whether an automaton will halt, Turing considered programming an automaton to halt if and only if it would foresee (predict) that it will *not* halt. As this situation leads to a paradox, the initial assumption must be wrong: automata that can predict halting cannot always exist. In order for such a proof to work, you have to be able to *specify* one automaton to another, which is what Turing figured out how to do.

BOX 1.2 The Halting Problem

machine can do. The idea of a computer simulating a computer (by being fed the information necessary to characterize that computer, i.e., the set of rules) led von Neumann to a formulation of the requirements for an automaton to *construct* another automaton. In analogy with Turing, he drew up what he considered to be a complete list of elementary parts to be used in building automata (a catalog of machine parts). The construction automaton is imagined to float in an infinite supply of these parts and, when furnished with the description of a particular automaton, proceeds to construct it. This is the universal constructive automaton, A. Let I then stand for the *description* of a particular automaton. (This is the analogue of the information tape in Turing machines.) As a special case, A could be given the description of itself, and therefore self-replicate. Further, consider automaton B, which receives instructions I and proceeds to make a copy of them (i.e., B is essentially a Xerox machine). Finally, combine automata A and B, and imagine a *control mechanism* C with the following function: When A is furnished with instructions I, C instructs A to construct the automaton described by I. Then C causes B to make a copy of I and insert it into the newly constructed automaton. Finally, the constructed automaton with instructions is separated from A and B. For the sake of being definite, let

$$D = (A, B, C) \ .$$

Clearly, D can only function if supplied with instructions I, to be inserted into A of D. Now consider the instructions I_D, which describe D rather than just A or B. Then the aggregate

$$E = (D, I_D)$$

is reproductive, and by construction, a universal self-reproducing automaton.

It is important to emphasize here the qualitative difference made between an automaton and the description of an automaton. According to von Neumann, these are objects in different hierarchical categories. One of the reasons for this distinction lies in the puzzling observation (as far as self-replicating automata are concerned) that automata have (as deduced from observing natural automata) the ability to construct automata with higher complexity, or complication, than their own. There was no evidence for this in construction automata of the time (mainly machines that build tools). Still, von Neumann argued that this could easily be understood as long as constructive automata are provided with instructions that are *more* complex than themselves. What is left open is where this extra information comes from. Von Neumann argued that there was a "level of complication" above which complication could increase (in later generations)—below which, however, self-reproduction was degenerative. This was very obviously inspired by the famous incompleteness theorem of Gödel, which made reference instead to a logical system that was complex enough (see also Box 1.2).

Von Neumann's idea clearly has to be amended at this point. In hindsight (knowing the structure of the genetic code), it is of course an increase in information (more complicated instructions) that is responsible for an increase in complexity of the organism. However, the information is written into I_D (which von Neumann identified with the genome in natural automata) by random mutations, and does not have to be present in I_D from the beginning. In this sense, the complete reproductive system must include the environment to which the system is adapting. We shall return to this point when we discuss such concepts as complexity and measurement from a computational point of view in Chapter 5. In the following section we retrace the first steps at implementing von Neumann's logical construction in a computational medium.

1.5 Cellular Automata

Cellular Automata (CA) were invented by von Neumann (according to common lore, upon a suggestion by the famous mathematician Stanisław Ulam) in order to construct a realization of the universal self-reproductive automaton while an engineering solution was infeasible. The idea followed closely the Turing construction of a universal machine. Each organ of the self-replicator was to be constructed as a pattern (in two dimensions) of cells in different states, while the state of each cell $X(t+1)$ at time-step $t+1$ is determined by the state of this cell's *neighboring* cells at the previous time t, where a neighborhood is defined as the four adjacent cells in the cardinal directions and the cell itself (see Fig. 1.10). In other

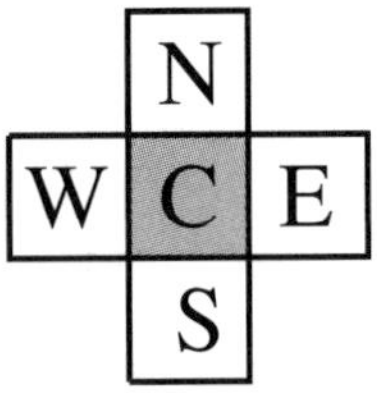

FIGURE 1.10 von Neumann neighborhood in his 2D cellular automaton. Each cell could take on 29 states, and the state of the central cell C at time $t+1$ depends on the state of the surrounding cells at time t via update rule f.

words,

$$C(t+1) = f[N(t), E(t), S(t), W(t), C(t)]\,, \tag{1.2}$$

where $C(t)$ denotes the state of the central cell at time t and so forth. The number of rules of the type (1.2) to implement von Neumann's self-reproducing automaton was enormous. Because von Neumann wanted his self-replicator to be *universal*, he designed it along the principles of the stored-program computers that he had helped design, which meant that the replicator had many different parts that played the role of pulsers, clocks, encoders, decoders, as well as pipelines that channeled the *description* of the automaton to the relevant organs. The complexity of this machinery precluded it from ever becoming functional, and von Neumann never got to finish the construction. Recently, attempts at reviving and completing the original design appear to have been successful [Pesavento, 1995], a feat that now carries only historical significance.

A simplified version of a cellular von Neumann automaton was designed by Codd in 1968 [Codd, 1968], involving only eight states per cell, but still insisting on *construction universality*, i.e., the automaton was supposed to construct *any* other automaton. Because of this requirement Codd's machine was still very complex, and as a consequence was never built. Work in this direction effectively stopped until 1984, when Chris Langton revived the ideas and put the field on a new level [Langton, 1984; Langton, 1986]. Essentially, Langton realized that in order to study aspects of living systems (such as self-reproduction) in a computational medium, one need not insist on providing the *sufficient* ingredients for self-replication, but only the *necessary* ones. An important element in this discussion is that of computational universality (rather than construction universality) of cellular automata. Thus, before describing Langton's self-replicating automata and the idea of *virtual state machines* (VSMs), we need to introduce CAs more formally and discuss their classification in terms of complexity classes.

Formal Definition of CAs

A cellular automaton is a lattice of sites, each of which can take on k values. Each site of the automaton is updated at discrete time-steps by a *finite state automaton* residing at each site, and assigned a value depending on the value of the sites around it. The neighborhood of the CA is crucial. In one dimension, each site has two nearest neighbors. In this case, the value of site i at time $t+1$, $a_i^{(t+1)}$, is determined according to

$$a_i^{(t+1)} = \phi(a_{i-1}^{(t)}, a_i^{(t)}, a_{i+1}^{(t)}), \tag{1.3}$$

where ϕ is a function of three k-valued variables. The function ϕ is called the CA rule. The idea of a neighborhood in 1D (one-dimensional) CAs can be extended to include the next two, or more, nearest neighbors. This defines the radius r of the automaton. Naturally, automata where only the nearest neighbors determine the state of the site at the next time-step are called $r = 1$ rules, whereas rules where the two leftmost and the two rightmost sites (in addition to the value of the site itself) are used to determine the next state are called $r = 2$ rules, and so forth. Finally, there is a special state for each site called the *quiescent* state (usually denoted as state 0). As we shall see later on, rules can be classified according to how many of the rules *return* to this quiescent state [Langton, 1992]. Note that the number of different rule tables for a CA can be quite large. For example, for a CA with $r = 2$ (five neighbors if you count the site itself) and $k = 8$ states, there are 32,768 possible neighborhood states. Each of those can go into eight different states; thus the total number of different transition functions is $8^{32,768}$. Usually, certain restrictions are applied to the possible rule-tables, such as requiring that all neighborhoods consisting entirely out of the quiescent state should return to the quiescent state. Another restriction imposes *spatial isotropy*, which requires that all planar rotations of a neighborhood should map to the same state. In one dimension this means that the rules must be *symmetric*, i.e., neighborhoods such as (1,0,0) and (0,0,1) must map to the same state. Rule tables that satisfy these two conditions are termed *legal* [Wolfram, 1983]. In two dimensions, the definition of a neighborhood depends also on the topology of the grid. For a regular grid, a neighborhood might consist of the north, south, east, and west squares ("von Neumann" neighborhood), or it might contain the diagonal squares as well (a nine-neighbor rule, "Moore" neighborhood). More sophisticated geometries can be designed, such as hexagonal grids (six neighbors). In the following, we shall attempt to classify the behavior of CAs (according to Wolfram, 1984) from a dynamical systems perspective.

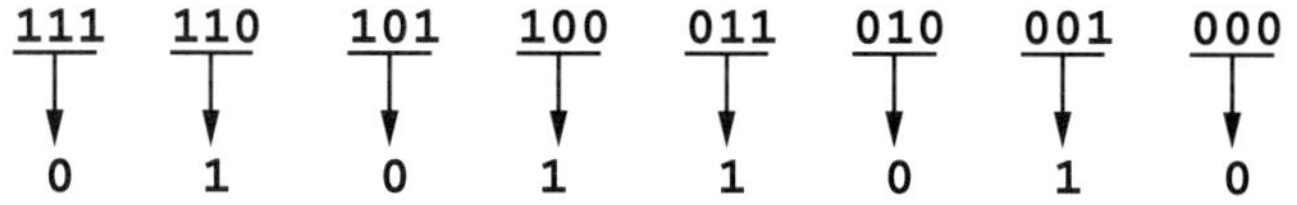

01011010 = 0x128 + 1x64 + 0x32 + 1x16 + 1x8 +0x4 + 1x2 + 0x1 = 90

FIGURE 1.11 Specifying the update rule for a 1D CA with radius $r = 1$ and two states. Shown here is rule 90.

Wolfram first investigated 1D CAs with $k = 2$ and $r = 1$, whose rule tables can be specified by the decimal equivalent of the binary number that results from concatenating the bits that result from the update of all possible neighborhoods (see Fig. 1.11). In the simplest case ($k = 2, r = 1$), there are eight different configurations for the neighboring sites, and the rules can be distinguished by specifying the value of a bit at the next update as a function of the different neighborhoods before the update. (Fig. 1.11 shows rule 90 as an example).

Of the 256 possible rule tables, only 32 turn out to be legal. Several other classifications of rule tables have been introduced. For example, certain rules are such that the value of a site at the next time-step only depends on the *sum* of the values in the neighborhood rather than the values of each site itself. Such rules are termed *totalistic,* i.e., they can be specified by functions of the form [cf. Eq.(1.3)]

$$a_i^{(t+1)} = \phi(a_{i-1}^{(t)} + a_i^{(t)} + a_{i+1}^{(t)}) \ . \tag{1.4}$$

You can easily convince yourself that of the 32 legal $k = 2$, $r = 1$ rules, only 8 are totalistic. Other rules depend only on the values of the sites to the left and right of the center site. Such rules are termed *peripheral.* (Rule 90 is an example of such a rule). Table 1.1 lists the 32 legal $k = 2$, $r = 1$ rules, and their classification in terms of totalistic or peripheral, and the Wolfram classes introduced below. Note that illegal rules are not necessarily uninteresting. A special $k = 2$, $r = 1$ rule is the well-studied Rule 110, which is nonsymmetric. A typical trace of this rule, started with a random initial condition (with about half zeros and half ones), is shown in Fig. 1.12.

Time in this case runs downward, and the asymmetry of the rule is clearly visible. The rules for 1D totalistic CAs with *two* neighbors ($r = 2$

TABLE 1.1 Table of $k = 2$, $r = 1$ legal rules and their classification: P = peripheral, T = totalistic. Roman numerals indicate the Wolfram class defined below.

Rule	Classif.	Rule	Classif.	Rule	Classif.	Rule	Classif.
0	T,P,I	72	I	128	T,I	200	II
4	II	76	II	132	I	204	II
18	III	90	P,III	146	III	218	II
22	T,III	94	II	150	T,III	222	II
32	I	104	T, I	160	P, I	232	T, II
36	II	108	II	164	II	236	II
50	I/II	122	III	178	II	250	P, I
54	III	126	T,III	182	III	254	T, I

rules) are described by a *code,* obtained by specifying the value of the site as a function of the possible sums of the sites of the five neighbors.

Since there are six different sums (0 to 5), a rule is specified by prescribing the value of a site for any of the six different sums. In Fig. 1.13 we show how the rule is obtained for code 20 and some example updates.

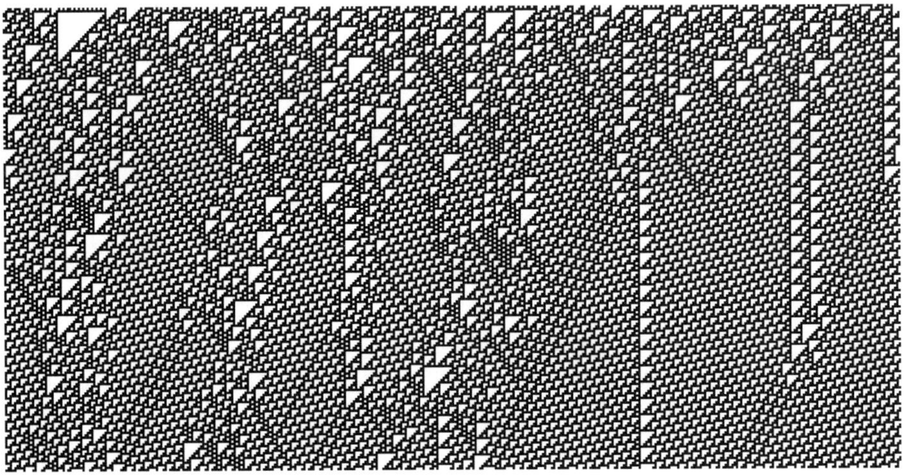

FIGURE 1.12 Time evolution of the asymmetric (and therefore illegal) Rule 110, showing complex interactions and patterns.

$\Sigma=5$	$\Sigma=4$	$\Sigma=3$	$\Sigma=2$	$\Sigma=1$	$\Sigma=0$		
0	1	0	1	0	0	=	20

$\Sigma=1$: 01**0**00 $\longrightarrow$ **0**

$\Sigma=2$: 10**0**01 $\longrightarrow$ **1**

$\Sigma=3$: 01**1**01 $\longrightarrow$ **0**

$\Sigma=4$: 11**1**01 $\longrightarrow$ **1**

$\Sigma=5$: 11**1**11 $\longrightarrow$ **0**

FIGURE 1.13 Specifying the update rule for a totalistic 1D CA with radius $r = 2$ and $k = 2$. Shown here is code 20, and some examples of transitions implied by the code.

Wolfram divided the rules into four distinct classes according to their long-time behavior. As $k = 2, r = 1$ CAs are too simple to show all possible behaviors, we concentrate instead on $k = 2, r = 2$ totalistic rules, of which there are exactly 32 [Wolfram, 1984]. The first class (Wolfram class I), displaying *limit point* behavior, is the simplest, as it evolves towards a homogeneous state (either all zeros or all ones) from almost all initial conditions in finite time (e.g., the code 60 rule shown in Fig. 1.14).

The second class (Wolfram class II), with *limit cycle* behavior, generates persistent structures in the long-time limit, i.e., we usually observe periodic behavior, or patterns shifting uniformly to the left or right (try code 56, shown in Fig. 1.15). Note that legal rules can show only trivial periodic behavior (cycle length 0 as in Fig. 1.15), as uniformly shifting patterns require a spatial inhomogeneity in the rule.

Class III CA rules almost always lead to aperiodic, *chaotic* states (Rule code 10, Fig. 1.16 is of that sort). In this class, the structures emerging may be ordered, but they show no obvious periodicity and no uniform

FIGURE 1.14 Temporal behavior of the $k = 2$, $r = 2$ code 60 totalistic rule (class I).

FIGURE 1.15 Temporal behavior of the $k = 2$, $r = 2$ code 56 totalistic rule of class II.

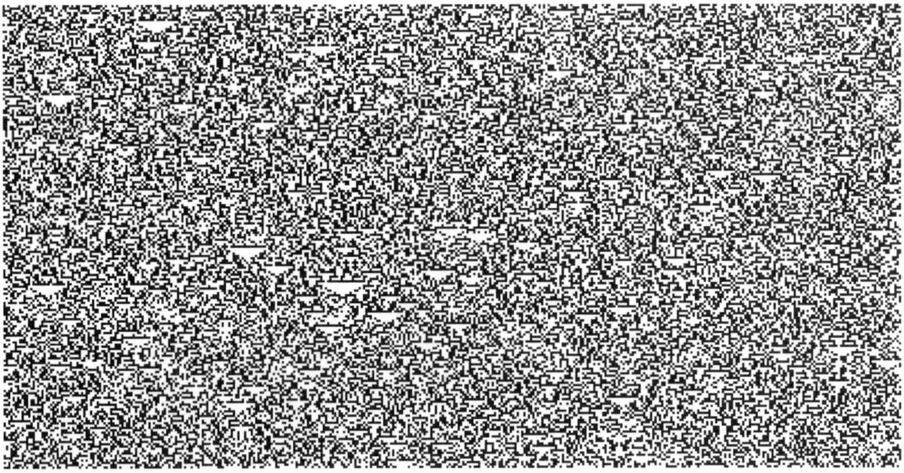

FIGURE 1.16 Temporal history of the $k = 2$, $r = 2$ code 10 totalistic rule, with class III behavior.

shifting of patterns. Naturally, for any finite CA (i.e., with a finite number of sites), any pattern must necessarily recur at some time, as for a CA with N sites there can be only k^N different patterns. The classification of the rules is thus meant to apply to the $N \to \infty$ limit. Note also that for some rules the behavior depends strongly on the initial pattern with which the CA is started. In that case the classification applies to average behavior over many different initial conditions.

Arguably the most important of the CA rules are those that fit into neither of these categories: the class IV automata. The temporal behavior of these CA is very complex, showing complicated patterns that seem to self-organize, and perform complicated operations (see the code 20 rule in Fig. 1.17). It has been argued that such automata are capable of universal computation, i.e., that the patterns are rich enough to serve as a universal Turing machine. It is of course for this reason that this class of automata is deemed to be so interesting.

For example, Langton (1992) has suggested that class IV automata maximize the information processing capacity for such automata, i.e., they allow for the maximal transmission of information from input to

FIGURE 1.17 Temporal history of the $k = 2, r = 2$ code 20 totalistic rule, displaying class IV behavior.

output. In a certain sense, he suggests that such automata are "perched at the edge of chaos." The "Game of Life" CA invented by the mathematician John Conway [Berlekamp et al., 1982] also seems to be in this complexity class, even though we will hardly be so overly impressed by its dynamics as to infer that the game has direct relevance to natural living systems. Still, we will be discovering an intriguing analogy in living systems to these dynamical systems perched at the edge of chaos, when we study aspects of critical self-organization in Chapter 6.

1.6 Overview

The field of Artificial Life covers a wide range of disciplines, from engineering over biochemistry to physics, biology, and computer science. At the root of the endeavor lies the desire to understand the general principles that govern the living state, and to construct life in an artificial medium so that its properties can be compared to the terrestrial life with which we are familiar, but also the conviction that modern technology can help us learn more about the living world around us by simulation and emulation. Artificial Life may yield experimental data that can be used to *falsify* theories about what constitutes the most general living state, just as evidence of extraterrestrial life in our solar system may one day do.

In order to properly place the approach followed in this book among others, highlights from different disciplines are introduced to convey the different *flavors* of Artificial Life that have emerged. The concepts of emulation and simulation are discussed with examples from robotic engineering, physics, and biology. The biochemistry of in vitro evolution and core-and-shell self-reproduction are mentioned as precursors to a carbon-based Artificial Life. Finally, the roots of the field are found in Turing's Automata Theory and the logical, self-reproducing automata of von Neumann. The latter's theories and ideas can be said to be the true foundations of the field. The cellular automata that have simplified our thinking and our approach to computation are introduced formally, and complexity classes are discussed as a means to classify the *computational chemistry* that is created.

Problems

1.1 The CA that is Conway's Game of Life is a two-state, nine-neighbor (Moore neighborhood) CA with three very simple rules that lead to an amazing amount of complex behavior, given the proper starting conditions. The rules are,

- A living cell (state "1") with 2 or 3 neighbors remains alive.
- A dead cell (state "0") with exactly 3 neighbors is born.
- All other cells die (from loneliness or overcrowding), or else remain dead.

Here are a few sample updates:

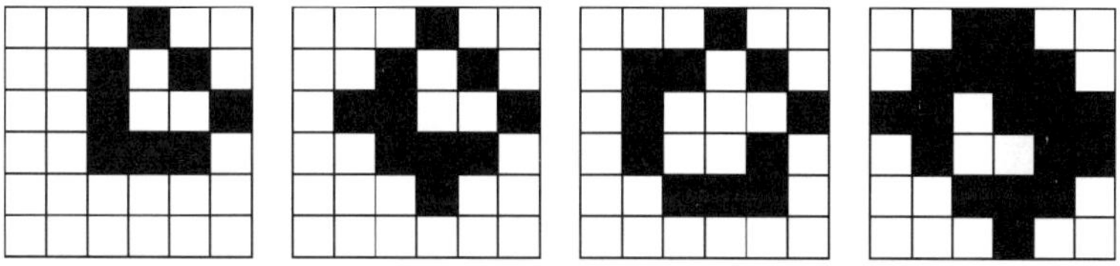

What is the smallest possible configuration (i.e., using the fewest number of living cells) that will completely die out in a single update? To make this question nontrivial, at least one cell must die from overcrowding (from having four or more neighbors). Here is an example that dies in two updates:

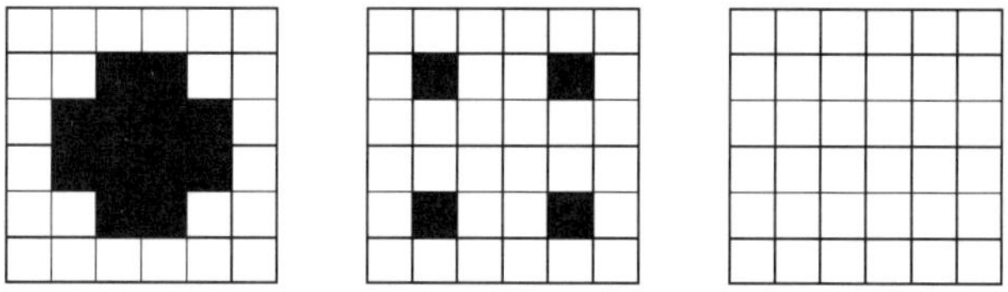

1.2 Classify the 32 legal totalistic rules for 1D CA with two states and a neighborhood of five cells ($k = 2, r = 2$) in terms of the Wolfram classes. A CA simulator can be found at `http://alife.santafe.edu/alife/topics/ca/caweb/`.

CHAPTER TWO

Artificial Chemistry and Self-Replicating Code

Das Sein ist ewig; denn Gesetze
Bewahren die lebend'gen Schätze,
Aus welchem sich das All geschmückt.[1]

J.W. von Goethe

2.1 Virtual Machines and Self-Reproducing CA

As mentioned before we entered the more formal discussion of CA, the appeal of CA as a substrate for Artificial Life lies in the suspicion that CA can constitute universal automata of the sort Turing had in mind, and thus that they could compute any function. While it is clear that this is not generally the case (certainly not for 1D automata with $k = 2$ and $r = 1$), some CA rules have been proven to possess the capability of universal computation (such as Conway's Game of Life mentioned earlier). Moreover, the global long-time behavior of a CA cannot, in gen-

[1]Being is eternal; for there are laws protecting the living treasures the Universe is dressed in.

eral, be predicted with standard mathematical analysis (precisely because of computation-universality), whereas the rules are easily implemented on a computer. As such, they lend themselves naturally to the simulation of complex 2D phenomena such as crystal growth, reaction-diffusion systems, turbulent flow patterns, and many more [Gaylord and Wellin, 1995; Gaylord and Nishidate, 1996].

In a paper that appears to have relaunched the field of Artificial Life [Langton, 1986], Langton discussed the possibility that certain aspects of living systems may be universal, in the sense that they are independent of the substrate upon which the rules that govern the interaction of the parts operate. From this point of view, he set out to explore the possibility of an *artificial chemistry* acting on *artificial molecules*. His medium of choice was the cellular automata we have been discussing, with the artificial chemistry specified by the update rules, and with *virtual automata* playing the role of molecules. Aware of the Wolfram classification introduced in the previous chapter, Langton ventured to discover which rules lend themselves most favorably to an artificial biochemistry, by investigating rule tables that were obtained *randomly*. Specifically, this was done for automata on a 2D regular lattice with eight states and a von Neumann neighborhood. Rather then trying out all possible $8^{32,768}$ rules, rules were sampled randomly, but keeping track of how many (of the 8^5 different) neighborhoods were mapped to the quiescent ("0") state. To this effect, Langton defined a parameter λ that represents the probability that a neighborhood in a particular rule is mapped to an active (i.e., nonquiescent) state. Naturally, rule tables with $\lambda = 0$ are boring: nothing ever happens. If λ is raised slightly, some regions on the array may hold some activity for a certain amount of time which, however, will eventually die out. This behavior is naturally reminiscent of class I rules. When λ is raised to about $\lambda = 0.2$ (20 percent of the neighborhoods map to any of the seven nonquiescent states), the behavior changes to a new pattern. Single active states are persisting, or propagating steadily in a fixed direction. Also, particular *cycles* of states can emerge that propagate across the array. Here, we recognize class II behavior in Wolfram's scheme. If λ is raised some more to about $\lambda = 0.3$, the activity in the array changes again dramatically to unpredictable, complex interactions of structures, as the mean distance between persistent active structures has decreased to such an extent that they interact and trigger each other. It is in this region where apparently the most interesting behavior takes place. For

λ values of the order 0.5 or more, the activity centers are so close to each other that chaos reigns and no complex structures can survive for long, a state of affairs best described by Wolfram's class III. Clearly, as λ approaches 1, the dynamics becomes less interesting again, as most rules result in active sites and the array does not display any localized structures or interactions. To a certain extent, the λ parameter seems to set the temperature of the computational world. Let us focus therefore on the region between $\lambda = 0.2$ and 0.4, where the more interesting dynamics seems to take place. Such rules seem to belong to the mysterious class IV, and are the most promising as far as computational universality is concerned.

First, we focus on propagating two-dimensional structures on the lattice. From an abstract point of view, such structures can be viewed as *virtual state machines* (VSMs), or more specifically, as virtual Turing automata. Trivially, if the quiescent background they propagate in is viewed as the *information tape* of the automaton, such a structure reads zero, advances, and writes zero, all through the lattice. However, more interesting structures can be found that change their state (such as the direction of travel), depending on the value of the cell they encounter, and that leave behind (write) cells in different states. Of course, such structures are exceedingly rare and usually have to be constructed by hand. Still, the important point here is that the computational medium allows for the *spontaneous emergence* of virtual Turing automata that are computation-universal. Note that the construction of such automata has enormous consequences. If we were able to implement a VSM with the ability to *self-reproduce*, we would have succeeded in constructing a *minimal von Neumann automaton* (compare Section 1.4). Indeed, if a VSM is capable of self-replication, we realize that because it is made from the same states that constitute the *information* tape, there exists the possibility that the construction automaton A, the tape-reproducing automaton B, the control mechanism C, and the tape describing all three I_{ABC}, are united in the same structure. In other words, the self-reproducing VSM could be machine and description of machine at the same time! Langton has implemented this idea by constructing a minimal vN (von Neumann) automaton with a CA of eight states and a vN neighborhood using 179 rules [Langton, 1986]. The self-replicating structure is a loop made out of a sheath (cells in state 2) within which circulates the information that is used to *construct* the loop (see Fig. 2.1). Left to self-replicate, the loops fill

	2	2	2	2	2	2	2	2					
2	**1**	**7**	**0**	**1**	**4**	**0**	**1**	**4**	2				
2	**0**	2	2	2	2	2	2	**0**	2				
2	**7**	2					2	**1**	2				
2	**1**	2					2	**1**	2				
2	**0**	2					2	**1**	2				
2	**7**	2					2	**1**	2				
2	**1**	2	2	2	2	2	2	**1**	2	2	2	2	2
2	**0**	**7**	**1**	**0**	**7**	**1**	**0**	**7**	**1**	**1**	**1**	**1**	**1**
	2	2	2	2	2	2	2	2	2	2	2	2	2

FIGURE 2.1 Langton's self-reproducing loop. Numbers 0–7 designate the eight different states of each cell. The sheath cells are in state 2. The information circulating through the loop is printed in bold.

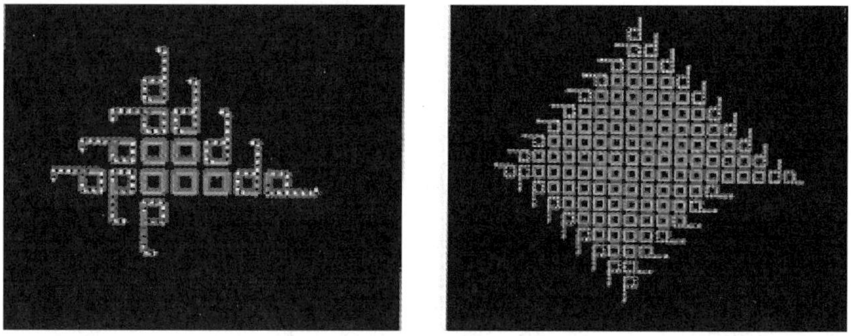

FIGURE 2.2 Colony of loops formed from a single self-replicating loop.

up the empty space (cells in the quiescent state) and form colonies with a regular lattice structure (see Fig. 2.2).

While Langton's experiment suggests that universal vN automata can exist in a computational medium, there is a clear limitation of the implementation that is its lack of adaptability. To a certain extent, fixing the update rules of the automaton fixes the chemistry (i.e., the world) that the automata live in, at which point only the information should be mutated, which is encoded in the value of the sites. Unfortunately (but by design), the automata themselves are embedded in the same medium, so that it becomes exceedingly difficult to mutate the information but not the automata. This is not a fundamental limitation, as we know that automata and information are identical also in RNA worlds. Still, the particular encodings used to date seem to lead to very brittle automata, implying that essentially all mutations of a self-replicator lead to a non-replicating structure. In principle, such a disadvantage may be overcome by searching for a world that reacts more kindly to noise. However, due to the number of possible chemistries in CA and the even larger number of possible automata roaming such a world, such a search appears, with current technology, hopeless. Nevertheless, work on self-replicating virtual automata has not stopped. The simplest self-reproducing CA found to date are 8-state CA with only 31 rules [Reggia et al., 1993], using a design that manages to get rid of the outer sheath in Langton's loops. In Fig. 2.3(a) we show a simple unsheathed loop that requires 177 rules and replicates in 150 updates, while in (b) we depict the replication sequence of the smallest replicator found to date, consisting of only five cells.

Due to the difficulty of finding an adequate computational chemistry in cellular automata, one may try to look for different computational paradigms that could harbor the promise of acting as a substrate for Artificial Life. In a manner of speaking, such a substrate manifested *itself*, without anybody looking for it, in the form of computer *viruses*. In the next section, we look towards self-replicating computer code to fulfill the promises awakened by the possibility of self-replicating virtual vN automata. In this case, as we know that our computers are computation-universal, we do not have to work hard to construct the artificial biochemistry.

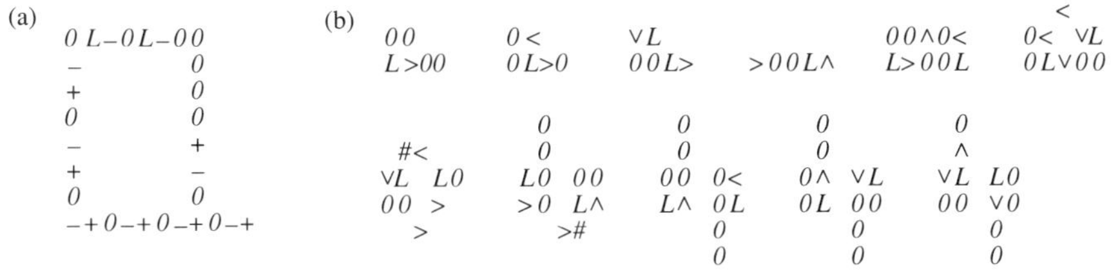

FIGURE 2.3 (a) Self-replicating unsheathed loop. The eight different states in this CA are denoted by a blank (the quiescent state) and $O\#L - *X+$. (b) Replication sequence for smallest self-replicating unsheathed loop. Here, the cells' states are denoted by $O\#L\vee > \wedge <$ (after Reggia et al., 1993).

2.2 Viruses and Core Worlds

The advent of computer *viruses* sparked the realization that computer code replicating in the core memory of PCs, laptops, and workstations could serve as a substrate for Artificial Life. In their simplest implementation, viruses do nothing other than replicate in the boot sector of the infected computer, although some viruses also attack device drivers or command interpreters (see, e.g., Spafford, 1994). In response to the effort of antivirus experts, virus programmers modified their creations to overcome more and more sophisticated antivirus programs using more and more sophisticated tricks. This evolutionary arms-race created whole phylogenies of viruses that are more or less distantly related, and their evolutionary tree can be analyzed using the same methods as are used for usual genetic kinship studies. In Fig. 2.4, we can see the genetic tree of the virus family "Stoned" that infects the DOS boot sector. The kinship analysis relies strictly on genetic rather than functional closeness of the viruses, and compares the genome of the variants with each other and with a copy of the DOS boot sector for reference. From the interaction between the programs and the environment they thrive in, particular species of viruses have adapted (by the actions of the programmers, not

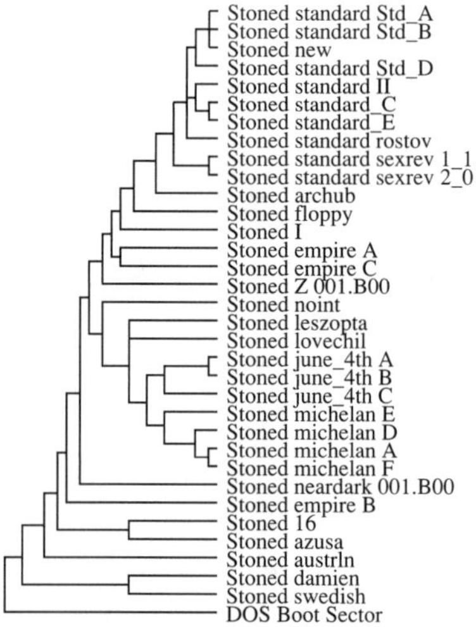

FIGURE 2.4 Phylogenetic tree for the family of boot-sector viruses "Stoned" (from Hull, 1995).

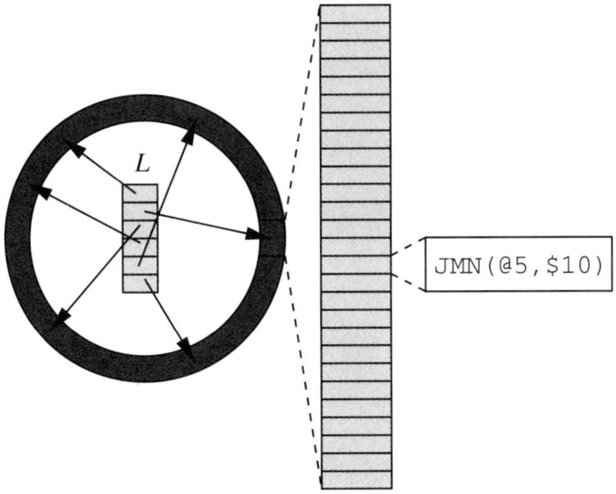

FIGURE 2.5 Central part of the Coreworld system, with circular memory of 3584 bytes and the execution queue of size L. One of the pointers in the execution queue is pointing to the address where the instruction `JMN (@5,$10)` is stored. (After Rasmussen et al., 1990.)

by random change!) to specific environments and niches. This kind of dynamic is all too reminiscent of very simple natural living systems, and it is this analogy that has given rise to the kind of Artificial Life we are dealing with in this book.

On another front, in the computer game "Core War" (see Dewdney, 1984) players are encouraged to write computer programs (in a virtual language) that fight for space in computer memory. The object of the game is to cause all of the opposing programs to terminate, leaving the winner's programs in sole possession of the machine. A typical strategy in this game includes self-replication, thus taking over all of the available space in this manner. Core War and the growing notoriety of computer viruses inspired a number of researchers to seriously think about computer programs executed on *virtual* processors (in order to avoid the pitfalls of programs attacking the real processor) and having the ability to self-replicate. The first serious attempt in this direction was undertaken by Steen Rasmussen and coworkers [Rasmussen et al., 1990] with their "Coreworld" program. Coreworld is a simulated computer core arranged in a circular manner, i.e., the linear address space wraps on itself, as indicated in Fig. 2.5. This space is seeded with assembly-like instructions of the original Core War language "Redcode," which has ten different instructions that take two addresses as arguments. Each argument is preceded by a mode identifier, which specifies whether the target value is interpreted directly, or as a pointer to another instruction, or as a pointer to a pointer. A typical instruction would be `MOV(#2354,$214)`, which moves the contents of memory block 2354 to the address pointed to in block 214. In this manner, Rasmussen's group could implement a low-level, Turing-complete rule system in a computational medium that was somewhere between Langton's cellular medium and a full-fledged von Neumann architecture. They termed it an *assembler automaton*. Finally, the core was subject to *noise* by allowing the `MOV` command, which copies an instruction from one place in the core to another, to be flawed with a certain probability. In this manner a random component was introduced into the system, enabling it to adapt.

The goal of Rasmussen was very close to Chris Langton's: to construct an artificial computational chemistry that could bear replicating artificial molecules, in the hope that these molecules would display *emergent* computational behavior. Unlike Langton, however, Rasmussen's starting point was not the design of a self-replicating program, and rather than searching for a set of rules that would allow universal computation, Coreworld settled on one such world that was *known* to be computation-universal. In fact, self-replicating programs can easily be written in Redcode. On the other hand, such programs cannot withstand a noisy environment (and indeed in the Core War game there is no explicit mutation), and as a consequence these programs *do not* represent a fixed point of the Coreworld dynamics in the presence of noise. Self-replicating programs introduced into the Coreworld do replicate for a while and introduce inhomogeneities into the core, but die out soon afterwards. Instead, Rasmussen was interested in the *true* fixed point (recurring stable patterns occurring at the end of a simulation) for this world. Thus, rather than hoping that self-replicating programs would just jump out of the random soup, his group counted on the possibility that programs would emerge that copy each *other*, so that maybe a network of replicating (but not necessarily *self*-replicating) programs might develop that self-organize in computer memory. Rasmussen's quest thus concerned the ever-intriguing question about the environmental conditions that gave rise to life: the mystery of its *origins*.

The dynamics in Coreworld turn out to be very intriguing indeed. For two rather different parameter settings, one corresponding to "desert" conditions in which survival requires a certain amount of hardship, and one termed "jungle" (where resources generally are abundant), very different evolutionary histories were found. While the adaptation in the desert regime usually did not lead to very interesting behavior, the jungle setting gave rise to complex adaptation, but very much dependent on random events in the early history of the core. Thus, the dynamics could take a number of very different avenues depending on chance events early in the development: a clear sign of *contingency* in the adaptive process. One particular outcome that occurred with reasonable frequency was a core that essentially consisted of the `MOV` instruction occurring together with the command `SPL` (which produces an additional pointer), as this mixture turned out to be quite stable. While no program with a fixed genotype ever emerges in this world, the overall core can evolve to be stable with cooperating program fragments. From a dynamical systems point of view, the system supports multiple fixed points with very delicate borders between the basins of attraction .

Still, Coreworld has a number of drawbacks (essentially the fragility under mutations) that are a direct consequence of using the Redcode scheme. Fortunately, it can be improved and as luck would have it, a young biologist named Tom Ray was visiting the Los Alamos National Laboratory (where Coreworld was developed) and looked over the shoulders of the Coreworld team as they analyzed their data. Impressed by their success and puzzled by the failures, Ray went back to the University of Delaware and started work on the tierra system. To a certain extent, tierra ushered in the second revolution in the still-burgeoning field of Artificial Life.

2.3 The tierra System

While the path that Rasmussen's team had chosen was fundamentally sound (namely to leave the CA world because the right artificial chemistry in that world was too hard to find, and to turn to the chemistry of actual von Neumann serial computers) their settling on Core War's Redcode as the default instruction set turned out to be too constricting. Ray took this seriously and started work on his own Coreworld, but with a completely new instruction set of his own design. To Ray, it was clear that the brittleness of Coreworld's replicating programs was due to the way instructions were paired with operands: any time a mutation takes place on an instruction, not only is the instruction changed but so are the *operands* (the addresses upon which the instruction operates). Instead, Ray decided to use a *pattern*-based addressing scheme, and a language where 32 instructions were coded in 5 bits (thus creating *codons* of length five) but with *no* operands. The pattern-based addressing scheme was loosely based on a biological paradigm. In genetic metabolisms, enzymes and proteins recognize the target of their catalytic activity not by being given an absolute address, but by recognizing a shape that is complementary to theirs and binding to it. Similarly, Ray decided to introduce instructions that do nothing when executed (so-called "No-ops"), but which carry a distinguishing mark. So, out of two such instructions (called `nop0` and `nop1`), Ray could fashion arbitrarily complicated patterns by stringing together `nop0`'s and `nop1`'s. Then, a pattern-matching algorithm would be able to search for the *complementary* pattern (where `nop0`'s and `nop1`'s are interchanged) and return its address when found. Otherwise, his language was loosely based on Intel i860 machine language.

To implement such an instruction set, Ray had to design a virtual CPU to go with it. His design involved an instruction pointer, four registers, a cyclic stack that could hold ten instructions, and a number of flags (this CPU is sketched in Fig. 2.6). Life and death in the tierra world is determined by the two queues that every program is placed in, irrespective of its position in the circular core. At birth, each program is entered in the *slicer* queue and is executed a fixed number of instructions (the *time slice*) if it is on top of the queue, and goes back to the bottom right after. Also at birth, it is entered at the bottom of the *reaper* queue, and moves up the queue as it "ages," i.e., every time it appears at the top of the slicer queue. If a program finds itself on top (or anywhere near the top) of the reaper queue, it will be eliminated any time empty space in the core becomes low, and a memory request by a mother program to place its offspring cannot be honored. At this point, it appears opportune to explain how programs can give "birth" to other programs in the first place!

The instruction set that Ray designed includes simple commands that allow the allocation of memory of a certain size by issuing a single command: `mal`. The CPU would then check register CX (see Fig. 2.6) and allocate a strip of memory (either close to the mother program or at an arbitrary location) of the size of the number contained in CX. Then, with

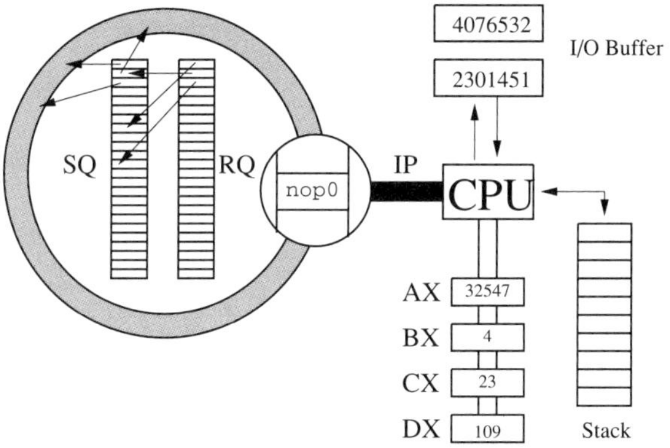

FIGURE 2.6 Virtual CPU of the tierra system. SQ is the slicer queue, which doles out time-slices to the programs, while RQ is the reaper queue, which decides which programs are slated for elimination.

the help of the `copy` command, the mother program can copy instructions from its own genome into the empty space it just allocated. The size of the space to be allocated can be determined by the mother program by finding the address of its own beginning and end and subtracting these two numbers. Of course the beginning and end of the program are tagged with patterns (templates) of `nop` instructions, and commands such as `adrf` and `adrb` followed by a complementary template will search forward and backward for the template and return the address if successful. After finishing the copy process, a successful program would issue the `divide` command, which removes write-access of the mother program to the daughter memory, hands the daughter program its own instruction pointer, and enters the cell into the relevant queues. In such a manner, computational birth has taken place! With these ideas, Ray set out to write an "ancestral" program (reproduced in Fig. 2.7) that replicates faithfully, and cautiously placed it into the new world.

The design turned out to be a stroke of genius, right off the bat. The program started filling the memory, and under the action of the cosmic-ray–like mutations that rained down on the core, started to diversify. Most of the mutations hitting the ancestor and its offspring would cause them to cease to be functional replicators, but the pattern-based addressing scheme was reaping its rewards: some of the mutations would leave the replicative abilities of the program intact, while some rare ones actually

```
nop1
nop1
nop1
nop1
zero
or1
shl
shl
mov_cd
adrb
nop0
nop0
nop0
nop0
sub_ac
mov_ab
adrf
nop0
nop0
nop0
nop1
inc_a
sub_ab
nop1
nop1
nop0
nop1
mal
call
nop0
nop0
nop1
nop1
divide
jmp
nop0
nop0
nop1
nop0
if_cz
nop1
nop1
nop0
nop0
push_ax
push_bx
push_cx
nop1
nop0
nop1
nop0
mov_iab
dec_c
if_cz
jmp
nop0
nop1
nop0
nop0
inc_a
inc_b
jmp
nop0
nop1
nop0
nop1
if_cz
nop1
nop0
nop1
nop1
pop_cx
pop_bx
pop_ax
ret
nop1
nop1
nop1
nop0
if_cz
```

FIGURE 2.7 Assembler source code for Ray's ancestral program (adapted from Ray, 1992).

improved it. Watching the dynamics unfold, Ray saw the programs' length slowly shrink. Figuring that the minimum length of the programs would be around 60 instructions long (the handwritten ancestor had 80 instructions), Ray was shocked to discover the sudden emergence of programs of size 45 that replicated virulently and seemed to be on the verge of overtaking the tierran world, as young as it was. Analyzing these creatures' genomes, he found them to be fragments of the dominating type of program, but devoid of the copy-loop that enabled the programs to replicate. What had happened was that mistakes in the calculation of their length (induced by mutations) had induced some programs to allocate only part of the space necessary for accurate self-replication, and thus to produce an incomplete fragment of themselves. Still, the beginning of the code was able to locate patterns, only now the pattern being searched for was missing from the program. If the command that searches for a pattern does not find it in its program, it has the possibility to simply search into the following program, as all programs are arranged linearly on the circular core. Jumping into the host program, the fragment's instruction pointer would now run into the host's copy-loop, and start copying its *own* instructions into space it had allocated, but using the copy-loop of the host! Thus, these short fragments turned out to be *parasites*, feeding on the reproductive machinery of their hosts. As parasites are much smaller in length, their gestation time (the number of instructions that need to be executed in order to produce an offspring program) normally is much less than the host's, which allows them to replicate much faster than the hosts. Thus, the parasites would take over the world until most of the hosts were driven into extinction. At this point, another curious dynamic started to unfold right under the astonished eyes of Ray, who was demoted to just *observe* the world he had created, a world that had taken life in its own hands rather quickly.

The parasites were dying out, partly due to having depleted the hosts, but it was noticeable that hosts were still abundant, and more and more so as the population of the parasites declined. Of course the question arose as to what had caused the *immunity* of these new self-replicators to the parasites that heretofore were so successful. An examination of these hosts quickly revealed that they operated with *different* templates of a form that could not be recognized by the parasites. Here then, we are witnessing in a computational medium a mechanism that is quite analogous to the development of immunity in, say, *E. coli* bacteria subjected to bacteriophages. The phage adsorb to *E. coli* by recognizing certain proteins in the cell walls, bind to it, and manage to enter the cell. After exposing a colony to such phage (which kill almost all *E. coli*), a few colonies will reappear after some time that are immune to the phage, mostly because they express different surface proteins that cannot be recognized by the phage. Whether such a mutation is prompted by the attack of the phage or whether it occurs randomly is a subject of much discussion [Foster and Cairns, 1992; Lenski and Mittler, 1993]. For us, it is immediately clear what happened, as there is no mechanism for directed mutation in tierra. The immune hosts are those that luckily had such irrecognizable templates, which due to the change in circumstances (i.e., the appearance of the phage) became all-important. We also note that the dominance of the new hosts *reflects* the presence of the parasites in the population. Programs are clearly *coevolving* in tierra.

What was unfolding before Ray's eyes was a veritable arms-race between hosts, parasites, immune hosts, immunity eschewing parasites, symbionts, "cheaters," "super-parasites," etc., all of it emergent, none of it anticipated. From a purely logical point of view, the barrier between life and artificial life seemed to have come down: the universality of life was proven. In hindsight we can see why Ray was successful where so many others had failed. Based on the work of Langton and Rasmussen, he put together a computation-universal pseudo-chemistry (the execution of programs) with a nonbrittle and *redundant* instruction set, stirred a little, and: *Voilá!*

2.4 avida, amoeba, and the Origin of Life

The success of tierra did not go unnoticed, and a number of researchers picked up the trail where Ray had ended up. From a scientific point of view, tierra has a number of features that are undesirable if it is to be used to understand the dynamics of natural living systems. While Ray moved on to find a "natural" habitat for his critters (the brave new world of the Internet [Ray, 1995]), the California Institute of Technology (Caltech) became the place where tierra was used for experiments in adaptation and the dynamics of evolution. (Results of some of these experiments are scattered throughout these lectures.) Subsequently, it was decided to retain only the paradigm of tierra's success (the idea of computation-universality and the redundant, pattern-based instruction scheme) and to start from anew. This marked the birth of the avida software (initially written by Titus Brown [Adami and Brown, 1994]). Charles Ofria wrote an independent version later, which is the software included in this book and introduced below. We also describe another mutation of the tierra system (also designed from scratch) that enables important insights about the origin of life, at least in this computational medium.

avida

In the design of the avida system, a number of important decisions were made very early on. First and foremost, the periodic-core structure of tierra and Coreworld was abandoned in favor of programs that live on a two-dimensional grid, with only eight nearest neighbors per program. The lack of *spatial* structure in tierra, while advantageous for some experiments, can have serious repercussions. Even more importantly, the global reaper queue in tierra implied that any program can, in principle, interact with any other one in the population. The most important interactions taking place in both tierra and avida arise from the birth of new

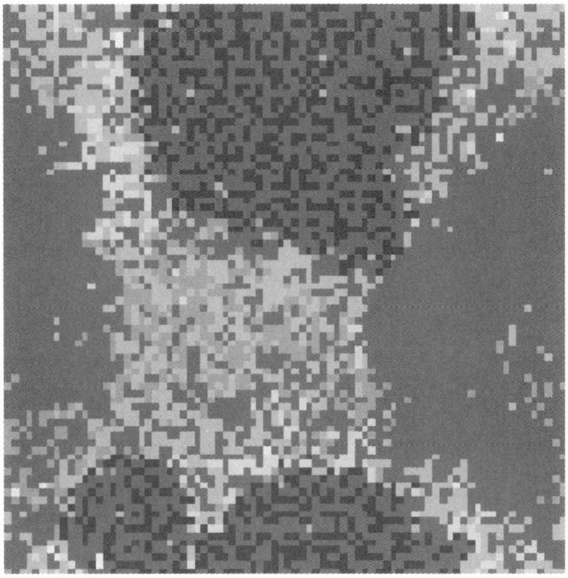

FIGURE 2.8 Snapshot of programs evolving in the two-dimensional avida world. The wavefronts of program clusters of the same genotype expanding over less fitter programs are clearly visible.

programs that replace older ones. In tierra, a request for placement of an offspring can entail the removal of *any* program in the soup (as only the position in the global reaper queue decides which program is removed). Thus there is no concept of locality in tierra, as programs arbitrarily far away in memory can still affect each other. By introducing a local geometry in avida, such effects are ruled out. In turn, there is a local reaper queue at each lattice location, which includes only the nine cells in the immediate neighborhood. Therefore, information can only propagate in a local manner, as we discuss in Chapter 10.

The locality of interactions in avida implies that, if the population is large enough, it will always be far from equilibrium (as we shall see in more detail in Chapter 10). In Fig. 2.8 we can see a snapshot of an 80×80 avida world and the programs inhabiting it. Different greyscales in this picture indicate different *genotypes* (or types of programs). Also, it is apparent in this figure that the two-dimensional avida world has a *toroidal* geometry in order to avoid boundary effects. Thus, the upper and lower, as well as the left and right boundaries of the world, are identified. Equilibrium can be said to be achieved if the information contained in one (usually the most successful) genome is able to propagate throughout the entire population. In that case, a certain *uniformity* of the population will be the result, which has a negative impact on the adaptability of the population. Indeed, a (genetically) homogeneous population can easily be trapped in a *metastable* state, which often prevents evolution to states of higher fitness. The more diversity in the population, the less chance of becoming trapped. Many experiments with avida demonstrated that local interactions do indeed prevent premature equilibration of the population, enhancing the adaptive powers. Note also that due to the local interactions, the population is much less vulnerable to infection by parasites. Metaphorically speaking, the global reaper queue in tierra entails intimate contact between *all* programs in the population, a behavior much discouraged in the presence of infectious diseases!

Concurrent with the change of geometry for the programs came a change in the way the instruction pointer (IP) behaves when it runs off the end of the program. In tierra, when an IP failed to be redirected to the beginning of a program, it simply ran into the following program, and so on. If the population is dominated by programs that have lost the ability to redirect the IP, instruction pointers would roam the entire core never to return to the programs that owns them. As far as natural living systems are concerned, this appears unsatisfactory. As we think of the execution of the program as an artificial chemistry, the Caltech designers deemed it much more appropriate that the chemistry of the program never be disabled. Thus, they decided to let the IP loop back to the beginning of the program *automatically* if it runs off the end, in effect making the code *circular*, much like the actual genome of most viruses and bacteria. Still, the ability to jump into the code of neighboring cells is maintained, *if* an appropriate (complementary) template is found there.

Since in avida a cell has *eight* neighbors (in the default mode), the designers also decided to introduce a *facing* for the programs. Thus, if a search for a template would be launched, the search would begin in the adjacent cell being faced by the program issuing the search, and continue clockwise if the search was unsuccessful. Finally, the replication process itself in avida is designed to be closer to *cell division*. If a program requests memory in avida, it is allocated at the *end* of the mother cell such that the genome is, during the replication process, generally doubled. The `divide` command, which otherwise is very similar to the `divide` command in tierra, then splits off the code at the end of the mother and places it in an adjacent spot, removing the cell that previously occupied it. Just how the cell to be replaced is chosen, and other details about the replication procedure and the structure of the virtual CPU of avida, can be found in the *User's Manual* (Appendix).

As avida will be the main platform for performing experiments in these lectures, we shall defer to Chapter 9 and the Appendix for a more detailed exposition of the design. Instead, we shall take a look at another tierra-inspired platform involving self-replicating computer programs. This particular system, dubbed amoeba by its creator Andrew Pargellis of Bell Laboratories (now Lucent Technologies), was designed specifically to investigate the issue of the *origin* of life.

amoeba

There is a great deal of interest in uncovering the mechanism that gave rise to the prokaryotic cell cycle, or more generally to self-replicating metabolisms. To a certain extent, the molecular ingredients for simple living systems are known, and amino acids form readily under conditions not unlike those generally assumed to have existed on a very early earth. Indeed, most if not all of the essential building blocks of proteins, carbohydrates, as well as nucleic acids and ATP (adenosine triphosphate, the carrier of chemical energy), can be readily produced under quite general primitive reducing (i.e., not oxidizing) conditions. Still, the origin of life appears to be one of the most fundamental biological problems, and naturally has attracted consideration from various incongruous points of view. Here, we shall follow only one of the hypotheses under discussion, namely that life can arise *spontaneously* from nonliving matter. Restricting the discussion to this view reveals a fundamental bias which of course is at the heart of this book. Only if life can emerge from nonliving materials is Artificial Life a worthwhile enterprise.

If we accept the premise that there are universal principles governing the dynamics of living systems, we may ask what it takes to produce Artificial Life from nonliving materials in a pseudo-chemistry. Systems such as tierra and avida seem to be ideal for the pursuit of this question. For example, we may ask whether self-replicating programs can emerge from a random core of instructions, and we may be interested in investigating the *path* by which such self-replicating programs arise from nonreplicating fragments. This is in principle the question that Rasmussen tried to answer with his Coreworld system, and this is the question that interested Pargellis when he designed amoeba [Pargellis, 1996a; Pargellis, 1996b]. Like avida, this system was inspired by Ray's tierra, but it is customized to investigate questions pertaining to the origin of life. The principle difference in Pargellis's amoeba as opposed to tierra is the choice of the instruction set. We have emphasized the importance of the right instruction set for evolution (which chooses the chemistry) earlier. Also, while cell death is governed by a global reaper queue much as in tierra, Pargellis's cells live on a two-dimensional grid that dictates which cells are nearest neighbors. Cells move about randomly on this grid so that the nearest neighbors change during a cell's lifetime.

Let us focus on the instruction set. In order to construct a set that will allow the *spontaneous* generation of self-replicating programs, Pargellis reduced the fragility of the language compared to that used in tierra or avida by restricting it to only *sixteen* different instructions. Furthermore, while in tierra and avida, for example, values in registers (and thus addresses) are not mutated directly (as opposed to Coreworld), there are many commands that affect the registers (such as any push and pop off the stack), as well as arithmetic operations on registers. Such instructions are absent in Pargellis's set, as is a stack. Once addresses are loaded into registers, they can only be changed by loading a different address. Also, the pattern-based addressing scheme is different in amoeba. There, *all* instructions can be used as patterns, and eight of the 16 instructions are the complement of another eight. The result is an instruction set that only requires *five* instructions to write a self-replicator. Note, however, that due to the simplicity of the virtual CPU, Pargellis's instruction set is *not* computation-universal. This is an important observation that we shall discuss later on. The obvious advantage of having such short self-replicators is of course that all possible replicators of length five can be written down, and as a consequence it is possible to calculate the *density* of self-replicators in the space of programs of length five:

$$P_5 = \frac{12}{16^5} \approx 10^{-5} \,. \tag{2.1}$$

Clearly this is a reasonable number, so that the *random generation* of a self-replicator becomes feasible. Pargellis's instruction set displays an even more important feature, however. The probability to find a self-replicator of size *six* within programs of length six (for example) is $P_6 \approx 2P_5$, i.e., longer programs have a higher probability to be self-replicators. As a consequence, the route to efficient self-replicating programs in amoeba usually involves the generation of *long* inefficient replicators *first*, followed by optimization to shorter ones.

In experiments conducted with a thousand cells, in which a constant percentage (4 percent in this case) of the cells were replaced by *random* code in order to maintain the influx of randomness, and where a cell's instructions would be mutated with a fixed small probability (10^{-4}–10^{-3}), self-replicating programs emerged spontaneously from cells with random code (the starting condition) in a typical fashion [Pargellis, 1996a]. First, short programs develop that completely lack the ability to replicate, either because they are too short (i.e., smaller than five instructions long) or because they lack one of the crucial commands necessary for replication (such as memory allocation, cell division, or copying). This is termed the *prebiotic* phase. Often, small parasitic programs emerge in this stage that manage to reduce the *entropy* of the population (see Chapter 3), presumably by using the copy instruction of other cells, or by copying each other. However, as these programs do not properly self-replicate, such a drop in disorder cannot be sustained. Quite abruptly, the *protobiotic* phase is entered if programs develop that copy themselves correctly *once*, but fail to do so a second time around. The emergence of such programs leads to a sustained drop in entropy, as can be seen in Fig. 2.9, which shows the entropy (as a measure of the disorder in the population) as a function of time (in generations, an arbitrary unit of time) for a specific, but representative, run.

While the first clumsy replicators lead to a sustained ordering of the population, these creatures are still fragile and can presumably sustain their genome only inside of a colony, which is a collection of programs of similar genotype that replicate each other if the instruction pointer is lost to a neighbor. However, from these protoreplicators, stable self-replicators do emerge (the *biotic* phase) that can copy themselves efficiently *and* correctly, at around 4000 generations in the run depicted in Fig. 2.9. Such a scenario appears to be typical in the amoeba world. Once true self-replicators are found, they tend to optimize and shed instructions, as smaller self-replicators are more effective than long ones in this setting.

Let us now analyze in which way the chemistry of the amoeba world is typical or atypical as far as realistic pseudo-chemistries are concerned. First off, irrespective on how realistic the path to self-replication is, it

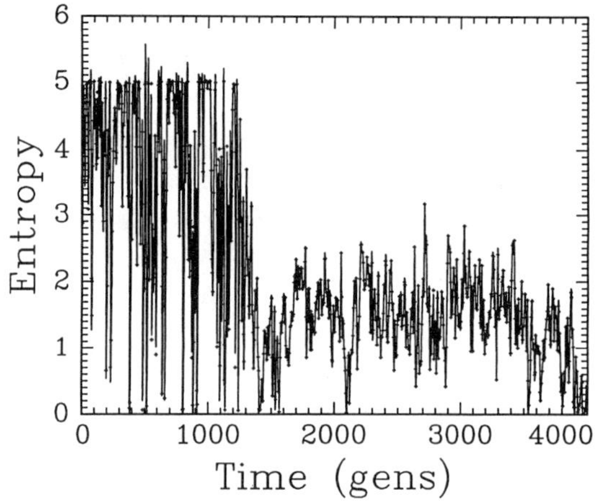

FIGURE 2.9 Entropy (disorder) as a function of time in the amoeba world. Worlds with random programs have an entropy around five. A robust self-replicator arose at generation 1274. (From Pargellis, 1996b, with permission).

certainly points out the *possibility* of a particular path, namely one in which self-replicators emerge in the limit of long sequences *first*. This is of course due to the special property of the amoeba chemistry, in which the density of self-replicators *rises* with sequence length. Pargellis has measured the probability of appearance of a self-replicator in amoeba, and the results do indicate a monotonous rise of the density of replicators with sequence length [Pargellis, 1996b]. (Because of the still-enormous number of possible sequences, the probabilities in Fig. 2.10 suffer from considerable sampling errors.) As we pointed out earlier, such a rise in the density of self-replicators is quite unlikely in either tierra or avida. Indeed, if P_n is the probability to find a self-replicator of length n in the latter worlds, P_{n+1} is almost certainly smaller than P_n, because of all the possible ways the additional instructions can modify the values in the registers of the virtual CPU. Such a modification will almost always lead to a nonreplicator. The higher fragility ("brittleness") inherent in the tierra and avida chemistries seems to be connected to the computation-universal CPU. We may thus *conjecture* that for a computation-universal chemistry, P_n *drops* monotonously with n, whereas this may be circumvented by using a nonuniversal instruction set. This does not rule out, however, life getting its start with a nonuniversal set, and then expanding it to

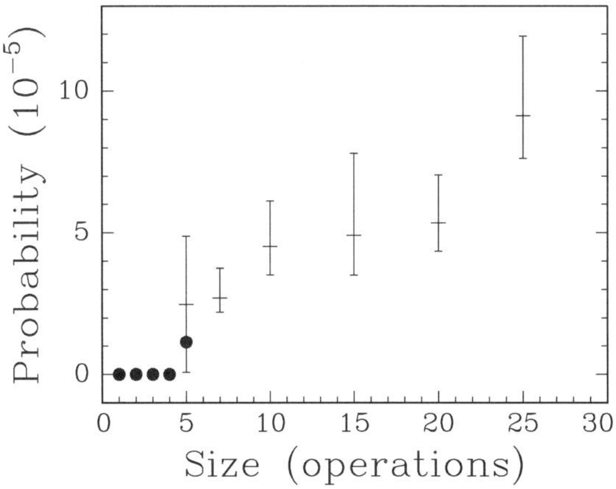

FIGURE 2.10 Probability P_n that a random sequence of n instructions is a self-replicator. The values $P_1 \cdots P_5$ are exact (indicated by solid dots). The points with error bars were obtained from running experiments with amoeba and counting spontaneously generated replicators of length n. (From Pargellis, 1996b, with permission.)

universality once a self-replicator arose. Indeed, one of the most important experiments in the (artificial) emergence of life would start out with *very many* possible instructions (say, for example, all the instructions of Coreworld, tierra, avida, and amoeba combined), letting the system decide which are the most appropriate. In amoeba, for example, it turns out that of the 16 instructions, most self-replicators use only six or seven. Thus, experiments with *variable* instruction sets may lead to important insights into how artificial life can emerge from artificial nonlife, and teach us something about the possible paths evolution in real systems may have taken.

2.5 Overview

In closing this chapter, we should take a step back and contemplate the value of experiments with artificial chemistries in the quest to understand the basic principles operating in living systems.

In a time span of almost fifty years, research on Artificial Life has come almost full circle. It was spawned by the brilliant ideas of von Neumann, who himself was inspired by the pioneering thoughts of Turing. From von Neumann's first steps in attempting to create a *universal* artificial chemistry, we have ended up using virtual CPUs modeled on those used in ordinary *von Neumann* computer architectures to achieve the goal. In so doing, we have given up more and more universality, while for the most part (with the exception of amoeba) retaining *computation* universality. Programs replicating in the tierra and avida worlds are, because of the remaining computation-universality of their virtual CPUs, isomorphic to populations of *minimal* vN automata. As a consequence, we feel compelled to speculate that such constructions constitute *minimal living systems*, and that they may play a similar role in the quest to understand the foundations of life as did, for example, the logistic map (see, e.g., Mandelbrot, 1977) in uncovering the principles of deterministic chaos.

In the following chapters, we prepare the ground for a thorough theoretical analysis of simple systems of self-replicating entities. Two important tools for this analysis are Shannon's theory of information [Shannon and Weaver, 1949], as well as the general framework of statistical mechanics and thermodynamics.

Problems

2.1 Calculate Langton's λ-parameter for Conway's Game of Life (see Problem 1.1). According to this value, in which of the complexity classes does this rule belong?

2.2 Estimate the number of self-replicating programs of size 11 instructions—the smallest self-replicator with the default instruction set in avida found to date—that can be written (consult the *User's Manual* in the Appendix on how to write programs with avida). From this ratio of replicators to non-replicators in avida, estimate the total time necessary to find *all* replicating programs of size 11 in avida (by examining all possible programs). Assume that one processor can perform 1 maips (1 million avida instructions per second), and that we have at our disposal a supercomputer with 10^4 processors. How can you improve on this limit?

CHAPTER THREE

Introduction to Information Theory

It is better not to speak about the "Entropy of the Universe."

L. Brillouin, 1949

The concept of information, even though it is used abundantly and intuitively in ordinary language, is nontrivial mathematically. In order to understand the dynamics of living systems, it is imperative that the notions of entropy and information are understood precisely rather than loosely. In fact, there is little doubt that the decades of confusion that have reigned over the treatment of living systems from the point of view of thermodynamics and information theory can be traced back to an imprecise understanding of these concepts. To a certain extent, this confusion is mirrored in Brillouin's quote at the beginning of this chapter. The discovery of the genetic code cemented the fact that information is the *central* pillar in any attempt to understand life, and the dynamics of information storage and acquisition that come with it. Let us then explore this theory.

3.1 Information Theory and Life

Clearly one of the most stunning aspects of living systems is the complexity they display so persistently in the presence of forces that work against the establishment of any structures, especially unlikely ones. This chapter introduces some of the elements that are necessary to unravel the connection between the complexity displayed by living systems and their ability to *store* information. This connection is a universal trait of *all* living systems, but it is especially simple in the artificial ones, and therefore easier to investigate.

The scheme Nature evolved to store and protect information from the deteriorating pull of thermodynamics (see Chapter 4) is a uniquely elaborate coding-decoding device. The question is whether this scheme is a result of historical contingency, (i.e., is just an accidental structure), or whether it reflects constraints on the system that make the scheme necessary, maybe even optimal. We will investigate later how this information acquisition and storage system can be responsible for the evolution of complexity. Some of these questions can be answered using the tools of *Information Theory*, a framework developed by Shannon in the mid-40s [Shannon and Weaver, 1949].

Information Theory deals strictly with messages, code, and the ability to transmit and receive messages accurately through more or less noisy channels. Initially, it was developed in order to gain a theoretical understanding of how well messages can be protected from corruption in noisy communication channels. Nowadays, it is used in very diverse environments ranging from computer science over physics to economy and, of course, engineering. Error-correcting codes that are the result of work in Information Theory and applied mathematics are now at work in every commercial CD player.

In order to appreciate the importance of Information Theory for living systems, let us sketch the basic communication channel: our main object of inquiry (See Fig. 3.1).

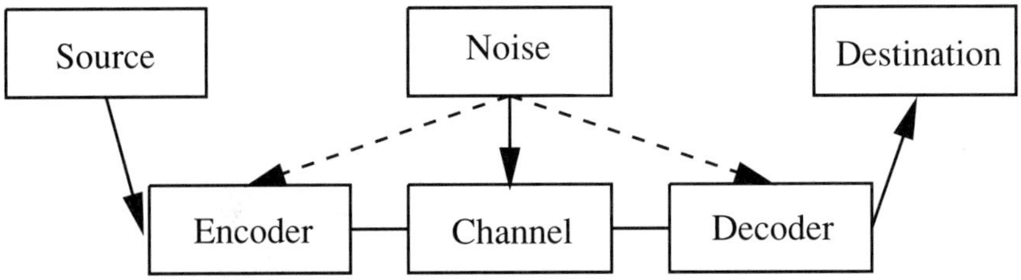

FIGURE 3.1 Communication channel.

Each of the units in this communication channel has a precise analogue in living systems. The source of messages is the environment in which the system under consideration lives and to which it adapts. In the process of adaptation, genomes become *correlated* to the environment: sequences of nucleotides (genes) *mirror* the environment's peculiarities. This information is encoded into the genome via a *stochastic* process: mutation. Once a sequence of nucleotides has been mutated in such a way that the host of the genome can take advantage of the properties and peculiarities of the environment, this information is duplicated and *transmitted* to the next generation. The channel for the information transmission is the DNA/RNA system that expresses the DNA into proteins. It is a noisy channel, as there are both replication errors (the incorrect copying of a base pair) and errors in the DNA due to external stress (be it heat, ultraviolet light, or viruses). All of these contribute to a deterioration of the information stored, and inhibit the correct transmission of the message to its destination: the expression of a protein that is advantageous to the host's functioning. The destination of these messages is thus the environment again: it is here where the organism needs to survive. Note that in living systems, noise obviously affects not only the channel itself, but also the encoding and decoding process. In Information Theory, the situation is simplified: the noise is thought only to affect the channel, and encoding and decoding can therefore be performed error-free. It turns out that such a simplification does not change the generality of the results. In order to make the above analogy more precise, we need to understand the main concepts in more detail and introduce a mathematical measure for information in order to track its fate from input to output.

3.2 Channels and Coding

Let us start by introducing the concepts of source, channel, and coding more rigorously, yet without trying to be mathematically precise. Imagine a source of symbols that produces messages by spewing out binary digits that are independent and uniformly distributed. Note that such strings of symbols by themselves do not constitute information. Only after such strings are *correlated* with certain messages that we want to send do they reflect information. (This will be made clearer in Section 3.5). Conversely,

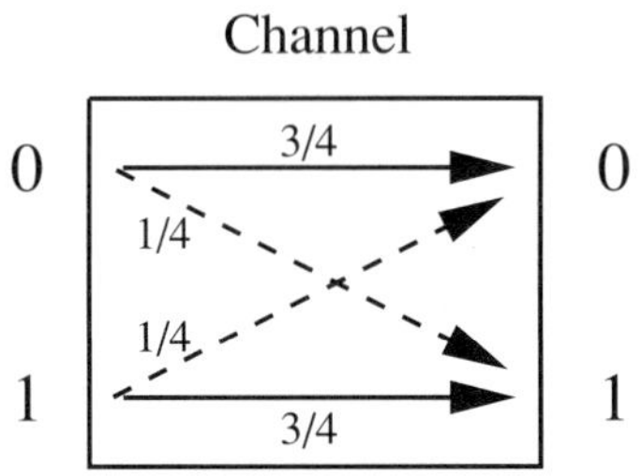

FIGURE 3.2 Binary symmetric channel with $p = \frac{1}{4}$.

we might want to be the source of the messages ourselves, in which case we translate our message into such binary bit strings for transmission. Note that in this case the symbols in the string will not (in general) be uniformly distributed. After this preparation, we would like to transmit the sequence through a channel, or medium. However, assume that the medium is noisy (like a bad connection over the telephone). For example, imagine that there is a 25 percent chance that the bit that is transmitted is flipped before it arrives. Imagine also that the bits are generated with a *rate* of 1 bit per second. The situation can be summarized by the diagram in Fig. 3.2.

In most cases, an error-rate of 25 percent is inadmissible, and ways need to be found to nevertheless transmit information accurately over such a channel. The central question of Information Theory thus becomes: "Is there a way to use a noisy channel and *encode* messages in such a way that it can be received and decoded with a reliability approaching 100 percent?" It is this question that Shannon set out to answer, when the general thinking at the time suggested that this could only be done at the expense of slowing the effective transmission rate to 0 bits per second. Indeed, the most straightforward scheme to reduce errors in transmission consists of repeating every bit of the message as many times as necessary, such that given a certain error rate, the message bit could be retrieved at the other side by a simple majority-rule. Imagine that you would like to send a 0 through a channel, but you know that with probability $\frac{1}{3}$, it is flipped to a 1. You might decide that, instead of sending 1 bit, you send a block of 3 (each bit identical in the block), and you inform the recipient of the message of this scheme. Then, while most such triplets have 1 bit flipped, a simple majority rule will allow the recipient to correct all 1-bit errors, i.e., the recipient will decode accurately any time zero or one error occurred in the transmission. Note also, however, that the rate of information transmission has slowed to $\frac{1}{3}$-bit per second. Also note that the transmission is *not* error-free. Indeed, some of the transmitted bits will still be decoded in error, and these are just those cases where more than 1 bit is flipped in the transmission. The probability of error, or conversely the *fidelity*, of the 3-bit redundancy code is the object of Problem 3.4. The redundant code used by Nature is outlined in Box 3.1.

Shannon's answer to the prevalent feeling that error-free transmission could only be achieved at the expense of vanishing transmission rate was his Fundamental Theorem. It asserts that in order to achieve *arbitrarily* accurate information transmission through a noisy channel, it is only necessary to reduce the transmission rate to the *capacity* of the channel. (Channel capacity is discussed in more detail in Section 3.7). The gain in reliability is always achieved via coding.

To calculate how much redundancy is needed for a particular channel, and what the *optimal* coding scheme is for such a channel, we need to mathematically define a unit, or a *measure*, of information. Conversely, a measure of information implies a measure of *uncertainty*, and both will be introduced below.

3.3 Uncertainty and Shannon Entropy

Let us define a source of symbols X, and treat X as a *random variable*. Thus, we assume that X can take on a finite number of states $x_1, \ldots, x_N$, each with probability $p_1, \ldots, p_N$. Also, let all probabilities lie between 0 and 1, and suppose that X can always be found in exactly one of the N states, i.e., $\sum_i^N p_i = 1$. We now may ask the question: "What is the uncertainty associated with X?" This question only has a precise answer if it is specified *who* is asking this question, i.e., how much is known about the random variable X. In general, *some* things are indeed known about X even though this is not always made explicit. For example, it may be known what the probability distribution for each x_i is. In the absence of any information, we must assume that each state is equally probable, i.e., all $p_i = 1/N$. (This is sometimes called the Principle of Insufficient Reason.)

As information preservation can always be viewed as a kind of transmission channel, we may ask what scheme Nature employed to preserve the information in the genetic code. Indeed, it is usually thought that early living systems developed in an environment much noisier than those prevalent today, on an earth that was hotter, had very frequent meteorite impacts, and was without an ozone layer. The simplest redundant encoding scheme we encountered earlier is not very efficient, and indeed is not used by Nature. Rather, more complex coding schemes are based on *codewords* that encode not single bits, but entire sequences. For example, we could invent an alphabet of letters to encode sequences of bits of length 5:

$$\begin{aligned} 00000 &\longleftrightarrow a \\ 00001 &\longleftrightarrow b \\ 00010 &\longleftrightarrow c \\ &\vdots \end{aligned}$$

In Nature, the message ultimately is the amino acid sequence of polypeptides, as it is those molecules that interact with the environment via chemistry. These sequences are formed from a 20-letter alphabet (let us take here the letters a through t). Each such letter is coded using a three-letter sequence, where the letters of the *code* are drawn from an alphabet of four (the nucleic acids cytosine, guanine, arginine, and thymine, denoted by CGAT), via

$$\begin{aligned} a &\longleftrightarrow CGA \\ b &\longleftrightarrow CUA \\ c &\longleftrightarrow UCA \\ &\vdots \end{aligned}$$

This scheme is redundant, since there are more codewords (64) than there are different amino acids. Indeed, in the translation scheme Nature has evolved, the last bit (A) of any of the above triplets can be replaced by any other nucleic acid without disturbing the capability to correctly transcribe. (However, this is not true for all triplets.) How well the natural code exploits redundancy will be examined in Box 3.2.

BOX 3.1 Information Coding in DNA

Let us go back to the general situation of a random variable X, where the outcomes x_i occur with probability p_i. Suppose we could associate an *uncertainty* $h(x_i)$ (rather than a probability) with each possible state of X. Then, clearly the uncertainty of X would be the average of the uncertainties $h(x_i)$ over all i:

$$H(X) = \sum_i^N p_i\, h(x_i)\ . \tag{3.1}$$

The uncertainty $H(X)$ that we would like to construct needs to conform to a number of *axioms*, which are mostly intuitive. The first is the axiom of monotonicity:

(i) monotonicity: The higher the number of (possible) different states in a system, the higher its uncertainty,

$$H_N(X) > H_{N'}(X) \qquad \text{if} \quad N > N'\,.$$

(ii) additivity: Uncertainty about two unrelated systems is equal to the sum of uncertainties of each,

$$H(X, Y) = H(X) + H(Y) \qquad \text{if } X \text{ and } Y \text{ uncorrelated.}$$

Interestingly, there is a unique function that satisfies these axioms, namely $h(p_i) = \log(1/p_i)$. With this, we have the Shannon entropy or uncertainty:

$$H(X) = -\,C\sum_i^N p_i \log p_i\ . \tag{3.2}$$

The constant C in front of this expression is related to the base chosen to express the logarithm. For simplicity, we shall take our logarithms to the base 2, which means $C = \log 2 = 1$. Let us now examine Shannon uncertainty more closely and try to understand intuitively what it measures.

For example, we may ask what the probability is of successfully predicting the state of a random variable X, armed only with the knowledge that X can be in N states, each with probability p_i. The answer is surprisingly simple in terms of the Shannon uncertainty. This probability turns out to be just

$$P = 2^{-H(X)}\ . \tag{3.3}$$

Note that in the absence of any information (the "insufficient reason" case mentioned earlier), we have $p_i = 1/N$, and therefore $H(X) = \log N$. As a consequence, $P = 1/N$, as we would expect. Note that we have discovered another important property of Shannon entropy:

$$H(X) = -\sum_i^N p_i \log p_i \leq \log N , \tag{3.4}$$

which expresses the idea that if we know that the probabilities for each state are *not* equal, then the uncertainty about this system is less than the maximal value. More precisely, $h(x_i)$ is the uncertainty *removed* from $H(X)$ if it is revealed that x_i in fact has zero probability to occur as one of the states of X. In other words, $H(X)$ can be viewed as the average uncertainty removed by revealing an average state x_i of X.

Another way of viewing uncertainty is as the "average number of yes/no questions needed to reveal the state of X precisely." Indeed, this is just a consequence of one of the formulas introduced above. As

$$P(X) = 2^{-H(X)} = \left(\frac{1}{2}\right)^{H(X)} , \tag{3.5}$$

we see that with each (judiciously chosen) question, the probability of predicting *incorrectly* is reduced by a factor 2. Thus, on average we need only ask $H(X)$ such questions to reveal X. This is explored further in Problem 3.1.

3.4 Joint and Conditional Uncertainty

After having treated the uncertainty associated with one variable (or more generally one system), we should consider the *joint* uncertainty of composite systems. This endeavor might seem trivial at first, on account of the additivity theorem introduced earlier. Note, however, that we insisted there that the two systems were *uncorrelated*. The insight gained by examining composite systems lies precisely in the correlations that might exist between the two. Imagine, then, two random variables X and Y, with X as before but Y having N' states, and define the probability for X to be in state x_i and Y *jointly* to be in state y_j:

$$P(X = x_i \text{ and } Y = y_j) = p(x_i, y_j) . \tag{3.6}$$

This allows us to define the *joint uncertainty*

$$H(X, Y) = -\sum_i^N \sum_j^{N'} p(x_i, y_j) \log p(x_i, y_j) . \tag{3.7}$$

It is easy to show the *subadditivity* character of this expression:

$$H(X, Y) \leq H(X) + H(Y) . \tag{3.8}$$

Earlier we saw that if two systems are uncorrelated, the joint entropy must be given by the sum of the individual entropies. Now we see that if this is *not* the case, there must be correlations present between X and Y. Let us illustrate correlations by a simple example.

Imagine a lattice of size $N \times N = M$, where each lattice point can harbor a ball labeled X or Y (see Fig. 3.3). Also, let us imagine that a ball is dropped randomly on the lattice such that each of the slots have an equal probability to be occupied (unless they are already occupied, in which case their probability to be occupied is zero). If we drop the X-ball or the Y-ball independently, the uncertainty of their position is $H(X) = H(Y) = \log(M)$, as there are M slots on the lattice. However, imagine now dropping the balls consecutively. Because the second ball dropped has one less slot to choose from, the joint uncertainty for the two balls is less than $2\log(M)$:

$$H(X, Y) = \log(M) + \log(M - 1) < 2\log(M) . \tag{3.9}$$

In this case the correlation is very small because both balls simply can't occupy the same slot. Much stronger correlations are introduced if we imagine tying a string between the balls in such a way that they *have* to occupy adjacent slots (Fig. 3.4). If we then drop this contraption onto the lattice, the uncertainty in the placement of the first ball (naturally, we now must drop the balls simultaneously) is still $\log(M)$. The second ball, attached to the first, however, has only four options: to the west, east, north, or south of the first. The joint uncertainty is then (neglecting edge effects)

$$H(X, Y) = \log(M) + \log(4) . \tag{3.10}$$

In this case the joint uncertainty is reduced considerably, and the locations of the balls are tightly correlated. Moreover, knowing the location

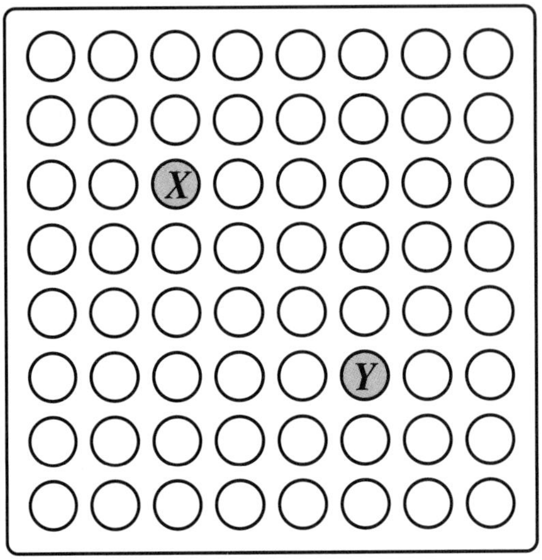

FIGURE 3.3 Lattice of M sites with two sites filled randomly.

of one ball allows us to predict the location of the other with reasonable certainty. This leads us to the concept of *conditional probability* and *conditional uncertainty*.

Define the probability to find Y in state y_j, *given* that X is in state x_i, as $p(y_j|x_i)$. The vertical bar in the probability is used to express this conditioning, and the conditional probability is usually read as "y knowing x," or "y given x." In the previous example, we clearly have

$$p(y_j|x_i) = 0 \quad \text{if } y_j \text{ not adjacent to } x_i, \tag{3.11}$$

$$p(y_j|x_i) = \frac{1}{4} \quad \text{if } y_j \text{ adjacent to } x_i\ . \tag{3.12}$$

Armed with such probabilities, we can define *conditional* uncertainties, expressing the uncertainty about a variable when some other variable is known. For example, we can define the uncertainty of Y given $X = x_i$:

$$H(Y|X = x_i) = -\sum_j^{N'} p(y_j|x_i)\log p(y_j|x_i)\ . \tag{3.13}$$

Note the particular construction of this uncertainty: it looks just like a conventional uncertainty but with the ordinary probabilities replaced by conditional probabilities. The quantity $H(Y|x_i)$ is read as the "uncertainty of Y given X takes on the particular value $X = x_i$." The conditional uncer-

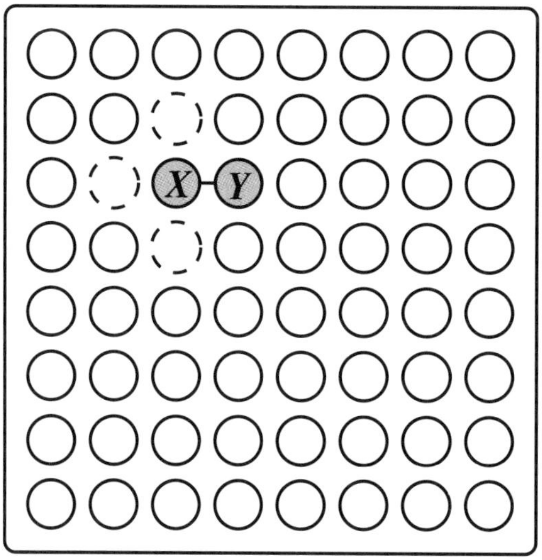

FIGURE 3.4 Lattice of M sites with two sites filled in a correlated manner.

tainty $H(Y|X)$, which is the uncertainty about Y knowing X (X taking on *any* value), is just the *average* of (3.13) over all the possible outcomes of X:

$$H(Y|X) = -\sum_i^N p(x_i)H(Y|x_i)$$

$$= -\sum_i^N \sum_j^{N'} p(x_i)p(y_j|x_i)\log p(y_j|x_i)\ . \tag{3.14}$$

The last expression can be simplified noting that

$$p(x_i)\,p(y_j|x_i) = p(x_i, y_j)\ , \tag{3.15}$$

i.e, that the conditional probability for y_j to occur given x_i occurred *multiplied* by the probability that x_i occurred is just the probability that both x_i and y_j occur simultaneously. This relation (also known as Bayes' Theorem) is used often and effectively in Information Theory. We can then write the expression for the conditional entropy in the usual manner:

$$H(Y|X) = -\sum_i^N \sum_j^{N'} p(x_i, y_j)\log p(y_j|x_i)\ . \tag{3.16}$$

Using (3.15) and its twin obtained by exchanging x and y, one can prove the relations between conditional and joint entropies:

$$H(X, Y) = H(X) + H(Y|X) \tag{3.17}$$

$$= H(Y) + H(X|Y) . \tag{3.18}$$

Before introducing the all-important concept of information, let us briefly mention a property of conditional entropies, namely that the revelation of the value of X does not ever *increase* the uncertainty about Y:

$$H(Y|X) \leq H(Y) . \tag{3.19}$$

It is easy to see that the *equality* holds if and only if the variables X and Y are independent, i.e., if $p(x_i, y_j) = p(x_i)\, p(y_j)$. Then, indeed, revealing the value of X does not change the uncertainty about Y. Let us now pinpoint this correlation between random variables, and see how it leads to the concept of information.

3.5 Information

Information is defined as the *correlation entropy* (or *mutual entropy*) between two random variables or two sets of random variables. As before, take two random variables X and Y, with joint entropy $H(X, Y)$. The information shared between the two (start keeping in mind from here on that information is *always* shared between two ensembles) is

$$I(X{:}Y) = H(X) + H(Y) - H(X, Y) . \tag{3.20}$$

Rewriting this as $H(X, Y) + I(X{:}Y) = H(X) + H(Y)$, we realize that the information is that piece of entropy that would have to be added to the joint entropy in order for the combined system to have the same entropy as the sum of the entropies of the subsystems X and Y. In other words, if the joint entropy is *not* equal to the sum of the entropies of its subsystems, there is a correlation between these subsystems, and this correlation gives rise to information.

We can also examine this from another point of view. If there are correlations between the subsystems, then by possessing knowledge about one of the systems, you can extract knowledge about the other. Indeed, using some of the formulas derived in the previous section, we find that

$$I(X{:}Y) = H(X) - H(X|Y)\ , \tag{3.21}$$

i.e., information is the difference between the entropy of X and the conditional entropy of X *given* Y. To distinguish the correlation entropy (or *mutual* entropy, or sometimes *mutual information*) $I(X{:}Y)$ from the other entropies introduced earlier, we put a colon (:) between the two parts, here X and Y. Note that mutual entropy is symmetric, i.e., $I(X{:}Y) = I(Y{:}X)$, unlike the conditional entropy. To show that $I(X{:}Y)$ has the functional form of an entropy, we can introduce a mutual probability

$$p(x_i : y_j) = \frac{p(x_i)p(y_j)}{p(x_i, y_j)}\ , \tag{3.22}$$

and find with ease

$$I(X{:}Y) = -\sum_i^N \sum_j^{N'} p(x_i, y_j) \log p(x_i : y_j)\ . \tag{3.23}$$

Before illustrating the concept of information with examples, let us consider a diagrammatic way of summarizing the relationship between all the entropies introduced so far. We shall think of entropies as *areas* of circles that can intersect with other circles (the entropy of other systems). If the intersection between the circles of X and Y, say, is zero, the variables are uncorrelated, while the envelope of the two circles represents the joint entropy of the two. Thus, for X and Y we can draw (in general) a diagram, as in Fig. 3.5. This can easily be extended to represent the relationships between the entropies of *three* systems X, Y, and Z, as in Figure 3.6. The notation introduced for mutual entropies *conditional*

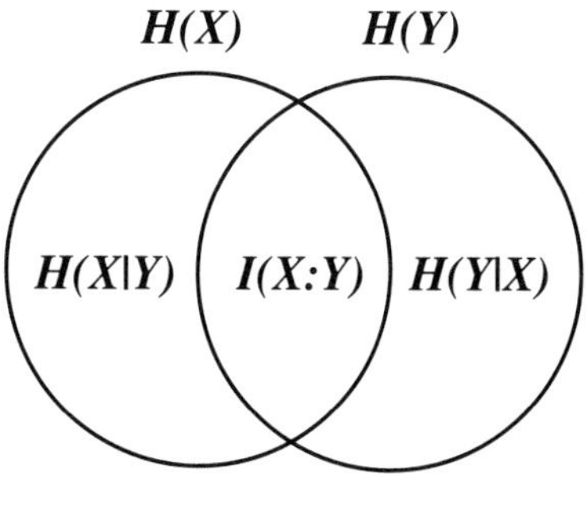

FIGURE 3.5 Entropy Venn diagram for two random variables X and Y.

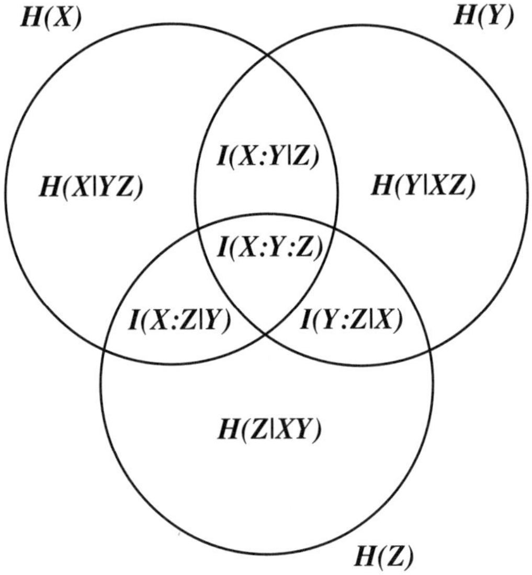

FIGURE 3.6 Entropy Venn diagram for three random variables X, Y, and Z.

on a third variable [$I(X{:}Y|Z)$, for example], or entropies conditional on joint variables [such as $H(X|YZ)$] should be self-explanatory. Note that all entropies introduced so far are *positive semi-definite*, i.e., they are strictly positive or zero, but never negative. The lone exception is the *ternary* mutual entropy $I(X{:}Y{:}Z)$, which can be negative (see Problem 3.2). This is so because it represents shared entropy between more than two systems. Note also that the positivity of some of the entropies introduced here (notably the conditional one) is not guaranteed if *quantum* variables are involved [Cerf and Adami, 1997] (see also Problem 3.3).

Let us now calculate the information for some of the examples introduced earlier. In the case where two balls labeled X and Y are dropped randomly on the lattice (Fig. 3.3), we found that the joint uncertainty (entropy) was just a little smaller than the sum of the uncertainties for each of them. Thus, we expect the amount of correlation, or information, to be small. Indeed,

$$\begin{aligned} I(X{:}Y) &= H(X) + H(Y) - H(X,Y) \\ &= \log(M) + \log(M) - \log[M(M-1)] \\ &= -\log\left(1 - \frac{1}{M}\right). \end{aligned} \tag{3.24}$$

Thus, in the limit of large M, $I(X{:}Y) \approx 1/M \to 0$. The situation is different when a string is tied between the two balls (Fig. 3.4). In that case

$$I(X{:}Y) = H(X) - H(X|Y) = \log\frac{M}{4}\,, \tag{3.25}$$

where we used the fact that the conditional entropy for the second ball is $\log 4 = 2$ bits. We could have calculated this directly using the conditional probability (3.12).

Finally, let us introduce a notation for the entropies showing that they are really the same as the corresponding entities in statistical mechanics (introduced in the next chapter):

$$H(X) = -\langle \log p(x)\rangle\,, \tag{3.26}$$

$$H(X|Y) = -\langle \log p(x|y)\rangle\,, \tag{3.27}$$

$$I(X{:}Y) = -\langle \log p(x:y)\rangle\,, \tag{3.28}$$

where the brackets, as usual, indicate statistical averages. Note, however, that in equilibrium statistical mechanics and thermodynamics, the conditional and mutual entropies play hardly any role, as all those correlations that are not *permanent* disappear once one has reached equilibrium (compare section 4.9). Thus, as $I(X{:}Y) = 0$ in equilibrium, conditional entropies $H(X|Y)$ and *unconditional* entropies $H(X)$ are then identical. Still, if there are permanent correlations (such as the walls of a container that do not allow a gas to escape), the entropies written down in thermodynamics really are conditional entropies, conditional on the boundary condition imposed by the fixed correlations. In this respect, the only unconditional entropy is that of a completely isolated system. Still, conditional and mutual entropies do play a role in *nonequilibrium* thermodynamics. We shall exploit this later with respect to the evolution of complexity and the concept of measurement.

3.6 Noiseless Coding

Armed with the concepts of entropy and information, we can now investigate how good or bad a certain *code* used to encode information is, and how much a code could be improved. Our first concern is to make messages as short as possible, as the longer messages are, the more target they offer to noise. To investigate this, we again use our random variable X that can be in states $x_1 \cdots x_N$ with probabilities $p_1 \cdots p_N$, and translate these messages into codewords formed from an alphabet $\{a_1 \cdots a_M\}$, e.g., $x_1 \to a_3a_1, x_4 \to a_2a_1a_3$, etc. If we associate to each message x_i a codeword of length n_i, we arrive to the *average codeword length*

$$\langle n \rangle = \sum_i^N p_i \, n_i \; . \tag{3.29}$$

We would like to find then, of course, the code that *minimizes* the average codeword length. First, we need to mention, however, that not just any assignment $x_i \to a_k a_l a_m \cdots$ corresponds to a viable code. Indeed, a code must be made in such a way that it is uniquely *decipherable*. An example of an undecipherable code is given below:

msg	code
a	0
b	010
c	01
d	10

Clearly, the messages *b*, *ca*, and *ad* are all coded as 010, preventing an unambiguous decoding. This is because in the code above, some codewords appear as *prefixes* to other codewords, like 0 and 01 are both prefixes to 010. Then, a necessary condition for a good code is that no codeword can be a prefix to another codeword. Such codes are called *instantaneous*, or *step-by-step decodable*. If we confine ourselves to codes of that nature, we can state the following theorem:

A code made from an alphabet of size D and codeword lengths n_i to encode messages exists if and only if

$$\sum_i^N D^{-n_i} \leq 1 \; . \tag{3.30}$$

We can quickly check if this works for DNA, as we *know* that this code exists. For DNA, the length of each codeword n_i is 3 (these codewords are also called *codons*). The size of the alphabet is, of course, $D = 4$, and the set of messages is the 20 different amino acids making up the polypeptides. Thus, $N = 20$, and we find

$$\sum_{i}^{N} D^{-n_i} = 20 \cdot 4^{-3} = \frac{20}{64} < 1 \,. \tag{3.31}$$

Thus, this (slightly trivial) example teaches us that with the genetic code, we could have coded up to 64 different amino acids. Nature has not chosen to do this, of course, but rather has opted for a certain amount of redundancy. Remember that the above theorem only guarantees the *existence* of a code, not how well it holds up under noise. Let us now answer the more difficult question of how small the average codeword length can be made to be. This is the content of Shannon's *Noiseless Coding Theorem*, which states that there is a *lower* bound to $\langle n \rangle$, essentially given by the entropy of the source

$$\langle n \rangle \geq \frac{H(X)}{\log D} = -\sum_{i}^{N} p_i \log_D p_i \,, \tag{3.32}$$

where we have indicated on the righthand side the logarithm to the base D, the size of the alphabet. The equality holds when the source is *equiprobable*, and $p_i = 1/D^{n_i}$. Rather than proving this theorem (as is done in all standard Information Theory textbooks), we discuss the average codeword length of DNA in Box 3.2 as an example.

An *optimal* code is an instantaneous code for which

$$\frac{H(X)}{\log D} \leq \langle n \rangle \leq \frac{H(X)}{\log D} + 1 \,. \tag{3.33}$$

With the average codeword length $\langle n \rangle = 2.105$ for human DNA derived in Box 3.2, we conclude that DNA is an optimal code!

3.7 Channel Capacity and Fundamental Theorem

This section covers communication channels whose inputs are subject to random disturbances in transmission. If we view *inheritance* of genomic information as just such a transmission channel, we can develop a formalism that allows us to judge how well Nature has mastered communication over noisy channels.

If X is the source of amino acids $x_1 \cdots x_{20}$, let us calculate the minimum codeword length under the assumption that all amino acids occur with the same probability $p_i = \frac{1}{20}$. This is, of course, not true in reality as the abundances vary between 1% and 10% (equiprobability would imply a uniform abundance of 5%). As we shall see, the assumption of equiprobability, however, is not too bad. In that case the entropy of the source is

$$H(X) = \log_2 20 = 4.322 , \tag{3.34}$$

such that

$$\frac{H(X)}{\log D} = \frac{4.322}{2} = 2.16 . \tag{3.35}$$

Thus, the minimum codeword length would be $\langle n \rangle \geq 2.16$. Seeing that not all amino acids occur with the same probability, we can cull the probabilities for specific families of organisms from tables (see, e.g., Brown, 1991 and Box 3.3). For humans, the abundances vary between 1.38% (trypsin) to 9.56% (leucine). There is also, in fact, a *stop* amino acid that signals the end of a string. As it occurs with a very small probability (0.24%), we can ignore it here. With the specific measured abundances, the entropy of the source is lowered as expected, and we find

$$H(X) = -\sum_{i=1}^{20} p_i \log_2 p_i = 4.2108 . \tag{3.36}$$

The minimum length, then, is $H(X)/\log D = 2.105$. Thus, a code of average length 2 can *almost* exist. If some amino acids were more dominant (exceeding 10%, for example), while a few others were correspondingly more rare, $\langle n \rangle = 2$ could easily be achieved. Note, however, that it would come at the expense of codes with varying codeword length, a scheme with logistics too daunting for Nature to adopt.

BOX 3.2 Noiseless Coding in DNA

A channel is a device that translates a coding $\{a_1, \ldots, a_n\}$ to a coding $\{b_1, \ldots, b_m\}$ with probability distributions P_n associated with the translation of each symbol. Thus, we can construct a *channel matrix*

$$a_{ij} = p(b_j|a_i) , \tag{3.37}$$

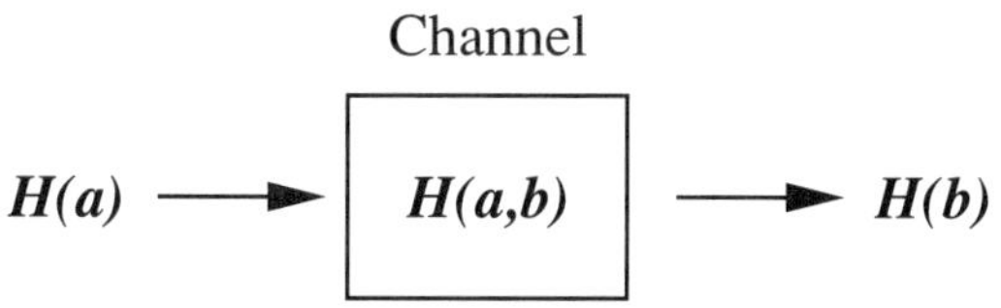

FIGURE 3.7 Noisy channel with channel matrix $p(b_j|a_i)$.

where $p(b_j|a_i)$ is the probability to obtain output b_j, given input a_i. If we enter a channel with a probability distribution on the input alphabet with uncertainty $H(a)$, we end up with a probability distribution on the output with entropy $H(b)$ (see Fig. 3.6). The noisiness of the channel can then be described by the joint uncertainty $H(a, b)$. If there is *no* correlation between a and b, i.e., between input and output, we are dealing with a *useless* channel, and

$$H(a, b) = H(a) + H(b) . \qquad (3.38)$$

On the other hand, if there is *complete* correlation, we are dealing with a *perfect* channel, and

$$H(a, b) = H(a) = H(b) . \qquad (3.39)$$

The previous can also be expressed in terms of the mutual entropy, or information $I(a{:}b)$. For a useless channel we have $I(a : b) = 0$, whereas for a perfect channel $I(a : b) = H(a) = H(b)$.

Having described these simple limiting cases, we can turn to the central question of Information Theory. What is the maximal amount of information that the channel can process (with arbitrary precision) in the presence of noise? To answer this question, Shannon introduced the concept of the channel *capacity*. In words, the channel capacity is the maximal mutual entropy between input and output distributions, where the maximization is performed over the input distribution:

$$C = \max_{p(a)} I(a : b) . \qquad (3.40)$$

Shannon's fundamental theorem (which we will not prove here) can now be stated in terms of this channel capacity: "It is possible to transmit information at a rate less or equal to the channel capacity with arbitrarily small probability of error." This is a somewhat surprising result, as this theorem does not involve the amount of noise in the channel. It contradicts the notion (widely held before Shannon) that as the noise goes up, the rate of impeccable information transmission must go to zero. Rather, Shannon stated that it is bounded from below by a constant, the channel capacity. This can be achieved by means of *redundancy* and error-correction.

An obvious strategy is suggested by the form of the channel capacity for *symmetric* channels, i.e., channels in which the error probability does not depend on which bit goes through the channel. In that case, $H(a|b)$ is independent of $p(a)$, and it is then sufficient to arrange for a coding in such a way that every symbol received at the output occurs with about equal probability. Then, $H(b)$ is maximized, and consequently also $I(a{:}b)$. Let us consider an example of coding that uses redundancy. Suppose we send messages of length 2 bits. Theoretically, we can use the alphabet $\{0, 1\}$ to code for the messages $\{a, b, c, d\}$:

$$a \to 00$$
$$b \to 01$$
$$c \to 10$$
$$d \to 11$$

However, any bit-flip in the coding will make the message irretrievable. A redundant code would use, for example,

$$a \to 00000$$
$$b \to 01101$$
$$c \to 11010$$
$$d \to 10111.$$

In this coding, 2 bit-flips have to occur before decoding becomes ambiguous. Error correction can be performed on such codes, as for example the erroneous code 11111 can unambiguously be traced back to the string 10111 (no other string could have given rise to 11111 if only one error occurred). In that case, a simple algorithm can flip back the bit to restore the string 10111. Redundancy is also used in genetic channels to protect against errors (see Box 3.3). More importantly, however, Nature uses the double-strandedness as a very effective way of protecting against errors. As nucleotides always have to be paired, an erroneous substitution can be detected as long as only one member of the pair is affected. A very sophisticated molecular error-correction mechanism can detect such errors and correct them.

The following table shows the translation between RNA codons and amino acids in humans. The relative abundance of the amino acids is indicated (in percent) behind the three-letter code for the amino acids.

Am.-acid ([%])	Codings
ARG (5.28)	CGA CGC CGG CGU AGA AGG
LEU (9.56)	CUA CUC CUG CUU UUA UUG
SER (7.25)	UCA UCC UCG UCU AGC AGU
THR (5.68)	ACA ACC ACG ACU
PRO (5.67)	CCA CCC CCG CCU
ALA (6.99)	GCA GCC GCG GCU
GLY (7.10)	GGA GGC GGG GGU
VAL (6.35)	GUA GUC GUG GUU
LYS (5.71)	AAA AAG
ASN (3.92)	AAC AAU
GLN (4.47)	CAA CAG
HIS (2.36)	CAC CAU
GLU (6.82)	GAA GAG
ASP (5.07)	GAC GAU
TYR (3.13)	UAC UAU
CYS (2.44)	UGC UGU
PHE (3.84)	UUC UUU
ILE (4.50)	AUA AUC AUU
MET (2.23)	AUG
TRP (1.38)	UGG
STOP (0.24)	UAA UAG UGA

Note that the most abundant amino acids usually have the most redundancy. For example, the rare amino acids MET and TRP are only represented by a unique codon. An exception is the twenty-first combination STOP, which signals the end of a protein sequence. It must be relatively rare, but it is extremely important for the functioning of the organism that the STOP sequence is read correctly. As a consequence, it has a much larger redundancy than its abundance would suggest.

BOX 3.3 Actual Coding in Human RNA

3.8 Information Transmission Capacity for Genomes

Let us now turn to self-replicating genomes and view them as information transmission channels. In essence, we consider a genome at time t as the *source* of messages, and the genome at time $t+1$ as the received message, where the unit of time is the time it takes to copy the genome once, i.e., the gestation time of the genome. Let the genome be of length ℓ, and the copy process be flawed with a rate R. Then, the fidelity F of the copy process is

$$F = (1 - R)^{\ell} . \tag{3.41}$$

The information processed by the channel is just the mutual entropy between the population of genomes at time t and at time $t+1$:

$$H(t+1 : t) = H(t+1) - H(t+1|t) , \tag{3.42}$$

where $H(t+1|t)$ is the conditional entropy of the population at $t+1$, given the population at time t. In extreme simplification, this can be calculated using the copy-fidelity F. The channel matrix (3.37) can be written in such a manner that the diagonal terms are just the conditional probabilities that the genome is copied correctly, whereas all other elements indicate the probability for copy errors. Let us imagine an equilibrated situation, where each genome can take on N states. Note that this idealizes the real process tremendously, as we keep the number of possible states N constant, and we treat the copy process simply as an error rate acting on the symbols. The channel matrix, then, is

$$\begin{pmatrix} F & \frac{1-F}{N-1} & \cdots & \frac{1-F}{N-1} \\ \frac{1-F}{N-1} & \ddots & \cdots & \frac{1-F}{N-1} \\ \vdots & \cdots & F & \vdots \\ \frac{1-F}{N-1} & \cdots & \frac{1-F}{N-1} & F \end{pmatrix} .$$

The conditional entropy is just the entropy of one of the rows (as this is a *symmetric* channel):

$$H(t+1|t) = -F \log F - (1-F) \log \frac{1-F}{N-1} . \tag{3.43}$$

The entropy of the population at any time $t+1$ can be estimated from the *equilibrium* distribution at time t, and using the conditional probabilities to obtain the distribution at time $t+1$. The entropy at time t, on the other hand, can be estimated from Eigen's model of self-replicating molecules (see Section 11.3), which indicates that almost all molecules have abundance

$$\rho_i = \frac{1-F}{N}\,, \tag{3.44}$$

whereas one (the quasi-species) has abundance F. The entropy $H(t+1)$ is then obtained with the probabilities

$$p_1 = F^2 + \frac{1-F}{N}\,, \tag{3.45}$$

$$p_i = \frac{(1-F)^2}{N-1} - \frac{1-F}{N(N-1)}\,, \tag{3.46}$$

so that in the limit $N \to \infty$,

$$H(t+1:t) = H_2[F^2] - H_2[F] + F(1-F)\log N\,. \tag{3.47}$$

For large N, this function is dominated by the last term, so that

$$H(t+1:t)/\log N \approx F(1-F) \approx e^{-R\ell}(1-e^{-R\ell})\,, \tag{3.48}$$

which is shown in Fig. 3.8 for different string lengths ranging from 30 to 150, and for mutation rates $0 < R < 0.1$. Note that the function peaks at $F = \frac{1}{2}$, i.e.,

$$R\ell \approx \log_e 2\,. \tag{3.49}$$

For fidelity $F \approx \frac{1}{2}$, i.e., mutation probabilities of about 0.7 times the inverse length of the sequence, the channel transmits the optimum amount of information. This capacity then acts as a bound on the rate of information acquisition in genomes. There can be no process that achieves a higher rate of arbitrarily accurate information processing in this channel. We shall show in Chapter 11 that for more realistic channels (where the probability to mutate to another sequence is *not* uniform), we find a different optimal rate, closer to $R\ell = 1$.

It is worth noting that such a channel is quite peculiar and unlike any channel usually encountered in information theory, as the entropy of the source ensemble is tied to the error rate $1-F$. Thus, if the copy process were perfect ($F = 1$), the channel still has no capacity here,

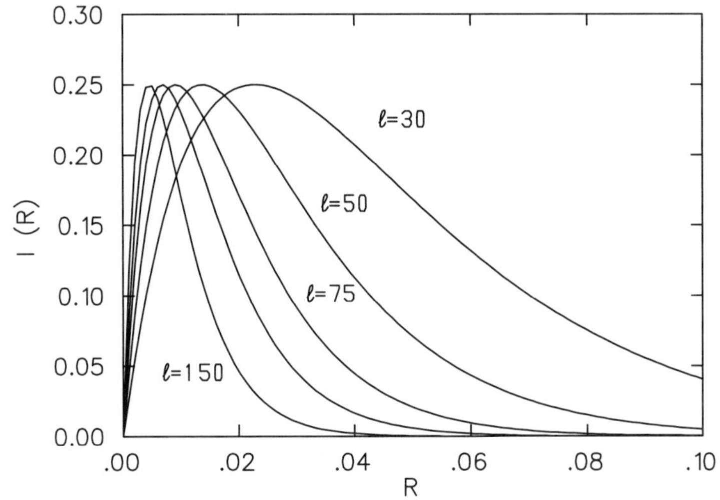

FIGURE 3.8 Mutual entropy calculated from Eq. (3.48) for $\ell = 30, 50, 70,$ 100, and 150.

because for $F = 1$ the population has (asymptotically) *zero* entropy, and thus cannot share any entropy with another system (see Problem 3.5 for the equivalent binary symmetric channel). The mutual entropy here is therefore *not* that amount of information that one genome passes along to the next generation, but rather the information *processed* by the channel, which in this case puts a limit on how much information the genome can *acquire and store*. Thus, it measures how much the genomes can *learn*. In order to calculate how much information is passed from generation to generation, we need to take into account the *redundancy* due to the many copies of the genome in the population. As this number is usually very large, the entire genome is usually passed on through every generation, a virtually perfect channel.

3.9 Overview

This chapter introduced the concepts of entropy and information in order to study the statistical aspects of information acquisition and storage of genetic systems. Information is *not* the description of an object in terms of bits, or the number of bits necessary to describe an object, but rather the mutual entropy between two ensembles. In other words, information measures the amount of *correlation* between two ensembles, which allows you to make predictions about one ensemble armed only with probabilities garnered from another. Limits on error correction and reliable information transmission were introduced, and finally, the genetic replication process was investigated in terms of an information transmission channel. There, we found that genetic channels are very different from the ordinary ones, because the process that corrupts the messages also controls how much entropy is being sent across the channel. Thus, a channel with a perfect copy-process turns out not be a channel at all, as all messages are identical in such a population (zero entropy ensemble), and the information transmission capacity vanishes. Thus, it turns out that there is an optimal error rate for genetic channels that allows the maximum rate of information processing. Such a rate is inversely proportional to the length of the code, a proposal that we shall test in Chapter 11.

Problems

3.1 (a) A fair coin is flipped until the first head occurs. Let X denote the number of flips needed until this happens. Thus, X is a random variable. Find the entropy $H(X)$ in bits.

(b) If a random variable is drawn from this distribution, find an efficient sequence of yes/no questions to determine the value of that variable. Compare $H(X)$ to the expected number of questions required to determine X.

3.2 (a) The mutual entropy between two random variables $H(X{:}Y)$ is positive semidefinite (always larger than or equal to zero). The *ternary* mutual entropy between *three* random variables can be negative, however (see Fig. 3.6). Find binary random variables X, Y, and Z such that $H(X{:}Y{:}Z) = -1$ bit. (Hint: think of logical operations.)

(b) Express (use Venn diagrams!) $H(X{:}Y{:}Z)$ in terms of the entropies of each of the random variables alone, mutual entropies between all pairs of variables, and the joint entropy of all three.

3.3 Two binary random variables are said to be 100% *uncorrelated* if the conditional probability matrix is

	y_1	y_2
x_1	1/2	1/2
x_2	1/2	1/2

,

whereas the variables are 100% *correlated* if

	y_1	y_2
x_1	1	0
x_2	0	1

(a) Draw the entropic Venn diagram for both situations.

(b) Draw the diagram for the *nonclassical* situation where the entropy of each random variable is 1 bit, but where the *joint* entropy of both *vanishes* (so-called EPR pairs).

3.4 For the binary symmetric channel with error rate p, calculate the probability of faulty transmission (1 minus the *fidelity*) of the 3-bit redundant code shown below:

$$0 \to 000$$
$$1 \to 111\,. \tag{3.50}$$

3.5 (a) Again for the binary symmetric channel, find the mutual entropy between messages and received symbols for the case where the probability to find 0 or 1 as a message is *equal* to the *noise* in the channel (the binary analogue to the genetic channel).

(b) Find the probability p that maximizes the mutual entropy.

CHAPTER FOUR

Statistical Mechanics and Thermodynamics

Tout se fait dans le monde par la matière et le mouvement.[1]

René Descartes

This chapter introduces basic concepts and methods from statistical mechanics and thermodynamics. Since our goal is to construct and analyze populations of self-replicating code in order to establish a baseline for minimal living systems, we need a theoretical framework to describe such populations without having to understand in detail the interactions between each and every member. This is precisely the object of statistical mechanics and thermodynamics: to describe the behavior of an aggregate, knowing only the forces between the microscopic constituents. Therefore, statistical physics must be the basis of any theory of complexity.

This chapter can be used in different ways. A reader unfamiliar with the basic constructions of thermodynamics can use it as a primer, but not as a substitute for a thorough study of the subject (see Landau and Lifshitz, 1980 for such a purpose). Alternatively, a reader who has had a basic course in statistical physics can browse through the topics and stop only at the shaded boxes, which contain applications of the theory to populations of self-replicating strings. Elements of statistical mechanics and thermodynamics are used throughout these lectures. The reader is thus

[1] Everything in the world is made from matter and movement.

invited to refresh his or her acquaintance with terms used throughout different chapters with the short expositions presented here.

In statistical physics, we study the specific laws that govern the behavior and properties of macroscopic bodies that are made up of very many microscopic particles (e.g., atoms, molecules, or any units). These laws emerge from a process of averaging over the possible configurations that the system can take on, and are more accurate the larger the number of microscopic particles. Rather than derive these laws (something that is done in most standard textbooks), we shall study their predictions as far as they concern living systems. Theoretically, there is another approach that would let us predict the dynamics of many particles: writing down the system of coupled differential equations that describe them, and solving that system numerically using the initial conditions for each particle. This is difficult for two reasons. On the one hand, the numerical solution of the differential equations would, with current computer technology, take longer than the age of the universe for decent-sized problems and satisfactory accuracy. On the other hand, it appears impossible to specify (or know) the initial conditions accurately enough for a system with a large number of components. Nevertheless, this computational approach to statistical systems has led to important insights in specialized systems, and is of course the only avenue open when such general laws as those of statistical mechanics and thermodynamics are unavailable, i.e., when the interactions are so complicated that a simple (mathematical) averaging over possible configurations is impossible. In the following, we introduce the basic concepts that allow us to describe systems with a large number of degrees of freedom, and try to understand what the circumstances are under which such a description is possible. This will be interspersed by application of these concepts to the statistics of self-replicating strings of code, such as DNA-strings or the bit-strings of our self-replicating computer programs.

4.1 Phase Space and Statistical Distribution Function

In order to accurately describe the behavior of an aggregate of particles (agents, units), we first need to find a characterization of the *essence* of each unit. In statistical mechanics, for example, we deal with point particles, each described by its position and its velocity. In more general systems, we may have to specify more attributes, e.g., a magnetic moment, or a minimum volume occupied by each particle. In any case, the determination of the basic *degrees of freedom* of each subunit is going to determine the accuracy of the general laws that are going to emerge. In the simplest situation, the specification of position and momentum is enough: the state of each particle is determined by the pair

$$(\vec{q}, \vec{p}) \, , \tag{4.1}$$

i.e., by six numbers (in three dimensions). Then, the state of a macroscopic number of particles is specified by those six numbers for each and every particle in the system, i.e., (q_i, p_i) for each particle i (we omit the vector notation from now on and think of each q_i or p_i as representing three numbers).

In order to follow the evolution of each particle's *state*, we construct a space in which every state is just one point: this is the *phase space*. For example, in one dimension, the temporal evolution of one particle can be described by a curve in (q, p) space, as depicted in Fig. 4.1. Thus, each particle traces out a trajectory in phase space given by the coordinates $[q(t), p(t)]$. Let us now imagine a two-particle system. Each state of the system is then specified by the numbers (q_1, p_1, q_2, p_2), and the trajectories could be visualized by plotting the pair (q_1, p_1) versus (q_2, p_2), in a four-dimensional space. The projection of an example trajectory onto two dimensions is depicted in Fig. 4.2. Note that there we draw the trajectory as a closed curve: often the possible trajectories recur, as for example

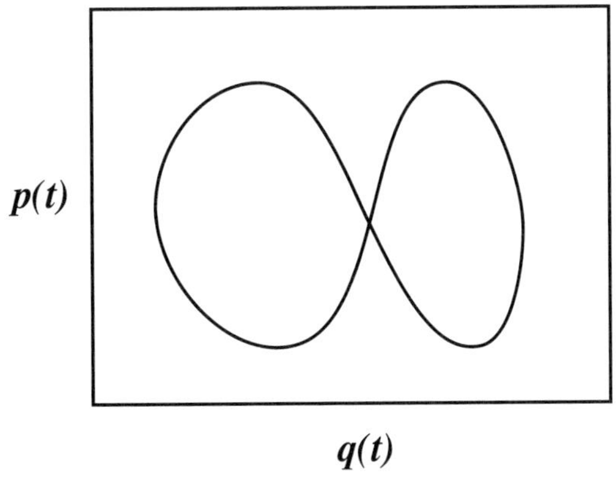

FIGURE 4.1 Trajectory in phase space for one particle.

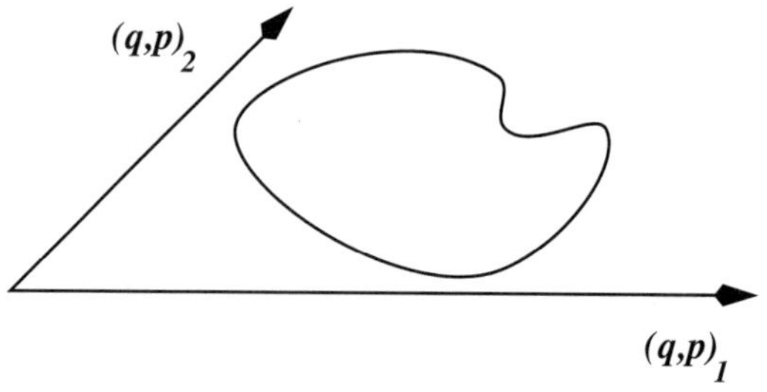

FIGURE 4.2 Trajectory in a two-particle phase space.

in periodic motion. The phase space, then, is a finite volume of points (if time is discretized, of course) that describes all possible states and all possible motions of the system. In general, the phase space for N particles is extremely high-dimensional, yet it is often imagined as a cloud of points (where each point represents *one* possible state of the system) and where the density is proportional to the system's probability to occupy this state (see Fig. 4.3).

Let us examine a small piece of this volume, $\Delta q\,\Delta p$, at location q, p. In essence, we are examining all those possible configurations of the system for which the positions are between q and $q + \Delta q$, and the momenta between p and $p + \Delta p$. Clearly, if the system is evolving in time, it will spend only a small amount of time in this little box. Let us define then the time Δt that it spends in this region if we observe the system for a very long time, T. In this case, we can say that the *probability* w for that system to be in a state with coordinates between q and $q + \Delta q$, and momenta between p and $p + \Delta p$, is

$$w = \lim_{T\to\infty} \frac{\Delta t}{T} \,. \tag{4.2}$$

Now we can take the limit to infinitesimal box sizes, and write for the volume element of each box,

$$dq\,dp = dq_1, \ldots, dq_N\,dp_1, \ldots, dp_N \,, \tag{4.3}$$

where we make explicit the fact that this volume element has very many dimensions, and write for the infinitesimal probability

$$dw = \rho(q_1, \ldots, q_N, p_1, \ldots, p_N)\,dq\,dp \,. \tag{4.4}$$

According to this formula, the probability dw to find the system in a state with coordinates between q and $q + dq$, and momenta between p and

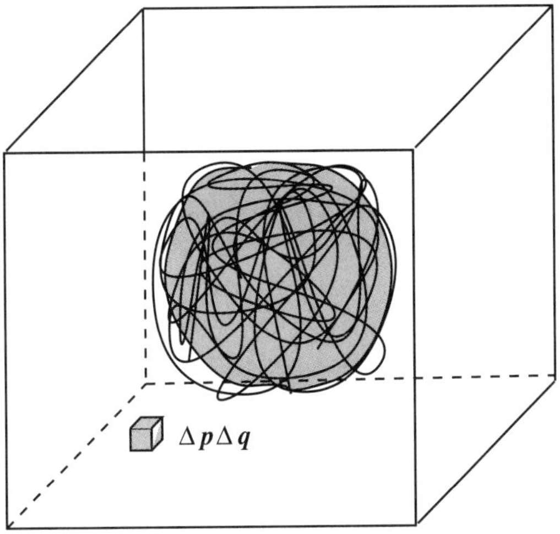

FIGURE 4.3 Density of trajectories in phase space.

$p + dp$, can be obtained from the density of states $\rho(q, p)$: the statistical distribution function. It describes the density of points in phase space that can be reached by trajectories. Areas with very high density are, on average, visited much more often than areas with low density. Therefore, areas of phase space that have zero density are strictly forbidden: the system can never be found in such a state, no matter how long you wait. Naturally, the statistical distribution function is normalized, meaning the probability to find the system in *any* state is 1:

$$\int \rho(q, p)\, dq\, dp = 1 \; . \tag{4.5}$$

It is important to note that, under conditions pinpointed in the next section (the *ergodicity* condition), the distribution function ρ does *not* depend on the initial state of the system, i.e., the probability to find the system in state (q, p) does not depend on which coordinates and momenta the system was started with initially. This is because under such conditions, the trajectories (which can be thought of as tracing out the density of the cloud in Fig. 4.3) traverse *all* possible points eventually, and thus each of these points could have been used as the system's starting condition.

4.2 Averages, Ergodicity, and the Ergodic Theorem

Now let us turn to the advantages of having a statistical distribution function for a statistical system. For example, suppose we wish to know the average of a certain *physical* quantity (such as the energy) that is a function of the coordinates and momenta, averaged over the many many particles that make up our statistical system. Rather than measuring this quantity for each particle in the system over time and averaging them, it turns out that we only need to average over all possible *states* that the system can take on, i.e., we need only to average over the statistical distribution function. Thus, the average for a quantity $f(q,p)$ is denoted by $\langle f \rangle$ and given by

$$\langle f \rangle = \int f(q,p)\,\rho(q,p)\,dq\,dp\,, \tag{4.6}$$

which can easily be obtained if we know the function $\rho(q,p)$. Because of the way we defined ρ, we could surmise that $\langle f \rangle$ could just as well have been obtained by averaging $f(q,p)$ over time:

$$\langle f \rangle = \lim_{T\to\infty} \frac{1}{T} \int_0^T f[q(t),p(t)]\,dt\,. \tag{4.7}$$

However, this is not always the case. Imagine that there is a region of phase space that *cannot* be accessed from every initial condition (maybe because there is a barrier of some sort blocking the development into that region). Then, the two averages (4.6) and (4.7) are not equal, and the system is called *nonergodic*. Thus, ergodicity is a property of systems whose statistical distribution function is traced out by a *single* trajectory. More mathematically, we can say that a system is ergodic if the trajectory of the system in phase space, observed at small enough time interval Δt, approaches the statistical distribution $\rho(q,p)$ arbitrarily well as $T \to \infty$. The idea of ergodicity will become very important when we consider the phase space of living systems: genotype space. Such a space was first introduced by Maynard Smith (1970) in the guise of a *protein space*. The idea is that all sequences can be assigned a point in a high-dimensional lattice, where neighboring sequences differ only at one position. Even though we may think that all possible genomes can be obtained by mutations from any other genome (ergodicity of genotype space), this is so only theoretically (see Box 4.1). Practically, such a process would take much longer than many many ages of our universe; thus such systems are practically

If we assume that evolution is a random process (random walk) in genetic space, is there a chance that the entire volume (all possible genomes) is sampled over the age of the universe? To answer this question, let us conduct a *thought experiment* and imagine that the DNA strings consist of 10^9 nucleotides. If we assume a replication rate of one nucleotide per second [Eigen, 1985], it follows that only about 10^{60} configurations can be sampled by all such genomes in parallel, during the age of the universe. On the other hand, for strings of length 10^9, there are 4^{10^9} possible arrangements, far outnumbering the number that can be tried (no matter what replication rate we assume). Thus, the process of evolution is *highly* nonergodic, and it can be said that only a *negligible* fraction of genetic space will ever be explored. Also, this means that an area in genetic space that has been abandoned (via extinction) will, with all probability, never be visited again. As a consequence, a statistical treatment of evolution can *never* make any predictions as far as the global outcome of the evolutionary process is concerned (e.g., What is the most probable genome after 4 billion years?). Still, the statistical method can be applied to the *subsystems*, i.e., small populations. We can thus imagine the evolutionary process as guiding many independent clouds of points through the vast genetic space, each cloud never dissipating enough to cover any significant amount of that space, but sometimes splitting never to merge again (the process of speciation). On small time-scales, the clouds seem to drift aimlessly around (performing a Gaussian random walk), while on large time-scales one can observe a jumpy behavior that is related to sudden bursts of invention. This phenomenon will be highlighted throughout the book from several different perspectives.

BOX 4.1 Ergodicity in Genetic Space?

nonergodic. Clearly, in nonergodic systems the *history* of the system is important, and the state of the system at any time will usually depend on this history. We will learn more about this and its consequences later. In the meantime, let us investigate the important concept of statistical and thermodynamical equilibrium.

4.3 Thermodynamical Equilibrium, Relaxation

From the previous discussion it is apparent that predictions obtained with the statistical distribution function are *probabilistic* in character, unlike the deterministic laws we are used to from classical mechanics. This is because we are interested in making predictions about the macroscopic behavior of a large number of units using only a limited amount of information. In this statistical treatment, the probabilistic nature of the laws we derive, however, is hardly noticeable: the macroscopic body will take on the most probable value with close to certainty.

Using the statistical distribution function $\rho(q,p)$, we can construct a probability distribution function for the quantity $f(q,p)$ that will tell us which of the values f is the most probable, as a function of the state (q,p). This distribution function turns out to be very strongly peaked around the average value $\langle f\rangle$, which is a consequence of the law of large numbers (or central limit theorem). In essence, the law of large numbers guarantees that the sum of a large number of independent *random* variables approaches a peaked, Gaussian, distribution. It is instructive to see this law at work by summing, for example, the output of a random number generator. Even though each number is identically distributed (the probability to obtain a number between 0 and 1 is the same), the sum of these random numbers converges toward a Gaussian distribution peaked at $\frac{1}{2}$. Indeed, the fact that the probability distribution functions of physical variables become so very strongly peaked ("delta-function–like") around their average value for large systems is really the reason why the statistical description is so accurate. Also, the deviations from this average (fluctuations) become exceedingly small in this limit:

$$\sigma^2(f) = \langle f^2\rangle - \langle f\rangle^2 \sim 10^{-N} . \tag{4.8}$$

Using this concept, and the concept of a *subsystem*, we can define what it means for a system to be in *equilibrium*. The general idea will be that a system is in equilibrium if in every one of its subsystems (an arbitrary contiguous piece of the full system), the mean value for its physical quantities is close to the mean value of these quantities for the *whole* system. The whole system is then said to be in *thermodynamical equilibrium*. Note that this does not hold for all possible measurable quantities. As stated earlier, we have to be able to assume that the contribution to the overall average from each subsystem is *independent*, and the variables we observe must be *additive* (i.e., the value of the observable for a combined system is just the sum of the variables in each of the systems). For this to be the case, the system may have to be observed for a sufficient amount of time so that the influence of any starting condition or the weak interaction between subsystems has had a chance to die out. This time is usually different for every system, and depends on the nature of the interactions. For each statistical system, there should therefore exist a characteristic *relaxation time* that determines how long it takes for a system to return to equilibrium after it has been perturbed.

The concepts of thermodynamical equilibrium and relaxation will also become important for our simple living systems, where we might ask, for example, whether the fitness of a subpopulation is the same as the average fitness of the whole population. We expect this to be the case after the system has equilibrated for a time longer than its relaxation time τ. Conversely, for systems that are always off-equilibrium, the relaxation time diverges, i.e., tends toward infinity. We shall find that if a (genetic) system is small enough, the time between perturbations can be smaller than the time it takes to equilibrate. Perturbations are usually introduced into the population as mutations that generate fitter genotypes. However, since the relaxation time in livings systems appears to increase dramatically as the population grows larger, large enough genetic systems tend to be off-equilibrium most of the time.

4.4 Energy

Let us now ask the question of whether we can find a functional form for the statistical distribution function, something that would clearly turn out to be very useful. In order to answer this question, let us state a general theorem (without proof) that will be of even more use later. *Liouville's Theorem* states that, if the statistical distribution function is a sum of observables (measurable physical attributes that are functions of q and p), then each of these observables must be *invariants*, i.e., quantities that must be conserved under time evolution. For the simple systems that we are considering here (particles at rest without angular momentum), there is in fact only one such invariant: the energy. As a consequence of Liouville's Theorem, we may write down the following formula for the statistical distribution function:

$$\log \rho(q,p) = \alpha + \beta\, E(q,p) , \tag{4.9}$$

where $E(q,p)$ is the energy of the state described by (q,p). Note that strictly speaking, for Liouville's Theorem to hold, one must guarantee that there is indeed a function (called the *Hamiltonian*) with certain characteristics (such as positivity), which guarantees the evaluation of the energy in the first place. Also note that in general, a system has a Hamiltonian only if its energy can be determined entirely from the degrees of freedom (here q and p) of the system. While this is always the case for the statistical mechanical systems we are considering here, this must be ascertained before we can extend such a theorem to the genetic realm. The logarithm of the distribution function arises because ρ is obtained from its subsystems by taking products: if subsystem 1 has distribution function ρ_1 and subsystem 2 has distribution function ρ_2, the statistical distribution function of the combined system is given by the product

$$\rho_{12} = \rho_1 \rho_2 , \tag{4.10}$$

such that $\log \rho_{12} = \log \rho_1 + \log \rho_2$. If a subsystem is described by E_1 or E_2, we see that

$$\log \rho_{12} \sim \beta(E_1 + E_2) . \tag{4.11}$$

Equation (4.9) will then allow us to calculate the statistical distribution function if the energy, as well as parameters α and β, are known. What these are will become clear in subsequent sections. We now turn to the important concept of *entropy*.

4.5 Entropy

Entropy is a measure of the *disorder* present in a system, or alternatively, a measure of our lack of knowledge about this system. Such a measure was introduced in the previous chapter for random variables, and the *same* measure is defined here for *physical* systems. At first sight, entropy seems like an odd definition in physics, as it seems to imply that it depends on the state of an observer. As we shall see, this is not the case. The general definition of entropy is such that if no knowledge is specified, the entropy, or disorder of a system, is always maximal, i.e., it is obtained by assuming that the system is in a superposition of all possible states with equal probability for each. In comparison, if an observer has some knowledge about the state of the system, for example, that it can only be found in two of its possible N states, the entropy (which is now conditional on the knowledge of the observer) is much smaller, much in the manner discussed in Chapter 3. Of course *conditional* entropies are *not* independent of the system to which they are coupled (typically an observer).

Let us estimate the number of states that a system can be in, in order to get a handle on such a measure. For this, let us return to our small piece of phase space volume $\Delta q \Delta p$ introduced in Section 4.1. Indeed, the number of states in this little box is proportional to the probability w [see

In our quest to treat genetic strings with the methods of statistical physics, let us attempt to define the equivalent of an energy variable for strings. The most important feature of the strings we are considering here is, of course, that they can self-replicate. Each string is uniquely identified by its *genotype*, the specific sequence of instructions. The genotype then plays the role of the coordinates and momenta in our general discussion, and the space of all genotypes is the equivalent of phase space. For each genotype i, let us define its *replication rate* ϵ_i, which measures number of offspring per unit time, and thus has dimensions of inverse time. A string that takes a long time to replicate has a small replication rate, while a fast-replicating string has a large replication rate. Clearly, in an interacting population whose total number is kept constant, the genotypes with the highest replication rate will survive, while the inferior ones will suffer extinction. Let us then, for each genotype, define a measure of *inferiority* I_i, which depends on the string's own replication rate ϵ_i, and the current highest replication rate $\epsilon_{\rm best}$:

$$I_i = \epsilon_{\rm best} - \epsilon_i \; . \qquad (4.12)$$

In the absence of mutations, the population will always tend towards a state of zero average inferiority, while with mutations present it will simply attempt to minimize the average inferiority. Thus, the inferiority seems to be the analogue of energy in these string systems. For an arbitrary string, $I = 0$ then denotes the *ground state*, while when hit by a mutation, the string will be left in a state with higher inferiority, only to be replaced by a (faster-replicating) string with lower inferiority subsequently—the analogue of decaying from an excited state to a lower energy state. Sometimes, we shall encounter replication rates that rise exponentially with fitness. In such cases, we shall define the inferiority as $I_i = \log \epsilon_{\rm best} - \log \epsilon_i$. *(continues)*

BOX 4.2 Energy for Genetic Strings

Note that the inferiority is defined with respect to a *global* constant (the replication rate of the best string), and in a sudden event where a new best string is discovered can change discontinuously. As we shall see later (see Box. 4.5), such a process can be described as a first-order phase transition. Even though we would be hard pressed to write down a function $I(i)$ that takes a genotype as an argument and returns its inferiority (this would be the analogue of a Hamiltonian), we can still be confident that such a function exists *in principle*, as long as the inferiority of a string depends only on its own genotype and the other genotypes in the population. Under such circumstances, such a fitness landscape might allow us to define a dynamic for the genotypes that involves the minimization of a Lyapunov function (here, the energy function), a cornerstone of any statistical treatment of many bodies. Consequently, an analogue of Liouville's Theorem might hold (at least in between large transitions), and it should be possible to write a statistical distribution function for genotypes that depends on the inferiority.

BOX 4.2 *(continued)*

Eq. (4.2)] that this box is visited by a trajectory traced out by the system. Thus, the number of states $\Delta\Gamma$ is

$$\Delta\Gamma = \frac{\Delta q \Delta p}{k} \,, \tag{4.13}$$

where k is a normalization factor that we shall encounter later. Clearly, the more states in a box, or cell, the higher its *statistical weight* $\Delta\Gamma$. The entropy of some subsystem described by a phase space cell $\Delta q \Delta p$ is then defined simply as the logarithm of its statistical weight:

$$S = \log \Delta\Gamma \,. \tag{4.14}$$

As is obvious from definition (4.13), the entropy is only defined up to a constant, a matter that does not need to worry us now. Roughly, the entropy is defined as the number of *different* states in which the system can possibly exist. If an observer gains knowledge about the system and thus determines that a number of states that were previously deemed probable are in fact unlikely, the entropy of the system (which now has turned into a *conditional* entropy), is lowered, simply because the number of different possible states is then lower. (Note that such a change in uncertainty is usually due to a *measurement*).

Clearly, the entropy can also depend on what we consider "different." For example, one may count states as different that differ by, at most, Δx in some observable x (for example, the color of a ball drawn from an ensemble of differently shaded balls in an urn). Such entropies are then called *fine-grained* (if Δx is small), or *coarse-grained* (if Δx is large) entropies. The difference in entropies depending on whether or not one knows the *distribution* of objects or particles in an ensemble, is exemplified in Box 4.3 for genetic systems. In the following, we try to find the entropy (given the distribution function) for general systems.

Let us then divide our system again into n subsystems, each being characterized by its total energy E_n. We saw in Section 4.4 that the statistical distribution function for each subsystem $w_n(q,p)$ can only be a function of the energy, because it must be a function of conserved quantities, and the energy is the only one available to us. Thus,

$$\log w_n = \alpha + \beta E_n \ . \tag{4.15}$$

Because of the linearity of this expression, we find for the superposition of the n systems:

$$\log w(\langle E\rangle) = \alpha + \beta\langle E\rangle = \langle \log w(E_i)\rangle \ . \tag{4.16}$$

Since the number of states in the combined system is just $\frac{1}{w(\langle E\rangle)}$, the entropy is

$$S = \log \Delta\Gamma = \log \frac{1}{w(\langle E\rangle)} = -\langle \log w_i\rangle \ . \tag{4.17}$$

The last average can be rewritten using the probability to be in subsystem i, w_i, as

$$S = -\sum_i^n w_i \log w_i \ . \tag{4.18}$$

(See also Exercise 4.1.) Note that the former analysis also entails that if we can define the entropy of a subsystem S_a, then the total entropy of the system is given by the sum

$$S = \sum_a S_a \ . \tag{4.19}$$

Quantities for which such an additive law holds are called *extensive*. We encountered another extensive quantity earlier: the energy. An example of a nonextensive (or intensive) quantity is pressure: the combined pressure of two systems at equilibrium, each with pressure P, also has pressure P rather than $2P$.

Let us define the entropy of a population in genetic space. Assume that the genetic phase-space is spanned by sequences of instructions that are numbered $i = 1, \ldots, N_g$, where N_g is the total number of different sequences in the population. If sequences are of length ℓ instructions (say), and each instruction can take on D values ($D = 4$ for DNA), the total number of different strings is D^ℓ. However, since the process of evolution is highly nonergodic, it is not useful to consider this vast space. Rather, let us consider the space that is currently occupied by the population of N strings, where $N \geq N_g$, of course. The volume of that corner in phase space is just the number of different strings (genotypes) in existence now, thus

$$\Delta\Gamma = N_g \; . \tag{4.20}$$

The entropy of the population is then

$$S = \log N_g \; . \tag{4.21}$$

If we have somewhat more knowledge about the population, this entropy can be smaller. Specifically, imagine that we were in possession of the knowledge of how many of each genotype i there are in the population. If n_i denotes the abundance of genotype i in the population, with $\sum_i^{N_g} n_i = N$, we can define the *genotype distribution function* $\rho_i = n_i/N$, and the entropy

$$S = -\sum_{i=1}^{N_g} \rho_i \log \rho_i \; . \tag{4.22}$$

Clearly, if this distribution is not known, we have to assume that all genotypes are present with the same probability $\rho_i = 1/N_g$. In that case, Eq. (4.22) reverts to (4.21).

BOX 4.3 Entropy in Genetic Space

4.6 Second Law of Thermodynamics

Let us now formulate one of the most fundamental observations of thermodynamics: the *second law*. While the first law generalizes the law of conservation of energy to include the possibility that energy is transformed to heat, the second law characterizes the manner in which a system approaches thermodynamic equilibrium.

If a system is away from thermodynamical equilibrium, it has an improbable statistical distribution, and will move towards a more probable state. This increase in probability happens very fast, in fact, exponentially fast, since from the above equations we can see that the change in the statistical distribution function is exponential in the entropy:

$$dw \sim e^S . \tag{4.23}$$

This also shows that, if in time the system moves from a less probable state to a more probable state, the entropy S has to increase. Generally, the law of increase of entropy can be formulated as follows: "If a *closed* system is in a macroscopic non-equilibrium state, the most probable change of the system will be towards higher entropy." Note that the words "most probable change" mean that, because of the large number of particles involved, the opposite would happen only with a probability smaller than anything that can possibly be observed, i.e., an astronomically small probability. Thus, for practical matters, we can formulate: "If at any time the entropy of a closed system is not equal to its maximal value, it will subsequently *not* decrease, but rather will stay constant or increase." Note also that processes for which the entropy stays constant are called *reversible*, while those for which the entropy increases are naturally termed *irreversible*. In classical physics, the term *isolated* system can be translated to *thermally* isolated, while this is not true in quantum mechanics, for example. Thus, the second law only holds for classical systems that are not in thermal contact with any other system. Such thermal interactions allow a decrease in entropy for a system, for example, by putting the system in question into contact with another one at a lower temperature. The ensuing thermal equilibration will lower the entropy of the system at a higher temperature, while increasing the entropy of the system at the lower temperature. For the combined system, though, the total entropy stays constant. Likewise, any kind of *measurement* performed on a system is excluded by the term "isolated." Thus, we can say that the second law is a statement about what can and cannot happen to a system if it is left completely alone.

4.7 Temperature

In general, the statistical distribution function is not enough to completely describe a macroscopic body, as more thermodynamical variables than just the energy are needed for that. Let us then start here with defining *temperature*.

The temperature of a thermodynamical system with entropy S_n and energy E_n is defined as the inverse rate of change of the entropy with energy:

$$\frac{1}{T} = \frac{dS_n}{dE_n} \,. \tag{4.24}$$

To have a better understanding of this definition, let us consider two systems $n = 1$ and $n = 2$ in thermal contact with each other. Then we can write down the total energy and the total entropy:

$$E = E_1 + E_2, \tag{4.25}$$

$$S = S_1 + S_2 \,. \tag{4.26}$$

To find out the temperature of the combined system, consider

$$\frac{dS}{dE_1} = \frac{dS_1}{dE_1} + \frac{dS_2}{dE_1} = \frac{dS_1}{dE_1} + \frac{dS_2}{dE_2}\frac{dE_2}{dE_1} \,. \tag{4.27}$$

But since $E_2 = E - E_1$, $dE_2/dE_1 = -1$, and

$$\frac{dS}{dE_1} = \frac{dS_1}{dE_1} - \frac{dS_2}{dE_2} \,. \tag{4.28}$$

Now assume that the system is in statistical equilibrium. Then S takes on its maximal value, and

$$\frac{dS}{dE} = \frac{dS}{dE_1} = 0, \tag{4.29}$$

and therefore

$$\frac{dS_1}{dE_1} = \frac{dS_2}{dE_2}, \tag{4.30}$$

or

$$\frac{1}{T_1} = \frac{1}{T_2} \,. \tag{4.31}$$

Having determined several statistical concepts for self-replicating strings that are analogous to classical thermodynamics, let us try to find the analogue of temperature. We suspect immediately from our previous considerations that the mutation rate will fill this place. Indeed, it is temperature that drives statistical systems towards higher entropy and equilibration, and mutation provides just this role in our genetic systems. Imagine, then, that our strings are subjected, in a Poisson-random manner, to mutations that replace instructions randomly by a different instruction from the set. In natural DNA or RNA systems, this is the replacement of a nucleic acid by one of the four possible ones. In general, there is another source of mutations unrelated to these cosmic-ray-type of mutations due to errors in the copy process. For the discussion here, they can simply be lumped together (although they play different roles in general). Thus, let us define a mean rate of mutational changes *per site* of a string, R, such that the probability for a string of length ℓ to be hit by a mutation is

$$p(\ell) = 1 - (1 - R)^{\ell} . \tag{4.32}$$

For copy mutations, an analysis of error thresholds (see Chapter 11) yields that

$$\log \tilde{\epsilon}_{\text{best}} - \log \langle \tilde{\epsilon} \rangle = R\ell, \tag{4.33}$$

where the replication rate $\tilde{\epsilon}$ is related to the one previously introduced by $\tilde{\epsilon} \approx 1 + \epsilon$. According to Eq. (4.12), the left-hand side of (4.33) is just the average inferiority, and we can then deduce that

$$\langle I \rangle \approx R\ell. \tag{4.34}$$

With the interpretation that I is the analogue of energy and $R\ell$ the analogue of temperature, Eq. (4.34) is just the analogue of the *Equipartition Theorem* of thermodynamics (see, e.g., Landau and Lifshitz, 1980).

BOX 4.4 Mutation Is Temperature

We are therefore able to say that if two systems are in thermal contact and in equilibrium, their temperatures must be equal. Since this way of reasoning can be generalized to any number of systems, we may conclude that *the temperature of equilibrated systems in thermal contact with each other is equal.*

4.8 The Gibbs Distribution

We have seen in previous sections that the total energy E_n is sufficient to describe the statistical distribution function of subsystem n, say, given certain parameters α and β:

$$\log \rho_n = \alpha + \beta E_n \ . \tag{4.35}$$

Inserting this into the formula for the entropy (4.17), we can write

$$S_n = -\langle \log \rho_n \rangle = -\langle \log \alpha + \beta E_n \rangle \tag{4.36}$$

and therefore identify one of the parameters (β) as related to the temperature, since [compare Eq. (4.24)]

$$\frac{dS_n}{dE_n} = -\beta = \frac{1}{T} \ . \tag{4.37}$$

In the following we shall try to identify the meaning of the constant α in expression (4.35), so that we can write down a general distribution function for statistical systems armed with only the knowledge of the energy. This very general form of the statistical distribution function is known as the *Gibbs* distribution. Even though we have omitted proofs for most of the equations that we have presented in this chapter, it is instructive to observe how this distribution can be derived.

Suppose we deal with a large system composed of many subsystems (an *ensemble*). Let each of the subsystems have n possible states, with energy E_n. (These are, then, systems with *discrete* energies, as is usual in quantum mechanics, for example.) Pick out one of these subsystems, and we may ask what the probability is to observe that system in state n with energy E_n (while averaging over all the other subsystems). The distribution function of the entire system is, as we determined earlier, just a product of the distribution functions of the subsystems

$$\rho_0 = \prod_\alpha \rho^{(\alpha)} \ , \tag{4.38}$$

and the energy of the combined system is just the sum of the energies of the subsystems

$$E_0 = \sum_\alpha E_\alpha \ . \tag{4.39}$$

On the other hand, the energy of each subsystem is an average of the energies that each state can take on:

$$E_\alpha = \sum_n \rho_n E_n^{(\alpha)} , \tag{4.40}$$

where we sum over all N states of the subsystem, and ρ_n is the probability to be in state n. We know from an earlier discussion (see Section 4.3) that, in equilibrium, the distribution function (as a function of the total energy) is strongly peaked around E_0:

$$\rho(E) \sim \delta(E - E_0) , \tag{4.41}$$

where the function $\delta(x - x_0)$ is a mathematical shorthand for a function that is infinitely large at x_0 and zero everywhere else, but in such a way that the integral of $\delta(x - x_0)$ over x is finite. We are interested in the distribution of energies in one particular system. Then let us divide the total energy E_0 into the energy of our subsystem, E_n, and the rest, with energy E':

$$E_0 = E' + E_n . \tag{4.42}$$

Let $\mathrm{d}\Gamma'$ be the number of states in the remaining system with energies between E' and $E' + \mathrm{d}E'$, so that

$$\mathrm{d}\rho(E_n) \sim \delta(E' + E_n - E_0)\, \mathrm{d}\Gamma' , \tag{4.43}$$

and as a consequence

$$\rho(E_n) \sim \int \delta(E' + E_n - E_0)\, \mathrm{d}\Gamma' . \tag{4.44}$$

The number of states $\mathrm{d}\Gamma'$ can be expressed in terms of the entropy of the primed system [see Eq. (4.14), with $\mathrm{d}\Gamma/\mathrm{dE} = \Delta\Gamma/\Delta E$]:

$$\mathrm{d}\Gamma' \sim e^{S'}\, \mathrm{d}E' , \tag{4.45}$$

such that

$$\begin{aligned} \rho(E_n) &\approx \int \delta(E' + E_n + E_0)\, e^{S'(E')}\, \mathrm{d}E' \\ &\approx e^{S(E_0 - E_n)} . \end{aligned} \tag{4.46}$$

In the last line we made use of the definition of the delta function:

$$\int_{-\infty}^{\infty} f(x)\, \delta(x - x_0)\, \mathrm{d}x = f(x_0) , \tag{4.47}$$

which holds for "reasonable" functions $f(x)$. Continuing with Eq. (4.46), let us expand the entropy $S(E_0 - E_n)$ around E_n:

$$S(E_0 - E_n) = S(E_0) - E_n \left.\frac{\partial S}{\partial E}\right|_{E=E_n} + \mathcal{O}\left(\frac{\partial^2 S}{\partial E^2}\right)$$
$$\approx S(E_0) - \frac{E_n}{T}, \tag{4.48}$$

where we have made use of the definition of temperature in the last line. Inserting this into (4.46), we obtain for the distribution function of energies in the system we picked,

$$\rho(E_n) \approx \exp\left[S(E_0) - \frac{E_n}{T}\right] \equiv A\, e^{-E_n/T}\ , \tag{4.49}$$

which is the Gibbs distribution. The constant A, which we see determines the coefficient α that we defined in Eq. (4.35), is thus related to the entropy of the entire system, but can also be tied to properties of just the system under investigation. Indeed, since $\rho(E_n)$ must be normalized,

$$\sum_n \rho(E_n) = 1\ , \tag{4.50}$$

we find

$$A = \frac{1}{\sum_n e^{-E_n/T}}\ . \tag{4.51}$$

Averages in each subsystem can now be written in terms of this quantity, and the corresponding averages are called *Gibbs-averages*:

$$\langle f \rangle = \sum \rho_n f_n = \frac{\sum_n e^{-E_n/T} f_n}{\sum_n e^{-E_n/T}}\ . \tag{4.52}$$

4.9 Nonequilibrium Thermodynamics

At this point the reader may ask whether it is possible to define the entropy of a system that is *not* in equilibrium, and whether the second law may be violated in such situations. Such an extension can be achieved borrowing some results of Information Theory (the subject of Chapter 3). The second law does not hold then, of course. However, one can show that the total entropy is still a conserved quantity, even off of equilibrium.

As we shall see later, living systems are notorious for very effectively staying away from the equilibrium regime. Quite obviously, this is of prime importance if specific patterns that represent information are to be conserved in time.

Let us start by wondering how it can be that the entropy of a system, in the complete absence of any knowledge (and at equilibrium) is maximal (the logarithm of the number of states), while it can be *smaller* if we write it in terms of a probability distribution function $p_i = e^{-E_i/T}$, for example. The difference is that such Gibbs probabilities are, in fact, *conditional probabilities*. The probability p_i written above is the probability to be in state i, *given* (while knowing) that the state has energy E_i. The corresponding entropy

$$S = -\sum_i p_i \log p_i \tag{4.53}$$

can, as a consequence, be lower than maximal, by an amount that reflects our knowledge about the system:

$$\Delta S = \log \Delta\Gamma - S\,. \tag{4.54}$$

Note that this quantity is often written *without* the constant first term $\log \Delta\Gamma$, and termed *negentropy*. The latter, i.e., $-S$, plays no special role in physics. Still, the second law applies to S as well as to $\log \Delta\Gamma$. As we have seen in the previous chapter, this entropy difference is not, strictly speaking, information (it is not a mutual entropy). However, the *average* ΔS, averaged over all possible systems with all possible energy distributions, is. Thanks to the concept of conditional probabilities, we are now in a position to specify entropies of systems away from equilibrium, *if* we know what kind of constraints are put on the probabilities. An example is the thermodynamics of measurement. This is a very general nonequilibrium situation, as entropies are *reduced* in the process of measurement since knowledge is acquired. In reality, all that happens in a measurement is that the probabilities that we use to compute the entropy, the p_i, change because we learn more about the system. For example, if we start out with probability

$$p_{i|E_i}\,, \tag{4.55}$$

the probability to be in state i, given its energy E_i (that's what the bar after the i means), then after making measurements j, k, and l, for example, the probability becomes

$$p_{i|E_i,(j,k,l)} \; . \tag{4.56}$$

For example, suppose that in a measurement we are able to obtain the position and momentum of each and every particle in a box. Before the measurement, let the system have entropy

$$S_0 = -\sum_i e^{-E_i/T} \log e^{-E_i/T} = \frac{\langle E \rangle}{T} \; . \tag{4.57}$$

After the measurement, the conditional probabilities have changed in such a manner that the $p_{i|\ldots}$ represent a *peaked* distribution (where the ... after the horizontal bar denote all the outcomes of the measurements): all the $p_{i|\ldots}$ are zero *except* one. Then, the entropy

$$S(p_{i|\ldots}) = -\sum_i p_{i|\ldots} \log p_{i|\ldots} \tag{4.58}$$

must be zero. However, entropy is still conserved, because Bayes' Theorem (see Chapter 3) tells us that the entropy S_0 is split in the following manner:

$$S_0 = S + I \; , \tag{4.59}$$

where I is the *entropy of correlation* (mutual entropy) and S is entropy of the system given the particular measurement outcomes: a *conditional* entropy. If we denote with $p(\ldots)$ the probability to obtain the measurement outcomes $(\ldots)$, then

$$I = -\sum p(\ldots) \log p(\ldots) \; . \tag{4.60}$$

Thus, entropy is conserved, even though the measurement operation was not an equilibrium process. Note that if we let the system evolve freely after the measurement, the entropy of correlation I will rapidly decrease, in fact exponentially fast. It is this quantity that decreases in the approach to equilibrium, whereas S approaches S_0 exponentially fast, according to the second law. Thus, we realize that in equilibrium thermodynamics, all correlation entropies vanish, and all conditional entropies are maximal, i.e., they are regular thermodynamical entropies. Away from equilibrium, we must use conditional probabilities and worry about correlations induced by the nonequilibrium processes.

4.10 First-Order Phase Transitions

One of the most recognizable aspects of evolving and adapting systems are the periods of extinction and innovation, collectively known as *evolutionary transitions*. While the period *between* such transitions can be described by the techniques of equilibrium thermodynamics, the transitions are decidedly *nonequilibrium* phenomena. As we shall see, those are best described in the context of *first-order phase transitions*.

The state of any homogeneous body in thermodynamical equilibrium is determined by the specification of *any* two thermodynamic variables. Examples for such pairs are the temperature and pressure, the temperature and volume, or the energy and the chemical potential. The chemical potential μ is a quantity that measures the *rate of change* of the energy with the number of particles (while keeping the entropy and the volume constant—this is indicated by the subscripts):

$$\mu = \left(\frac{\partial E}{\partial N}\right)_{S,V} . \tag{4.61}$$

On the other hand, specifying two of these variables does *not* guarantee the existence of a homogeneous body with these parameters in thermodynamic equilibrium. Rather, such a state could in principle *separate* into two *phases* that are each described by different pairs.

Different phases of a system can coexist under very special circumstances, namely

$$T_1 = T_2 \qquad \text{equality of temperatures} \tag{4.62}$$

$$P_1 = P_2 \qquad \text{equality of pressures} \tag{4.63}$$

$$\mu_1 = \mu_2 \qquad \text{equality of chemical potentials} \tag{4.64}$$

It is impossible for two phases to be in equilibrium at any possible pair of temperature and pressure. Rather, specifying one determines the other. On a diagram of any two thermodynamic variables, then, the phase boundary can be drawn as a line, as in Fig. 4.4. At any point on the line in Fig. 4.4, the two phases can coexist. If the parameters are tuned in such a way that the system's state point is off the line (while still two phases are present), a *phase transition* will take place that will result in the system to be in one homogeneous phase. When this happens, a certain amount of *heat* is produced or absorbed by each subsystem. This can be thought of as the energy necessary to melt a crystal or evaporate a liquid,

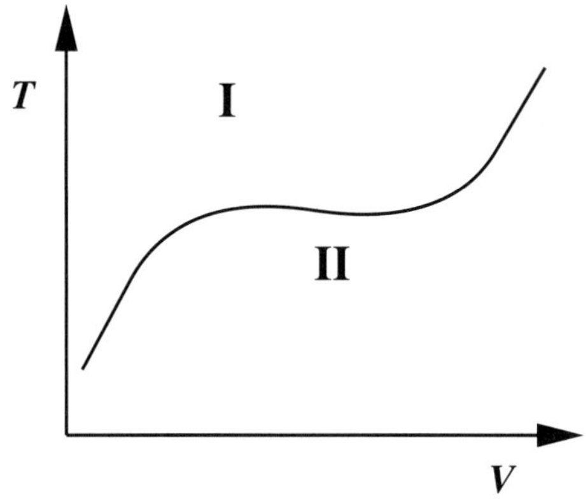

FIGURE 4.4 Phase diagram for two different phases I and II.

or, in the reverse direction, the amount of heat liberated when steam condenses or a liquid solidifies. In general, this amount of heat (per unit) is called the transition heat, or *latent heat*.

When a system undergoes a phase transition from phase 1 to phase 2, often the entropies of the two phases are different. (They certainly are in the examples mentioned above!) Such transitions are termed *first-order*. In transitions of second order the entropy is continuous. For first-order transitions, let us then define the latent heat as

$$L = \int_{S_1}^{S_2} T\,\mathrm{d}S\,, \tag{4.65}$$

where we integrate over the entropy from its initial value S_1 to its final value S_2. Then, we find

$$L = T(S_2 - S_1) = T\Delta S\,. \tag{4.66}$$

The time it takes to go from one phase into the other is usually related to the relaxation time of the system.

Phase transitions of this kind seem to occur in many systems that have metastable states, and where each metastable state defines a phase. An analysis of evolutionary transitions in terms of the language of first-order phase transitions can be found in Box 4.5.

In our string systems, a phase is characterized by a *global* variable, the replication rate of the best string, ϵ_{best}. A mutation that introduces a better string *renormalizes* the inferiority (the equivalent of energy, see Box 4.2) of every string in the population. If

$$\epsilon_{\text{best}}^{\text{new}} = \epsilon_{\text{best}}^{\text{old}} + \Delta\epsilon \;, \tag{4.67}$$

a string that was previously in the *ground state* ($I_i = 0$) suddenly finds itself in an *excited state*: $I \to I + \Delta\epsilon$. In the language of first-order phase transitions, this is due to the discovery of a *new vacuum*, and a phase transition with latent heat $L = \Delta\epsilon$ has to occur (see Fig. 4.5). The moment the new ϵ_{best} is discovered, the old vacuum defined by $I = 0$ becomes a false vacuum, and therefore metastable. The new phase has a lower entropy, and the difference in entropy between the old phase and the new one is given by

$$\Delta S \approx \frac{\Delta\epsilon}{R\ell} \;. \tag{4.68}$$

We can check this explicitly in artificial genetic systems such as avida. In Fig. 4.6, we can see a transition triggered by a self-replicating program that discovered how to perform a type of logical operation on inputs (see Section 9.3 for more details), thereby increasing its replication rate by an amount of $\Delta\epsilon$, indicated in the figure. As a consequence, the average inferiority (energy) of the population suddenly changes by an amount $\Delta\epsilon$, ushering in the phase transition. During the transition, the latent heat $\Delta\epsilon$ has to be dissipated. This has happened once the new genotype has wiped out all the inferior ones, and the energy in the lower part of Figure 4.6 has returned to its starting value. Note that in equilibrium, this value is determined by the mutation probability, as discussed in Box 4.2.

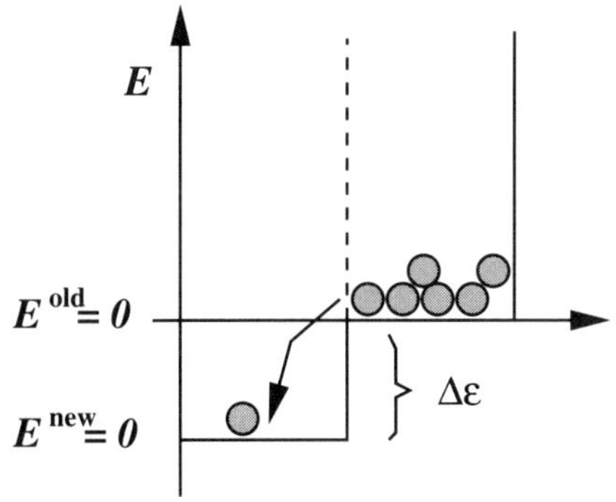

FIGURE 4.5 Phase transition engendered by the discovery of a new vacuum.

BOX 4.5 First-Order Transitions in Genetic Systems

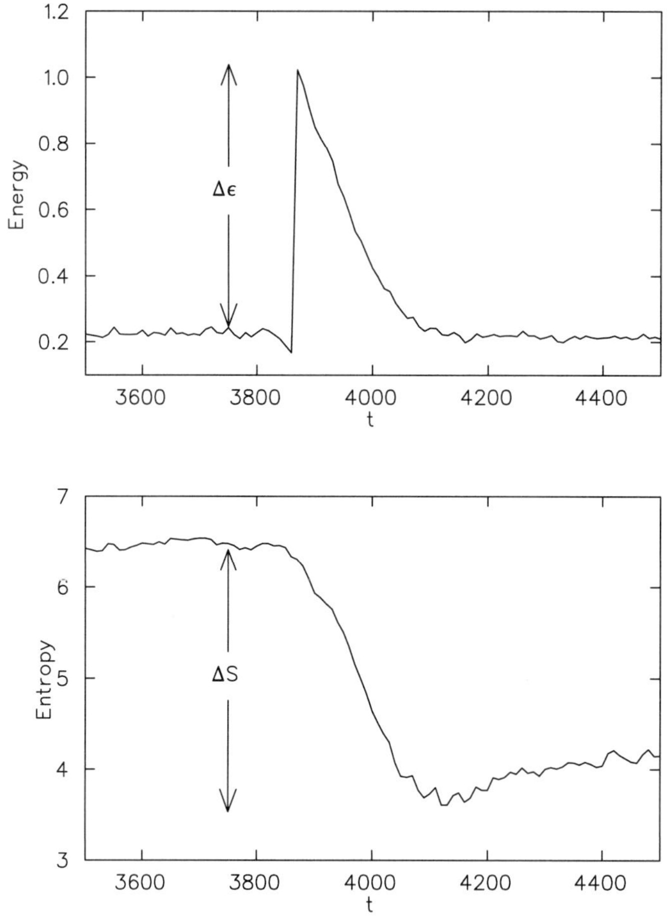

FIGURE 4.6 Phase transition in avida (triggered by the discovery of `~A or B` in Fig. 9.3). The entropy drops for as long as it takes to dissipate the latent heat (lower panel). After that, the entropy slowly starts to rise again as mutations of the new best program start to reinstate diversity. In the upper panel the abrupt change of the average inferiority (Energy) $\langle I\rangle = \log \epsilon_{\rm best} - \log\langle\epsilon\rangle$ reflects the discovery of a new best string. Note that in avida experiments, we designate by $\epsilon_{\rm best}$ the replication rate of the *most abundant* member, rather than the overall highest, as the former is easier to measure. They are usually equal in equlibrium, but differ just after the transition takes place. As a consequence, the measured energy dips below the average value right at the beginning of the transition for as long as the new best replicator is not also the most abundant.

4.11 Overview

The methods of statistical mechanics and thermodynamics are singularly well-suited to investigate the macroscopic properties of an aggregation of very many units interacting by simple laws. In the usual description of this theory, it is assumed that the system is *ergodic*, meaning that all possible states of the system can and will be taken on eventually. As a result, the dynamics of such systems do *not* depend on the *initial* state. Living systems can, to a certain extent, also be described by the methods of statistical physics, albeit minus the ergodicity assumption. As a consequence, we must bow to the principle of contingency: Everything we observe in living systems will necessarily depend, more or less strongly, on the ancestral genotype giving rise to all life on this planet. Furthermore, the evolution of a population is an intrinsically *nonequilibrium* process, except for the times *between* discoveries. These periods can, if the population is uniform enough such that its members can be considered the same species, approximately be described by equilibrium dynamics. To describe the monumental upheavals that occur during the major evolutionary transitions, the language of first-order phase transitions seems to be well-suited. Still, statistical theory will never be able to predict *when* a transition is going to occur, nor what the emerging genotype is going to be like.

Problems

4.1 Entropy in thermodynamic systems can be viewed as the logarithm of the total number of *possible* states in which a system can exist. Without appealing to *equilibrium*, the formula (4.18) can be obtained using the *microcanonical* ensemble. Suppose we have N systems that can be in states $1, 2, 3\ldots$, and that of these systems, N_1 is in state 1, N_2 is in state 2, etc., where $N = N_1 + N_2 + N_3 + \cdots$. Show that you can obtain the entropy *per system*

$$S = -k \sum_i p_i \log p_i \,, \tag{4.69}$$

where $p_i = N_i/N$, by writing S as k times the number of possible ways to realize a given set of occupation numbers $N_1, N_2, \cdots$. In (4.69), k is Boltzmann's constant that converts temperatures into energies (we have set this constant equal to one in the previous chapter). Hint: Use Sterling's formula $\log N! \approx N \log N$.

4.2 *Nernst's Theorem* (also known as the *third law* of thermodynamics) states that the entropy of any physical system at absolute zero temperature is

$$S(0) = k \log M(E_0) , \tag{4.70}$$

where $M(E_0)$ is the *degeneracy* of the ground state (i.e., the number of *different* states that have the lowest energy). It is conjectured that in physics the ground state is *always non-degenerate*, in which case $S(0) = 0$. In avida, start a run with a mutation rate of your choice, and set an *event* to change the mutation rate to zero at a time when you think the population will be reasonably equilibrated. Watch the entropy of the system approach (4.70) (with $k = 1$), and try to determine $M(0)$ from the limit $t \to \infty$. Consult the *User's Manual* in the Appendix about how to change the mutation rate during a run.

CHAPTER FIVE

Complexity of Simple Living Systems

The computing process [...] *is closely akin to a measurement.*
R. Landauer, 1961

That the complexity of life is a mystery as far as physics is concerned has been articulated often and forcefully, most notably by Schrödinger (1945). Today, however, we have reached a far better understanding of the dynamics that give rise to the hitherto mysterious ordering process, and we are ready to examine the evolution of complexity in living systems armed only with the tools of thermodynamics and Information Theory. The latter teaches us how to treat systems that are *away* from thermodynamic equilibrium, a state of affairs important to the physics of living systems.

While we are contemplating this matter, geneticists all over the world are sequencing genomes (including our own) at an unprecedented pace, and there remains little doubt that the complexity of living systems is just a reflection of the information stored in the genome. How, exactly, this conclusion can be made firm is the subject of this chapter. Artificial living systems can help tremendously in understanding the mechanism that allows information to be transferred from the environment into the genome. Furthermore, they can help us in forging a *measure* of complexity, a concept with a history of long debate. Apart from complexity measures based on automata theory (which we will encounter in a few pages), there are many others that will not be reviewed here; for an introduction, see Badii and Peliti, 1997.

5.1 Complexity and Information

In the following, we are going to investigate the basic process that allows the *stochastic* transfer of information from an environment into the genome. In the end, the main agent in this transfer is Darwin's principle of survival of the fittest. However, we abstract the mechanism down to its simplest form, operating on self-replicating binary strings.

Imagine such a string self-replicating in an environment consisting of other strings and physical principles that affect the manner in which a particular string replicates (the chemistry). Also, imagine that there is an agent of *noise* that induces errors (bit-flips) in the strings by either or both of two mechanisms: a random bit-flip acting on a string (akin to cosmic-ray mutations), or bit-flips resulting from the incorrect copying of instructions in the self-replication process. Note that a self-replicating string already contains quite an amount of information, namely the information necessary to self-replicate in its given environment. How such information can enter a *non-replicating* string in the first place is the subject of an entirely different discussion (the origin-of-life question), touched upon in Chapter 2. We should reflect at this juncture already about the fact that this information that allows the string to self-replicate is *context-dependent*, rather than absolute. If the particular string were placed into an environment with a different chemistry, it would most likely *cease to self-replicate*. The arrangement of bits that was so powerful in one environment can be quite meaningless in another: what is information given the chemistry the string evolved in, is just randomness given a foreign one. This will be our guiding principle in examining complexity and its relation to information: information is context-dependent. Of course the theory we are going to use to make this explicit is Information Theory, as outlined in Chapter 3. Let us now follow the evolution of a string in its native environment, and let us also assume that the environment is complex, i.e., that there are many things to be discovered, many ways to improve your fitness. In short, let it be replete with potential information.

The first observation we make is that it appears that the mutation probability (which is due to the noise) seems to be nonuniform across the bits of the self-replicating string. If the population is examined *post mortem*, i.e., its evolution interrupted, some positions (sites, loci) are highly variable across the population, while others are strongly conserved. Is this a reflection of a site-dependent mutation rate? We observe here that it is not. Rather, bit positions that are essential for self-replication are impervious to mutation owing to the simplest of all mechanisms: if they *are* flipped, the bearer loses the capability to self-replicate and as a consequence is incapable of promulgating the "tainted" code. Such a string will be replaced quickly by one whose information was not corrupted, in a sense *reversing* the lethal mutation. Bit positions that are inessential for the continued self-replication of the string, on the other hand, will not be corrected in this way, and such bits will take on all possible values throughout the population as time goes on. Thus, the nonuniform rate of substitution is just a reflection of important versus unimportant bits. In the following, we call bits that are highly variable across the population *hot* bits, while those that are strongly conserved will be termed *cold*. The mutations of hot bits, correspondingly, are in general neutral mutations (they do not affect the fitness of the bearer), while a mutation of a cold bit is usually lethal.

Imagine now a random mutation that, by chance, actually increases the rate of replication of the bearer. Such a mutation could have taken place on a bit that was previously hot, which, however, together with a mutation taking place somewhere *else* on the string, suddenly means something. In other words, the string with these mutations makes better use of the environment. Not only will this random mutation be passed on to the offspring in this case (like any other neutral mutation is), but it will be *amplified* in the population due to the superior replication rate of its bearer. Thus, offspring of the string with the beneficial mutation will increase in numbers relative to those strings that do not carry it, and in due time (in a world that is of finite size), *all* strings will carry this *allele*, i.e., this particular value of the bit. In other words, a position has reverted from hot to cold; it has been *frozen*.

The simple fact that the adapted string is more successful means that it is exploiting the environment in some manner. In order to do this, it must contain information about the environment. Any time, then, that a string becomes better adapted to the environment, we conclude that information about the environment has been written into the genome by the process just described: previously "blank" tape (sections of genome with hot bits that can be mutated with impunity) are written over with stable code: information. Note that this is a rather idealized picture: most of the time positions in the genome are neither totally volatile nor completely fixed. Also, in the adaptive process positions that were previously fixed can be overwritten with new code if the new information is at least as successful in exploiting the environment as the previous was. We will deal with the variable temperature of bits by considering the entropy of the population later. For the moment, we are only interested in elucidating the main mechanism for information transfer from the environment into the population. Information slowly and inexorably trickles into the genome as the population takes advantage of more and more features of the environment: the genomes become *correlated* to the environment. A well-adapted string can then be viewed as a "book" about the environment in which it evolved. By analogy, if we had the ability to decipher our own genome accurately by inferring the function of each and every protein and enzyme coded for in our DNA, we would be reading a book about the environment in which *we* evolved: earth. This book would include such simple things as chemical abundances, the temperature of the water and air, the composition of the ecosystem on which we thrive, as well as any other item relevant to the human metabolism.

Intuitively we surmise that life evolves towards higher and higher complexity. Such a statement is rather empty if complexity is only loosely defined. Here, we posit that life evolves in such a manner as to increase the amount of information about its environment coded into the genome. Also, we shall see that our intuitive notion of complexity in fact corresponds to the notion of information in physics. In that sense, then, life indeed evolves towards higher and higher complexity.

5.2 The Maxwell Demon

Using the concepts of Information Theory introduced earlier, let us quantify what happens at the instant information enters the genome. In the following, it will be useful to view this event as a *measurement* performed by the genome on the environment. In a measurement, the correlation between the measurement device and the measured system increases, and the conditional entropy of the measured system and that of the measured device decreases. Yet the measurement that the population performs on the environment is not a purposeful act; rather, measurements are performed spontaneously and randomly. Still, once information is acquired in this spontaneous manner, it is not released just as spontaneously, but is used to *lower* the entropy of the population instead. This is the prototype behavior of a Maxwell demon. Before describing such a beast, let us consider in more detail the physics of measurement.

In the following, we denote the environment or measured system by U, the "universe". The total amount of information that can be accessed by any measurement is then bounded by the entropy of the universe, $H(U)$. Entropy can thus be viewed as *potential information*. Let us also think of the universe as an ensemble of random variables X_i, correlated in arbitrary ways. Typically, the experiment reveals the value of one or more of the variables X_i of U. After the measurement, the entropy of U *given* X is all-important: it now becomes what is termed the *remaining* entropy of U. Concurrently, information $I(U:X)$ is gained. This process is expressed in the fundamental equation, encountered before in the chapter on Information Theory:

$$H(U) = H(U|X) + I(U:X)\ . \tag{5.1}$$

Remember from that chapter that information is a *mutual* entropy: it is the amount of entropy shared between two systems. In the measurement process the information gained is just a measure for the amount of correlation between measured system and measurement device introduced by the measurement, here between X and U. The measurement just described is slightly awkward from the point of view of our usual experience because the system performing the measurement (X) is *part* of the universe (see Fig. 5.1). Yet it is a useful example that teaches us that the entropy of the closed (isolated) system $H(U)$ does not decrease during measurement. In our everyday experience, a system S is measured by a device M that is *not* part of S, resulting in the correlation of some of M's variables with S. In this case, some of the entropy of the measurement device $[H(M|S)]$ is *not* correlated with the measured system (Fig. 5.2), a situation that we will encounter below. In both cases, however, we see that information is a quantity that needs the specification of *two* ensembles: what is being measured and by what. Thus, information is never context-free: information is always information *about* something.

Let us now consider the consequences of the second law of thermodynamics discussed in Chapter 4 more closely. As mentioned above, the

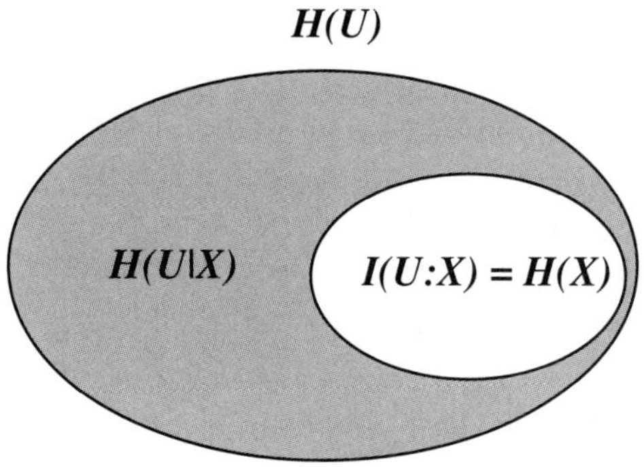

FIGURE 5.1 Measurement (correlation) of variables X in U.

entropy of the closed system (the universe in Fig. 5.1, or the combined systems M and S in Fig. 5.2) is unchanged through the measurement. This is true irrespective of whether the measurement process is a thermodynamic equilibrium process or not: the entropy of a closed system is *always* conserved. Note that this modifies the way the second law should be stated. According to this law, a system in thermodynamic nonequilibrium evolves in such a way as to increase its entropy. For example, the entropy of molecules stashed in the corner of a box is going to increase until the entropy reaches its maximum value, at equilibrium. Here we see that it is the *conditional* entropy that increases during this process (not the unconditional one, as standard thermodynamics teaches). The unconditional (or *marginal*) entropy—given by conditional entropy *plus* mutual entropy $H(S|M) + I(M : S)$ in Fig. 5.2—stays *constant*.

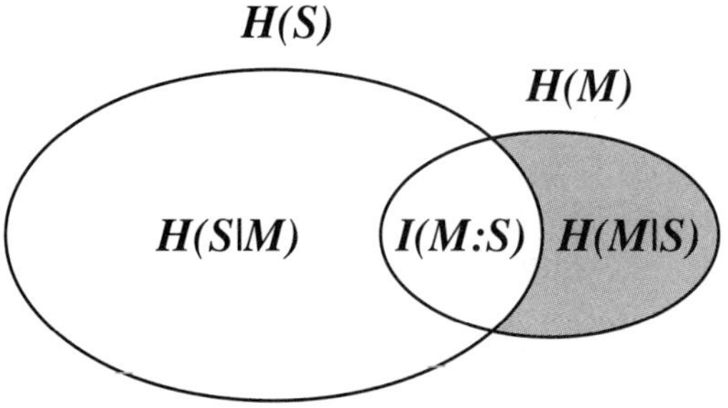

FIGURE 5.2 Measurement of variables of S by M.

The inverse (nonequilibrium) process that confines all the molecules in a corner of the box is a *measurement*, thereby creating information and reducing the conditional entropy. We may therefore formulate an extended second law as follows:

> If a *closed* system is in a macroscopic nonequilibrium *or* equilibrium state, the *marginal* entropy of the system is a *constant*.

The usual formulation (involving the *increase* of entropy) is sufficient for most purposes, however, if we just keep in mind that conditional entropies may decrease without violation of any law. The decrease in entropy due to the measurement

$$\Delta S = H(U) - H(U|X) \tag{5.2}$$

for the measurement situation of Fig. 5.1, or

$$\Delta S = H(S) - H(S|M) \tag{5.3}$$

in Fig. 5.2 is just the information gained in the measurement. Let us investigate whether it can be used to perform *work*, which would be an indication that the second law is violated. This leads us to the Maxwell demon paradox.

The Maxwell demon first appeared in Maxwell's book *Theory of Heat* [Maxwell, 1871]. There, he argued that a being armed only with the intelligence to make measurement decisions could change the pressure in one side of a container that is divided into two halves (A and B, say) by selectively opening and shutting a door connecting the two halves depending on the speed of the molecules at the door (see Fig. 5.3). So, for example, he could let fast molecules enter through the door from A to B, while denying the same thing to slower molecules. Meanwhile, he would allow slow molecules from B to enter A, but not fast ones. After a while, B would be filled with fast molecules, while A is left with the slow ones. As a consequence, this demon would have raised the temperature of B and lowered that of A, without the expenditure of work (assuming that the work necessary to operate the shutter is negligible), in violation of the second law. Up until Landauer's seminal work in 1961 and Bennett's in 1973, attempts at revealing flaws in Maxwell's argumentation were centered on showing that the measurement operation of operating the shutter by the demon could not be performed without the expenditure of work (see Leff and Rex, 1990, for a history and a thorough guide through the literature

FIGURE 5.3 Maxwell's Demon at work (reprinted from *Fundamentals of Cybernetics*, A.Y. Lerner, [Plenum Pub. Corp., NY, 1975], p. 257).

on the Maxwell demon). Landauer's key realization was that information was an object of physics, not an ethereal quantity. While Bennett showed that the measurement process could be achieved *without* the expenditure of work, Landauer showed that the acquired information could not be stored without such expenditure. Before we discuss this in more detail, let us illustrate how the realization that "information is physical" [Landauer, 1991] saves the day for the second law.

For this purpose, let us illustrate the Maxwell demon paradox in its simplest incarnation, the *Szilard engine* [Szilard, 1929]. Imagine a single molecule enclosed in a box that is in thermal contact with a heat bath at temperature T (Fig. 5.4). A piston can be inserted into the box separating it into two volumes. After determining on which side of the piston the molecule is located (the measurement), the piston is moved in a direction determined by the outcome of the experiment, namely in such a way that the kinetic pressure of the molecule assists the movement of the piston. As this operation proceeds isothermally (i.e., so slowly that all components remain in thermodynamical equilibrium), the reduction

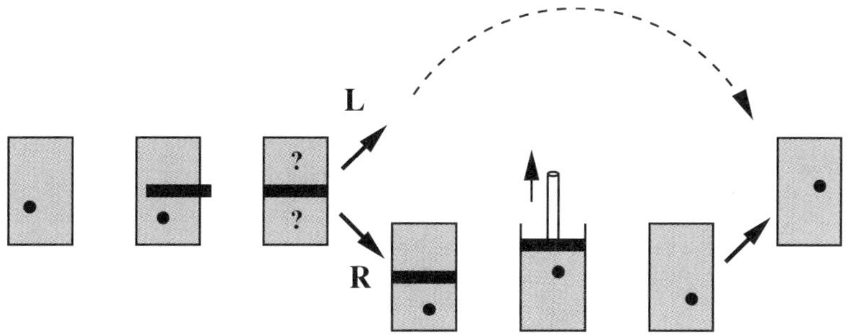

FIGURE 5.4 Szilard's one-molecule engine, designed to extract $\Delta Q = kT$ of work per cycle via measurement of the position of the molecule (adapted from Zurek 1990).

of entropy (achieved through knowledge of the position of the molecule) effects the extraction of useful work (via the movement of the piston). The extracted information, which is just the difference between the entropy before the measurement, $H(U) = k$,[1] and after, $H(U|X) = 0$, is used to perform work $\Delta Q = kT$ according to the relation $\Delta Q = T\Delta S$, where naturally $\Delta S = H(U) - H(U|X)$. Apparently, this violates the second law. However, we note immediately that the box with piston is *not* a closed system. Indeed, the information obtained via the measurement process has to be *recorded*, or equivalently, the information recorded in the previous cycle has to be replaced. This is where the physicality of information comes in. What does it mean to record information? To examine this issue, let us follow Landauer in imagining a very simple information storage device: a double-well potential (Fig. 5.5). A particle residing in either the left or right minimum of the potential denotes the value of the stored bit. If the state of the memory cell is known, a force can be applied to the particle that flips the bit. Theoretically, the energy needed for the particle to cross the hump can be recovered by breaking the particle after it crossed the barrier (the computer analogue of regenerative breaking used in electrical cars). Thus, a memory device can be switched without loss of energy *if and only if* the state of the device is known before the switching. Assume, however, that we want to write a specific bit into a

[1]In thermodynamics the entropy is given by the Shannon entropy (in bits) multiplied by k, where k is Boltzmann's constant. See Problem 4.1.

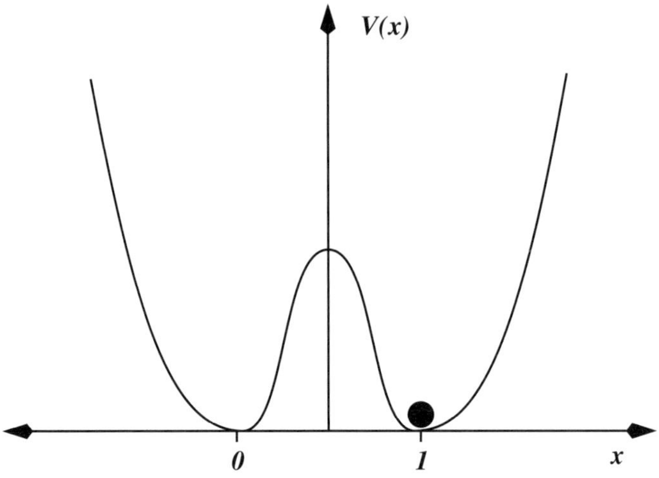

FIGURE 5.5 Landauer's double-well information storage device.

register (say, a zero) *without* knowing the state of the memory device prior to the operation. We must then apply a force to the particle in the well that will force it into the minimum $x = 0$ *regardless* of the initial state of the device. This can only be done by applying a *dissipative* (nonconservative) force, which slowly drags the particle from $x = 1$ to $x = 0$ if the initial state were $x = 1$, and otherwise leaves the state unchanged if it were $x = 0$. Then, as Landauer showed, a restore-to-zero operation on a binary memory device is necessarily associated with the dissipation of an amount of heat equal to $\Delta Q = kT$. This, it turns out, is *exactly* the amount of work that can be performed by exploiting the gain of entropy $I = \Delta S = k$. Thus, no net energy can be gained by the measurement, and the second law is inviolate.

This situation can be generalized as follows. A measurement involves a system to be measured S and an observer (or measurement device) X. The observer X is characterized by a recording device, which can be taken to be a tape that records the binary values 0 and 1. If the tape is *blank* (all zeros, say), the entropy of the observer vanishes (the state of the recording device is perfectly known). Consequently, a reduction of entropy in the measured system, having come into contact with a *zero temperature device* (zero entropy implies zero temperature, according to Nernst's Theorem, see Problem 4.2), does not violate the second law, because the entropy of a system in thermodynamic equilibrium at temperature T that comes into thermal contact with a system at a lower temperature naturally will lose entropy until the two systems are at equilibrium (see Section 4.7). If the observer's tape is in equilibrium on the contrary (random bits with equal probabilities for a 0 or a 1), the recording of information is associated with heat dissipation (the energy necessary to force a bit into a definite state), and the drop in entropy ΔS is cancelled by the heat dissipated in the recording of the information I. This is the content of the fundamental equation of measurement, mentioned before but worth repeating,

$$\Delta S = H(S) - H(S|M) = I(S:M), \tag{5.4}$$

illustrated by Fig. 5.2 and cemented by Landauer's analysis of Szilard's engine.

The importance of this observation for living systems and the evolution of complexity will be apparent in a short while. First, we need to dig even deeper into the logic of the measurement process. This will lead us, quite unexpectedly, back to the Turing machines introduced in Chapter 1.

5.3 Kolmogorov Complexity

As we have seen, there is no concept in thermodynamics that allows a determination of the entropy of *one* string. Entropy (and, as a consequence, information) is an entirely statistical concept. Nevertheless, it appears intuitive that not all bit-strings are alike; some appear more regular than others. The concept of *Kolmogorov complexity* [Kolmogorov, 1965]—sometimes called Chaitin complexity [Chaitin, 1966]—addresses this issue.

In short, the Kolmogorov complexity of a string is low if it can easily be obtained by a computation, whereas it will be high if it is difficult to obtain it. This difficulty is measured by the length of the shortest program that computes the string on a universal Turing machine. The use of Turing machines to determine the length of the shortest program that computes a particular bit-string is intuitive: Since a universal Turing machine can simulate any other Turing machine, the length of the program computing string s, say, on Turing machine T, can only differ from the program computing the same string on Turing machine T' by a finite length $u(T, T')$, the length of the prefix code necessary to simulate T on T'. As this difference is constant (for each string s), the length of the shortest program to compute string s on a universal Turing machine is constant in the limit of infinitely long strings s, and we correspondingly define the *algorithmic complexity* (Kolmogorov complexity) of string s as

$$K(s) = \min\{|p| : s = C_T(p)\}\ , \tag{5.5}$$

where $|p|$ stands for the length of program p, and $C_T(p)$ represents the result of running program p on Turing machine T. Let us illustrate this measure by a few examples. A blank tape (the string with all zeros) is clearly a highly regular string, and correspondingly its Kolmogorov complexity will be low. Indeed, the program needed to produce this string can be very short: `print zero, advance, repeat`. The same is true, of course, for every string with a repetitive pattern. Another way of viewing algorithmic regularity is by saying that an algorithmically regular string can be compressed to a much smaller size: the size of the smallest program that computes it. More interesting is the regularity of a string that can be obtained by the application of a finite but nontrivial algorithm, such as the calculation of the transcendental number π. The string representing the binary equivalent of π certainly *appears* completely random, yet the minimal program necessary to compute it is finite. Thus, such a string is also classified as algorithmically regular (though not quite as regular as the blank tape). Kolmogorov complexity also provides a means to define *randomness* in this context. According to the Kolmogorov measure, a string r is declared random if the size of the smallest program to compute r is as long as r itself, i.e.,

$$K(r) \approx |r|\ . \tag{5.6}$$

Thus, algorithmically random strings cannot be compressed in any way. From an intuitive point of view, algorithmic complexity does not seem to be a good measure for the *physical* complexity of a string. First, random strings should not be assigned maximum complexity, as we do not feel that they are very complicated. On the other hand, the regularity of a string does not reveal how complex the object or information is that this string represents. For example, it is possible to create an (admittedly insane) coding scheme in which the blank tape represents all of "The Brothers Karamazov." Again we see that for a true measure of physical complexity, *context* is of utmost importance. But let us see if complexity can still be defined in terms of Automata Theory.

5.4 Physical Complexity and the Natural Maxwell Demon

The basic flaw in the Kolmogorov construction (as far as *physical* complexity is concerned) is the absence of a context. This is easily rectified by providing the Turing machine with a tape u, which represents the physical "universe", while the Turing machine with u as input computes various strings *from* u. First, let us recall the definition of the *conditional* complexity of a string s as the length of the shortest program that computes s *given* string u [Kolmogorov, 1983]:

$$K(s|u) = \min\{|p| : s = C_T(p, u)\} \,, \tag{5.7}$$

where we introduced the notation $C_T(p, u)$ as the result of the computation running p on Turing machine T with u as input tape. The conditional complexity measures the remaining randomness in string s, i.e., it counts those bits that are *not* correlated with bits in u. In other words, the program p is the maximally compressed string containing those bits that cannot be computed from u, as well as the instructions necessary to compute those bits of s that *can* be obtained from u. The latter part of the program is of *vanishing* length in the limit of infinitely long strings, which implies that the program p mainly contains the remaining randomness of s. Of course we can then immediately define the *mutual* complexity

$$K(s : u) = K(s) - K(s|u), \tag{5.8}$$

which clearly just measures the number of bits that mean something in the universe u. This will be our measure of *physical* complexity. In Fig. 5.6 we give an example of a string where those bits that are obtained by computation from u are shown in gray, whereas those bits that cannot be obtained by a computation are white. By rearranging the bits on the tape we can see that each string s can be divided into two sections, one of length $K(s|u)$ and one of length $K(s : u)$.

This construction solves one immediate puzzle that Kolmogorov complexity raises. What does it mean that a *random* string is computed? Since

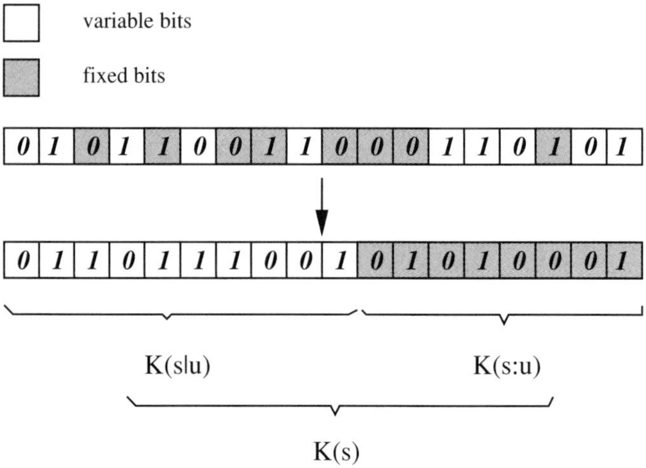

FIGURE 5.6 Division of a string s into the pieces $K(s|u)$ and $K(s:u)$.

these Turing machines are not probabilistic, random strings simply should be uncomputable. The way Kolmogorov circumvents this is by *defining* a string to be random if it can only be obtained by including the string *verbatim* on the program tape p (making it obvious that then $K(r) \geq |r|$). Note, however, that this defeats the purpose: while the string cannot be compressed to a length smaller than $|r|$, the combined tape (p, r) *can*. Thus, in the presence of p (which contains r), r *is* compressible. Indeed, the simple program `copy, repeat` started at the replica of r suffices to produce r. This reasoning solely relies on the fact that a Turing machine has essentially only *one* tape, so which pieces are program, input, or output tape is purely conventional. As suggested earlier, we can circumvent this problem by explicitly defining Turing machines that come with a tape that defines their universe. We then *only* ask whether a string s can be obtained via computation *given* a certain tape u. Random strings, on the other hand, *cannot* be obtained in such a manner. Rather, a string is defined to be random *with respect to u* if it cannot be obtained as a result of a computation on u. Randomness, just like information, is then only defined with respect to another system. Note, however, that this raises an interesting problem related to Turing's Decidability Theorem (see Box 1.1 on page 5). Indeed, it is generally *undecidable* whether a string s can be obtained by computation from u until it *has* been obtained. In other words, a string is random (with respect to u) until proven otherwise, and it can (if u is infinite) never be *proven* to be random. The reason is that it can never be the result of a computation that a string is uncomputable.

Let us now consider the physical complexity $K_P(s) = K(s:u)$ in more detail. Its meaning becomes clearer if instead of considering a string s obtained by Turing machine T with universe u, we consider the *ensemble* of strings S that can be obtained from a universe u with T. This ensemble can be thought of as a probabilistic mixture subject to random bit-flips. In other words, we imagine the output tapes to be connected to a heat bath. In that case, we can associate an entropy with the ensemble of strings S, $H(S)$. Now consider a Turing machine operating on u, a specific universe. Obtaining s from u then constitutes a *measurement* on the universe U, and consequently not only reduces the conditional entropy of S given u, but also the conditional entropy of U given s. Note that the universe is assumed here to be fully known, i.e., there is only one tape u in the ensemble U. While this must not strictly be so, here it is convenient to assume that there is no randomness in the universe. Also, the length of the smallest program that computes s from u, averaged over the possible realizations of s, then just equals the *conditional entropy* of S given u. It is known that the average Kolmogorov complexity over an ensemble of strings just equals the entropy of the ensemble. Then

$$H(S|u) = \langle K(s|u)\rangle_S = -\sum_s p(s|u) \log p(s|u) , \tag{5.9}$$

and

$$I(S:u) = \langle K(s) - K(s|u)\rangle_S = H(S) - H(S|u) . \tag{5.10}$$

Note that (5.9) is not strictly a conditional entropy, as no average over different realizations of the universe takes place. Indeed, it looks just like a conventional Shannon entropy only with all probabilities being probabilities *conditional* on u. Similarly, the physical complexity (5.10) is not strictly a correlation entropy, but represents information *conditional* on a specific universe u. In general, then, every computation, i.e., every measurement, reduces the conditional entropy of S and increases the information about u contained in the ensemble of strings S. Let us see how such a process can proceed naturally.

Imagine a string s that shares some information with u, but some bits are in fact random, i.e., it is undecided whether they correspond to anything in u. Mutations are constantly changing these bits, and the Turing machine operating on u is constantly trying if any of those bits that are "undecided" (formally random) can be obtained from u by a computation. If this happens, a bit that was previously random now represents *information*; in other words, the mutual entropy between the string and the universe has increased. Thus, we can view evolution much like it is depicted in Fig. 5.7, where we see the progression from $S \to S' \to S'' \to S'''$ as one where the mutual entropy $I(S : u)$ constantly increases (see [Adami and Cerf, 1996]). As we have decided, for the interpretation of physical complexity K_P, to consider ensembles of strings rather than single strings, we still must discuss what it means to calculate the entropy of string s. As we can separate the entropy of each ensemble of strings into a mutual and a conditional piece, $H(S) = H(S|u) + I(S : u)$, the idea is to consider *species* of strings determined by their common mutual entropy with the environment u. Such strings do differ, of course, in those bits that are not shared with u. Naturally, these are just the hot bits we considered earlier, while those bits that constitute information (and that are thus shared with the environment) are cold (see Fig. 5.6). Of course the entropy of S *unconditional* on anything is just the length of the string (in units of instructions).

We have thus been able to couch the intuitive discussion from the beginning of this chapter into a more mathematically precise language. We have determined that complexity is a *statistical* measure that can

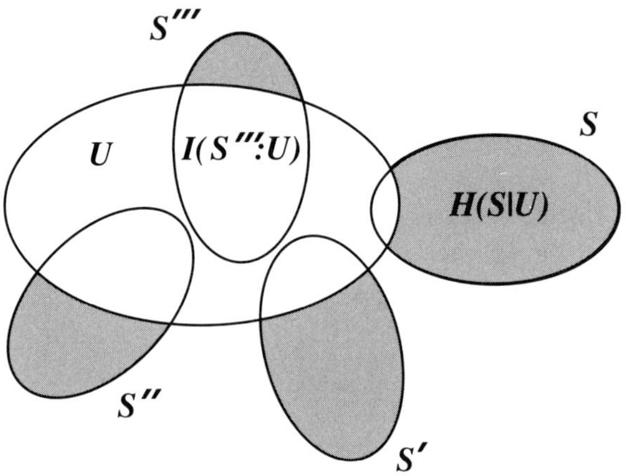

FIGURE 5.7 Evolution of complexity by increasing mutual entropy from $S \to S'''$.

only be applied to ensembles of strings, and that this complexity is just the information contained in the population of strings *conditional* on the environment to which the strings pertain. At the same time, because this information is the same among all strings that differ only in the hot bits, we can measure the complexity of *any* representative of this group by just counting the number of its cold bits. Evolution proceeds by a simple mechanism in which Turing machines constantly attempt to compute on the strings, trying to obtain from the random parts (hot bits) strings that also pertain to u. This can be viewed as ongoing, spontaneous *measurements* that are being performed on S. Any time a measurement succeeds (or in other words a Turing machine was able to derive some bits of s from u), the conditional entropy of S decreases as the mutual entropy of S with u, $I(S : u)$, [compare Eq. (5.10)] increases. At the same time, a cold bit usually does not revert back to a hot bit (information is not lost) because of the mechanism of survival of the fittest. Interestingly, this is just the Maxwell demon phenomenon: measurements are being performed that allow a reduction of the entropy, but the reverse process (the loss of information) is not occurring because such strings do not prosper. This is equivalent to the demon allowing slow molecules into one part of the box (thus reducing its entropy), while not allowing the fast molecules the same. In a sense, the demon provides a *semipermeable* membrane between the two halves of the box. The same can be said for information in the process of evolution: information can enter the genome, but it cannot leave. As a consequence, genomes are doomed to accumulate more and more information and grow longer and longer as a consequence. Thus, it seems as if it is indeed information that is "that which increases when a self-organizing system organizes itself" [Bennett, 1995].

In the next section, we shall try to see how these observations can be applied to real systems, whether natural or artificial.

5.5 Complexity of tRNA

If information is acquired by natural populations as described above, we expect to see its telltale sign: fixed positions that code for information and volatile positions that do not, in DNA or RNA. To reiterate the obvious, it is of course impossible to tell which nucleotide is fixed and which is not if we are given only one specimen of the DNA we are considering. Furthermore, if given an ensemble of such strings, we must make sure that enough time has elapsed since the last major evolutionary event (the last phase transition) so that the population has returned to equilibrium. In practice, this may be a very difficult condition to fulfill.

To illustrate how important this equilibrium condition is for estimating the complexity of a bit-string, imagine a piece of DNA that is completely random, and that as a consequence is not expressed. If given enough time for equilibration, each nucleotide should have a 25 percent chance of being found at any of the locations of the random string. According to the rules, each site is then declared volatile, i.e., uncorrelated with the physical world in which the population is evolving, and thus carries no information or complexity. Imagine further that shortly after we checked that this section is random, somewhere else in the DNA of this organism a mutation takes place that makes the host of this mutation far superior to those that do not carry this mutation. In principle, if the resources for the population are finite, this new strain is going to drive all the other strains into extinction, and shortly thereafter all strings will reflect the new genotype. Now, because the replication of the new genotype took place so rapidly and all the old equilibrated genotypes have vanished, we shall find close to 100 percent probability of finding one particular nucleotide at the positions in the region of DNA that we previously determined to be random, as opposed to the 25 percent measured before the transition. Of course, this particular nucleotide is inherited from the string that introduced the innovation. Still, we cannot conclude that the mutation that took place elsewhere on the string also reverted the random section to information. Rather, we have to wait until equilibrium is reached again, i.e., until the forces of mutation have had a chance to randomize the (previously) random section under consideration. Note that this can take, depending on the rate of nucleotide substitution and the length of the sequence, several million to hundreds of millions of years. It is therefore safest, when estimating the complexity of DNA, to consider only sections that are known to be old. Such is the DNA that codes for the well-known transfer RNA molecule (tRNA).

tRNA is the RNA molecule that translates a DNA sequence into its respective protein. This micromachine has two distinct ends: one where it attaches to a codon of DNA (the anti-codon region indicated by a curved

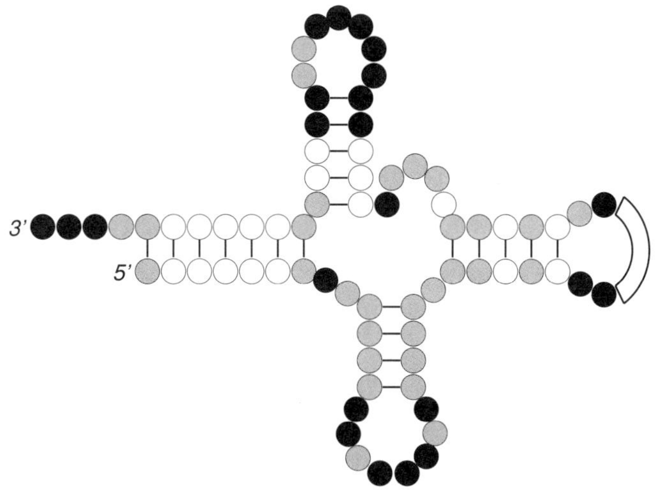

FIGURE 5.8 Secondary structure of tRNA with 76 common positions, of which 52 are independent and thus useful in the determination of its complexity. Fixed positions are black, moderately diverged positions are gray, and highly volatile positions are white (adapted from Eigen et al., 1989).

box in Fig. 5.8) and another where the respective amino acid is attached (the 3′ end). Thus, there are many different such micromachines (each specified by its anticodon). Clearly, these molecules with a distinct cloverleaf structure (see Fig. 5.8) must have formed around the time when the genetic code was fixed, i.e., around four billion years ago. Indeed, all life on earth makes use of these molecules, and we can therefore sample the genetic code for tRNA from a large ensemble indeed: all living things. The equilibration bottleneck that occurs any time a new species is created is avoided by determining the variation inside the tRNA of a specific species that is known to have evolved very early on, for example, bacteria.

For determining the complexity of tRNA, let us first make the simplification from a quaternary alphabet to a binary one. The four nucleotides of RNA can be divided into two groups: the purines (A or G), and the pyrimidines (U or C). Thus, we can give each nucleotide the purine or pyrimidine tag, and convert the quaternary string into a binary one. Furthermore, the RNA molecule is folded in such a manner that certain nucleotides are *paired*. Thus, only one member of the pair should be counted when counting either fixed or volatile bits, as the other one is

100 percent correlated to its partner. If we subtract from the 76 positions in the tRNA sequence in Fig. 5.8 the 21 paired positions, and the three positions determining the anticodon, we are left with 52 significant positions. A rough estimate of the complexity of tRNA of the bacterium *Bacillus subtilis* can be obtained by determining the volatility of positions from a pool of *B. subtilis* sequences. Manfred Eigen and his collaborators at the Max-Planck Institute in Göttingen [Eigen et al., 1989] determined by comparison of 28 sequences that 21 positions (those filled in black in Fig. 5.8) are identical through the pool, whereas 21 positions were moderately volatile (gray circles) and 10 positions are fully diverged (white circles). Thus, because we can assume that Nature has managed to compress the information contained in this tRNA to the smallest sequence possible, we conclude that the complexity of the tRNA of *B. subtilis* is somewhere between 21 and 42 bits. In general, a more precise result could be obtained by estimating the entropy of the *population* of tRNA strings, rather than of the individual bits. This is a difficult task, as the probability for each genotype, P_i, can only be estimated using its relative frequency in the sample, n_i/N, where N is the total number of strings in the sample. Due to the sampling error that this introduces, such an estimate will only be meaningful in the large population limit (populations of the order D^ℓ, where D is the size of the alphabet and ℓ is the average length of the strings). Indeed, it can be shown (see Basharin, 1959) that estimating the entropy of a variable

$$H = -\sum_{i=1}^{M} P_i \log P_i \tag{5.11}$$

by estimating the P_i from a *finite* sample N leads to a *sampling error* (to first order in $\frac{1}{N}$

$$\Delta H = \frac{M-1}{2N}, \tag{5.12}$$

where M is the number of possible values of the variable. Since the number of different values that a string made of ℓ instructions taken from an alphabet of size D can take on is $M = D^\ell$, we need a sample size of the order $N \approx D^\ell$ for the correction to be of the order 1.

Quite generally, for finite populations, the entropy per string $H(S|u)$ (the remaining entropy given the universe string u) is *not* just the sum of the *per site* entropies $H(s_j|u)$. The latter, the entropy for site j,

$$H(s_j|u) = -\sum_i^D p(i|u)\log p(i|u), \tag{5.13}$$

where $p(i|u)$ is the probability to find the ith instruction at site j (given universe u), measures the entropy per site *without* regard to correlations to other sites. As an example, consider a string of three sites that we label a, b, and c. Then the entropy per string $H(abc)$ is related to the per-site entropies $H(a)$, $H(b)$, and $H(c)$ by

$$\begin{aligned} H(abc) &= H(a) + H(b) + H(c) \\ &\quad - [H(a:b) + H(a:c) + H(b:c) - H(a:b:c)] \\ &= H(a) + H(b) + H(c) - H_{\rm corr}(abc)\ . \end{aligned} \tag{5.14}$$

While for independent sites the correlation entropy $H_{\rm corr}$ in general vanishes, this is not so for self-replicating systems. On the contrary, the correlations between sites (described by conditional probabilities such as $p_{a|b}$) are extremely important, and reduce the entropy $H(abc\cdots)$ to a value much smaller than the simple sum $\sum_j^\ell H(a_j)$. In general, for a string of length ℓ, the correlation entropy can have up to $2^\ell - 1$ nonvanishing terms.

As a consequence, the estimate based on counting the number of preserved sites in the string should be considered a *lower* limit [Adami and Cerf, 1997] on the amount of mutual entropy contained in it, as we may overestimate the number of volatile bits (if some are correlated with each other). Clearly, we have tried to avoid such correlations in the *fixed* sites by not counting as one site all base pairs (since if one of the sites is conserved its partner will be too), but in general the possibility of more correlations cannot be ruled out, and can only be detected in large samples.

5.6 Complexity in Artificial Life

In the following, let us attempt to estimate the information content, or complexity, of a population of self-replicating computer programs along the lines outlined in this chapter. Note that, with the avida system that accompanies this book on the CD ROM, we can monitor the information content from the beginning of evolution for as long as we want. To run this experiment, we seed the world with a single ancestor that has the capability to self-replicate. Clearly, this ancestor already contains some information about its world, namely the ability to self-replicate. However, since this program was written by humans, it certainly does not code this information in the most concise manner. If we let the population equilibrate after the computer memory is filled with the ancestor's offspring, we may begin to extract the information content.

In the limit of infinite population size, the information content of the population is, as we determined earlier, the maximal entropy of the population $H(S)$ minus the conditional, or remaining entropy $H(S|u)$. Note that if we take logarithms to the base of the alphabet size D, the maximal entropy $H(S)$ is just the average length of the strings in the population, and the conditional entropy is obtained as

$$H(S|u) = -\sum_i P_i \log_D P_i \ , \tag{5.15}$$

where P_i is the probability of finding genotype i when randomly sampling the population, i.e., *given* the particular environment u. For finite populations, we would like to approximate P_i by the number of times we find genotype i in the finite population of N strings, divided by N,

$$P_i = \frac{n_i}{N} \ . \tag{5.16}$$

Because we need populations of the order D^ℓ in order to accurately estimate the probability P_i from the samples n_i, entropies such as (5.15) cannot be used to estimate the true entropy of the population [see also Eq. (5.12)]. Instead, we revert to the counting of volatile instructions, i.e., we sum up the *per-site* entropies

$$\tilde{H}(\ell) = \sum_j^\ell H(x_j) = -\sum_j^\ell \sum_x^D p(x_j) \log p(x_j) \ , \tag{5.17}$$

where $p(x_j)$ is the probability to find instruction x at the jth site in the population, while ignoring the correlations. Again, this leads us to *overestimate* the entropy $H(S|u)$, and thus we *underestimate* the complexity

$$C(S) = \ell - H(S|u) \gtrsim \ell - \tilde{H}(\ell) \ . \tag{5.18}$$

Still, the hope is that the error committed is roughly constant for different ℓ, in which case we can monitor the increase in complexity during an evolution such as depicted in Fig. 5.7. As a preliminary step toward such a measure, let us assume that $H(S|u)$, i.e., the number of hot bits per string, is *approximately* constant during a run. Then we can monitor the evolution of complexity just by viewing ℓ as a function of time. Of course this is different from the physical complexity by at least a constant $H(S|u)$. In Fig. 5.9 we show the development of the average length of programs from the beginning of a run (where the information stored in the ancestral string only refers to self-replication) through the adaptive process in which the strings learn a number of logical and arithmetic tasks (upper curve, solid line). Since the information needs to be stored in the genotype, the length of the code grows constantly. However, even when the length (which can fluctuate) decreases, the amount of information about the environment that is stored in it should never decrease. For comparison we also show what happens if no information (except how to self-replicate) is extant in the environment. In that case the length of the code shrinks, and the complexity stays constant (lower curve, dashed

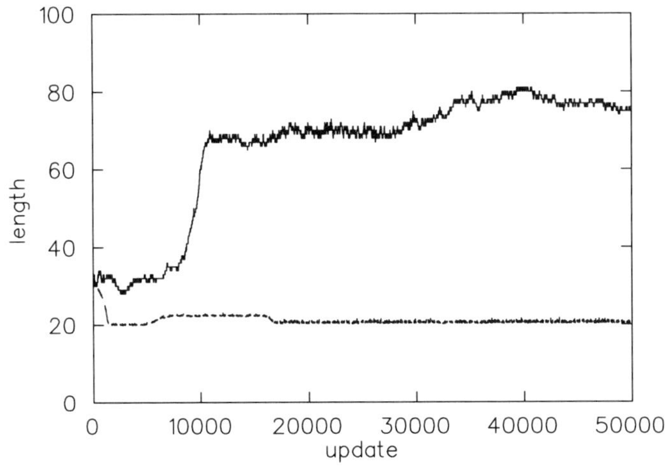

FIGURE 5.9 Evolution of average length in avida for complex landscapes (upper curve) and flat landscapes without information (lower curve, dashed line).

line in Fig. 5.9). Note also that attaching a random piece of code to the programs does not change the complexity either.

Finally, for a run where insert and delete mutations are turned off (such that all the strings in the population usually are the same length, except typically for size-doubling) the physical complexity (5.18) can be obtained, and its evolution compared to the development of fitness. In Fig. 5.10, we can see how each fitness increase (lower panel) is accompanied by an increase in physical complexity (upper panel), reflecting how much information was acquired per string during the transition. Naturally, the complexity first overshoots its equilibrium value during the transition (as explained earlier), but returns to equilibrium in a time

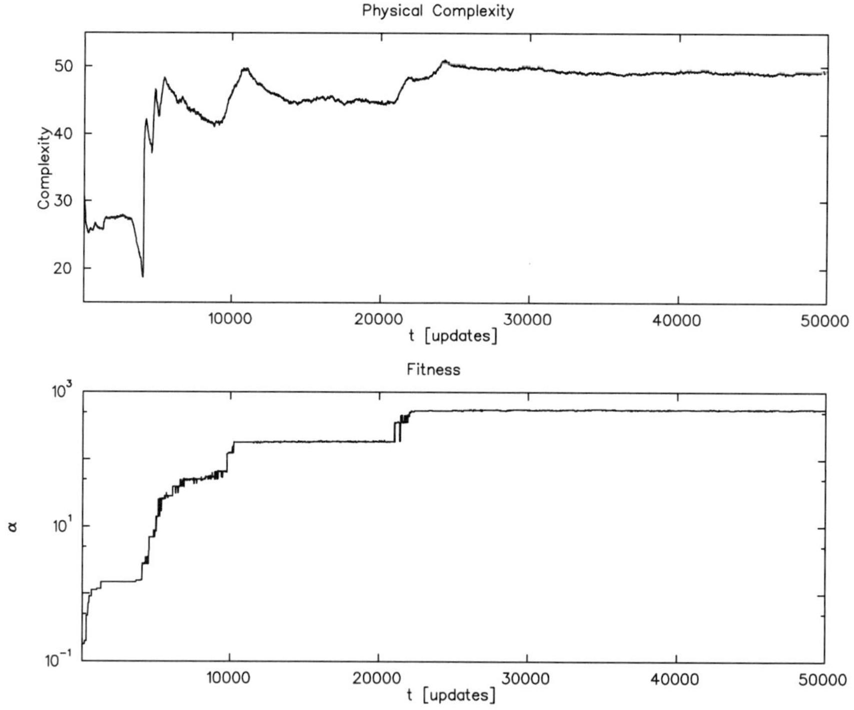

FIGURE 5.10 Evolution of physical complexity (upper panel) in avida in a complex landscape, compared with evolution of fitness (lower panel), in a population of 3,600 strings.

of the order of the relaxation time of the system[2]. Once the population has stopped learning, the complexity stays constant and is independent of any size changes. Note that the logarithm in Eq. (5.17) is taken to the base of the alphabet size, so that the complexity is expressed in units "instructions".

5.7 Overview

The complexity of a symbolic sequence is determined by the amount of entropy it shares with the environment within which it is to be interpreted. This has two consequences: complexity is *conditional* on the environment, and the complexity of a single sequence cannot be determined without reference to its environment because its mutual entropy cannot be determined.

This complexity can also be formulated in terms of automata theory. In this language, complexity is the *mutual Kolmogorov complexity* of a sequence s with the tape of a Turing machine u. Similarly, the tape of the Turing machine is the universe, or environment, within which the sequence s is to be interpreted. In other words, the complexity of a symbolic string is the length of the string minus the number of bits that mean *nothing* as far as the Turing machine is concerned, i.e., it is the length of the string minus the smallest length program that can compute the string from the machine's tape u. This smallest program represents the *randomness*, with respect to u, of the symbolic string under consideration, as any nonrandomness can be computed from u by a program of *vanishing* length in the limit of infinitely long sequences. For practical purposes, as universal Turing machines are not always at hand (and their behavior cannot always be predicted), we have to rely on the *average mutual*

---

[2]The decrease in complexity seen early in the run reflects the nonergodicity of the system: the ancestor we start with is not well adapted, but is difficult to mutate also. The true complexity of the ancestor is therefore much lower than indicated at the start of the run, but reaches this value after about 5000 updates (and some major structural changes). Similar structural changes and a corresponding drop in *essential* length is also seen in the run depicted in Fig. 5.9, lower (dashed) curve.

Kolmogorov complexity, averaged over sequences s from an ensemble S. This reduces to the mutual entropy considered above.

This measure of *physical complexity* can be used to estimate the complexity of genomes. For this endeavor, a lower bound on the complexity of the genome can be obtained by counting the number of *conserved* nucleotides within an ensemble of similar sequences that has had the time to equilibrate.

Problems

5.1 Consider a very small special-purpose computer with three binary elements p, q, and r. A machine cycle replaces p by r, replaces q by r, and r by $p \cdot q$. There are eight possible initial states, and in thermal equilibrium they will occur with equal probability. What is the *minimum* amount of dissipation that has to go on in one machine cycle [Landauer, 1961]?

5.2 An arbitrarily long sequence is formed by a flawed random number generator which, instead of generating a string of symbols taken from the alphabet $0 \ldots 9$ with equal probability for each symbol, will follow each zero by another zero, and then return to normal operation. The resulting sequence, as a consequence, will display some structure: it will be complex. What is its complexity *per symbol*? (Take your logarithms to the base 10 for convenience).

CHAPTER SIX

Self-Organization to Criticality

Die Lösung des Problems des Lebens erkennt man am Verschwinden dieses Problems.[1]

L. Wittgenstein, 1922

The idea that living systems tend to self-organize has been around for a quarter of a century [Eigen, 1971]. The concept was introduced to understand the apparent chicken-and-the-egg problem of what came first: proteins or nucleic acids. The main tenet is that biological systems are organized by the information present in them, and that the information in turn originates in the self-organized state by means of selection. As a consequence, one witnesses the establishment of structure or order in such a way that the entropy of the system is *not* maximal, i.e., the system is not in equilibrium. Many people believe that self-organization is one of the hallmarks of living systems.

Rather than reviewing the considerable amount of literature devoted to self-organizing systems, we will confine our attention here to theories of *self-organized criticality* (SOC), a concept introduced by Bak and cowork-

[1]The solution to the problem of Life is apparent as the problem vanishes.

ers [Bak et al., 1987]. The concept will allow us to take a very different look at the evolution of self-replicating systems, from the point of view of statistical theories that admit critical points, i.e., a branch of physics usually concerned with phase transitions in condensed matter systems. The idea is to abstract the interaction between the elements in the system (the self-replicating strings) to such a degree that they can be described by simple theoretical models. It is then the task of the theorist to isolate those characteristics of the models that carry over to living systems from those that are just an artifact of the abstraction.

6.1 Self-Organization and Sandpiles

The paradigm for the self-organized critical state is the *sandpile* that Bak, Tang, and Wiesenfeld (BTW) introduced, and that we will describe below. Its usefulness lies partly in the fact that it is so simple, yet displays some of the uncanny traits of natural systems such as power law behavior and self-organization (see Bak, 1996 for a nontechnical introduction to SOC in Nature). Power law behavior is seen in many physical systems, such as thermal noise in electronic devices (shot noise), the flashing of fireflies, turbulent fluid flow, activity patterns in neural networks, the distribution of earthquake sizes (the Gutenberg-Richter law), the distribution of solar flares and sunspots, the intensity fluctuations of quasars, and the size distribution of initial masses of stars (the Salpeter law), to name a few. Specifically, it was the frequency distribution of noise in many physical systems known as $1/f$ noise that prompted the idea of SOC. In general we distinguish between three main types of power laws in physical systems. First, we have the power spectral density distribution (such as $1/f$ noise), where

$$P(f) \sim \frac{1}{f^\alpha}\,, \tag{6.1}$$

where f is the frequency and $P(f)$ is the power at that frequency. In general, this function describes which frequency is the most dominant in the temporal behavior of the system under consideration: the power spectral density is just the square of the Fourier transform of the signal under consideration (see also Problem 6.2). The exponent does not necessarily have to be $\alpha = 1$ such as in $1/f$ noise, but it is in general a small real number. Another kind of power law appears in size distributions:

$$N(s) \sim \frac{1}{s^{\tau}}, \tag{6.2}$$

which reflects a distribution of frequency of events $N(s)$ as a function of the size of events. This is the kind of distribution observed in the Gutenberg-Richter law. Finally, we distinguish a power law in the *temporal* distribution of events, where τ is either the duration of an event (as in the sandpile experiments that we will meet below), or better, the time *between* events (also known as inter-event-interval distribution):

$$N(\tau) \sim \frac{1}{\tau^{\gamma}}. \tag{6.3}$$

Without trying to define SOC, let us immediately describe the sandpile. In the simplest, one-dimensional model, imagine a linear lattice of L sites, on which we can distribute grains of sand one at a time. Let the number of grains deposited on site j be denoted by $h(x_j)$ (the height of the pile at site x_j, as shown in Fig. 6.1).

The rules of the game are now such that grains of sand are free to accumulate as long as the height *difference* between adjacent sites does not exceed two (as for example is the case between the eighth and ninth site in Fig. 6.1). Such a situation is unstable, and a grain has to tumble from site j to site $j+1$. Should site $j+1$ become unstable due to this process, the tumbling continues up until there are no more sites

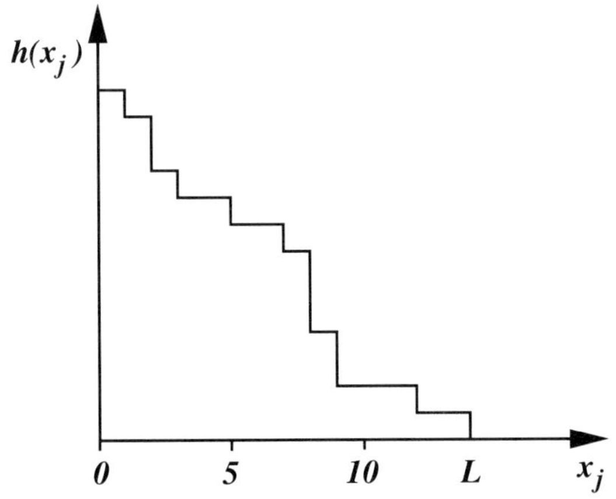

FIGURE 6.1 Sandpile on a linear lattice with L sites.

with a height difference between them exceeding two. This is called the *minimally stable state*. Indeed, dropping a grain randomly anywhere on the pile may result in either no transport of grains, or in a cascade involving any number of sites. Let us write down the update rules for the sandpile. They can be implemented in a straightforward manner as a simple cellular automaton updating the slopes $z_j = h(x_j) - h(x_{j+1})$. Here, we write rules in a general manner, where ν grains topple if the critical slope z_c is exceeded (a *supercritical* site):

Rule (i) — "adding of sand":

$$z_j \to z_j + 1 \,, \tag{6.4}$$

$$z_{j+1} \to z_{j+1} - 1 \,. \tag{6.5}$$

Rule (ii) — "tumbling of grains": (see Fig. 6.2)

$$\text{For } z_j > z_c : \quad z_j \to z_j - 2\nu \,, \tag{6.6}$$

$$z_{j\pm1} \to z_{j\pm1} + \nu \,. \tag{6.7}$$

The lattice is furthermore subject to the boundary condition $h(x_L) = 0$, which means that sand slides off the pile at that end.

We can easily imagine what happens if we run such a sandpile for an extended amount of time, dropping a grain of sand on a random site, updating the pile such that no more sliding occurs, and repeating. Surely we will witness many events where only one grain topples, a good number of events where a few grains are involved in an avalanche, and a few rare events where *all* the sites are involved in the avalanche. We may

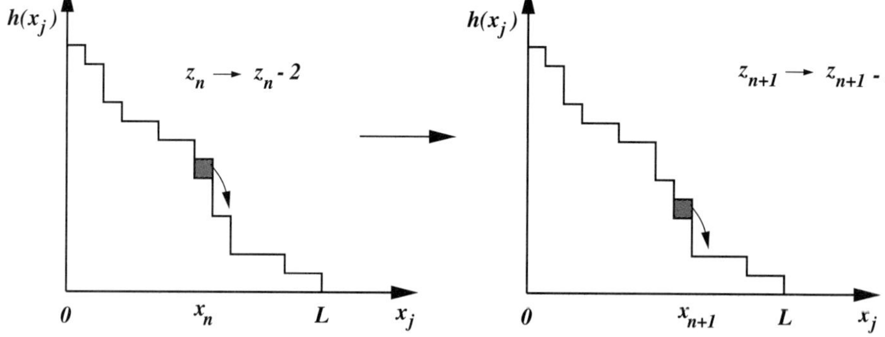

FIGURE 6.2 Tumbling of supercritical site x_n, with $\nu = 1$.

then ask what is the *distribution* of sizes of events. While this distribution can be obtained quite easily by simply performing the experiment [see Problem 6.2(b)], it is surprisingly difficult to arrive at the law from first principles, i.e., treating the model that we outlined above analytically. In Fig. 6.3 we show the abundance distribution of avalanches of size s for a one-dimensional sandpile of linear dimensions $L = 32, 64, 128$ and $\nu = 2$ [Kadanoff et al., 1989]. The case $\nu = 1$ turns out to be trivial, i.e., the abundance distribution is a constant. We plotted the function on a logarithmic scale for both the size s and the abundance $N(s)$, to highlight the power law behavior. Indeed, if the functional form is $N(s) = Cs^{-\tau}$, plotting the logarithm of $N(s)$ against the logarithm of s will result in the functional dependence:

$$\log N(s) = \log C - \tau \log s\,, \tag{6.8}$$

i.e., a *linear* law with slope $-\tau$. For the one-dimensional sandpile, we can fit the function with an exponent of $\tau = 1.0 \pm 0.1$ at small sizes. Note the strong dependence on the system size, however. The slope of the power law is best analyzed using finite-size scaling relations for this system [Kadanoff et al., 1989].

The experiment that yielded Fig. 6.3 was performed in such a way as to first fill the empty lattice with sand up to the point where adding a

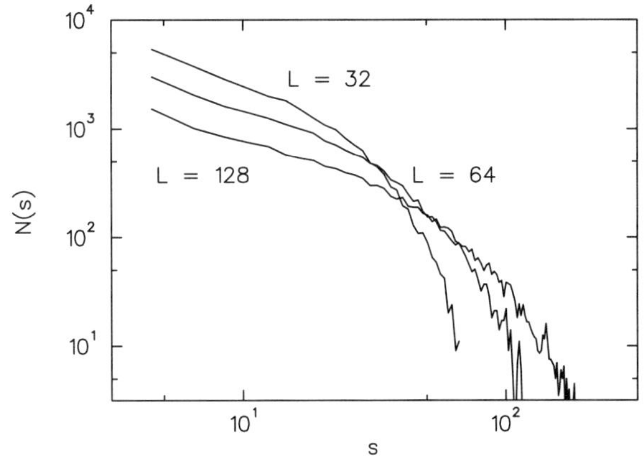

FIGURE 6.3 Abundance distribution $N(s)$ of avalanches of size s for a one-dimensional sandpile with $\nu = 2$ (the $\nu = 2$ "LL" model of Kadanoff et al., 1989) for sizes $L = 32$, 64, and 128.

single grain will almost always produce a sliding event. This means, of course, that the left side of the lattice (x_0) is filled to roughly its maximal height given $h(x_L) = 0$, and *almost all* of the slopes are at the critical point $z_j = 2$. This is called the *critical*, or minimally stable, state. Now, dropping a grain may result in an avalanche that disturbs the entire pile. Note, however, that while this critical state is disturbed by the avalanche, after the event the pile will *return* to the critical state due to the continuing dropping of the grains. Thus, the *average slope* of the pile represents a fixed point of the system: the pile will always return to it, but the state itself is highly unstable, and any small perturbation will result in another runaway event. This gives us a first indication of why sandpiles may be a good place to look for an analogy to the dynamics of living systems. There is a certain *robustness* to the pile, and it responds to perturbation by events of *all* sizes that may even be catastrophic, but still will always return to an equilibrium state which, in turn, is singularly prone to another perturbation. While the analogy is clearly very rough at this point, we will continue to refine the model to make it closer to natural systems, and find that the pile, as abstract as it is, may embody some of the same principles that govern simple systems of self-replicating entities, albeit in an entirely different medium.

Before examining the dynamics of the sandpile in a more rigorous manner, let us first generalize it to two dimensions. Rather than first constructing a lattice with a height distribution $h(x, y)$ and then calculating the slopes, let us immediately work with the slopes $z(x, y)$, defined at the coordinates x, y of a two-dimensional lattice. The rules for updating the lattice are again very simple. Adding a grain at site (x, y) results in

$$z(x-1, y) \to z(x-1, y) - 1\,, \tag{6.9}$$
$$z(x, y-1) \to z(x, y-1) - 1\,, \tag{6.10}$$
$$z(x, y) \to z(x, y) + 2\,, \tag{6.11}$$

while a toppling event takes place if

$$z(x, y) > z_c: \quad z(x, y) \to z(x, y) - 4\,, \tag{6.12}$$
$$z(x, y \pm 1) \to z(x, y \pm 1) + 1\,, \tag{6.13}$$
$$z(x \pm 1, y) \to z(x \pm 1, y) + 1\,, \tag{6.14}$$

i.e., if the slope at one site is supercritical, it is distributed evenly among its four neighbors to the north, east, south, and west. In the examples we

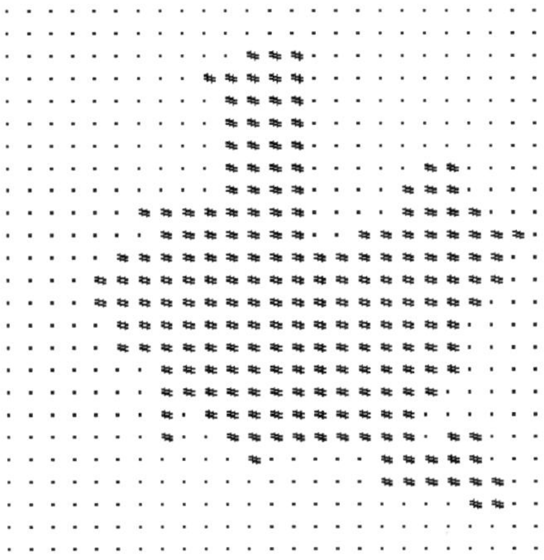

FIGURE 6.4 Two-dimensional lattice of size 25×25 with avalanche of 207 sites.

show here we choose $z_c = 4$, but this is not necessary. If the lattice is filled with sand again like in the one-dimensional case (for example by dumping an excessive amount of sand on the table and updating until no toppling takes place anymore), we can witness avalanches of all sizes by randomly dropping grains on the lattice at an arbitrary site and updating until no more site is affected. An example of such an avalanche for a 25×25 lattice is shown in Fig. 6.4. We can repeat this experiment many times and ask again about the distribution of avalanche sizes. Again, we will find power law behavior. However, this time the exponent of the power law is slightly different. In Fig. 6.5 we show the (binned) abundance distribution of 20,000 avalanches on a 50×50 lattice, fitted with an exponent $\tau_2 = 1.12 \pm 0.05$.

Clearly, the power of decay of the size distribution depends on the geometry of the system. For a three-dimensional lattice, BTW found an exponent $\tau_3 \approx 1.37$ [Bak et al., 1988]. The important point is that the law of decay does *not* involve any *scale*, as would be the case for an exponential law (such as radioactive decay, where the scale is the half-life) or a logarithmic law, which also requires a scale. Indeed, the power laws occur precisely in the case where there *is* no scale to set an average. More precisely, there is no scale of the same order as the range where we observe power law behavior. This is an important point that needs

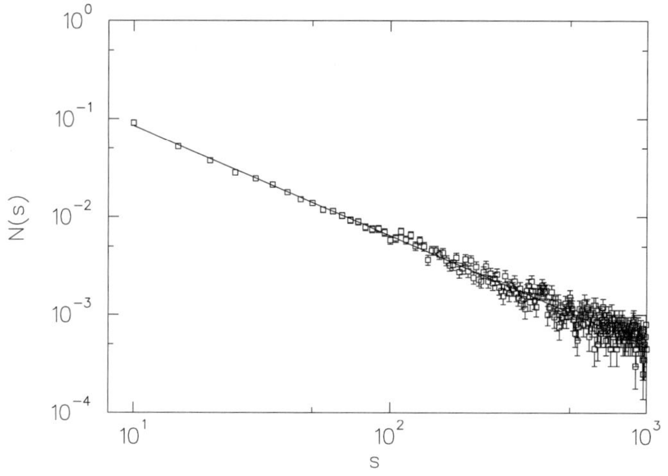

FIGURE 6.5 Size distribution of avalanches obtained from 20,000 avalanches on a 50×50 lattice.

to be stressed, as for any system that we investigate experimentally or numerically, there *is* one fundamental scale: the size of the system under investigation, as we have seen in the one-dimensional case. Of course for the sandpiles introduced here, this is the size of the lattice. Clearly, we cannot find avalanches that involve more sites than are in the lattice, so we cannot expect the power law to be valid close to this limit. In general, therefore, we must be sure to study finite size effects by fitting the power law in such a way that the finite size is taken into account. In Fig. 6.5 above, this was unnecessary, as we stopped the fit at an avalanche size 10^3, whereas the ultimate cutoff is of course 2500, the maximal number of sites.

Let us try to summarize the main requirements that must be met in order to observe a self-organized critical system. Note that this list does not specify if any of the conditions are necessary or sufficient. Because as of yet, there is no universal theory of SOC, the list looks somewhat like a shopping list, but gives a good idea about where to look to find SOC: It appears that we want

- A dissipative dynamical system with (locally) interacting degrees of freedom.

- Propagation of fluctuations described by a dissipative transport equation.
- Noise that can propagate through the *entire* system.
- An infinitesimal *driving* rate.

Let us go through these one by one. Dissipation plays an important role in self-organizing systems for the simple reason that it would be almost impossible to transmit signals in a system that is both noisy *and* nondissipative. There is a very general theorem of thermodynamics that states that fluctuations, which as we shall see essentially represent the signal, are *always* accompanied by dissipation. Without going into the mathematical details of this theorem, we can intuitively understand that it would be hard to encode and decode signals into the fluctuations if there were no way to damp them. Local interactions are obviously an important ingredient in self-organization, even though it is in principle possible to imagine systems that self-organize but where each part of the system is in direct contact with any other. Therefore, the locality may not be a necessary condition, even though most physical systems have that property.

The second condition is almost a consequence of the first: only in very awkward systems that are dissipative and have locally interacting degrees of freedom is the transport of fluctuations *not* described by a dissipative transport equation (such as a reaction-diffusion equation). However, it is important to require that the *entire* system is accessible to the fluctuations and therefore to noise (third condition). Indeed, we cannot imagine self-organization to occur in a system that has parts that are not connected to the rest. It is, after all, the fluctuations that provide for the communication between all parts of the system, which results in self-organization.

Finally, and most importantly, the system has to be driven at an infinitesimal rate. This most important condition will reappear in many different guises throughout this chapter. First, any system that is to self-organize has to be *driven*, i.e., there must be a force that is responsible for providing the fluctuations that may or may not result in catastrophically big events. For the sandpile, this is of course the dropping of the grains, and in the protocol outlined earlier we made sure to specify that one waits until the avalanche is over before dropping another grain. This is not, strictly speaking, a physical situation. Since the system is driven by the dropping of the grains, we ought to define a constant driving rate: the number of grains dropped per unit time [Becker et al., 1995] (see also Schmoltzi and Schuster, 1995, for introducing a time scale in the Bak-Sneppen model [Bak and Sneppen, 1993]). However, even if this is only specified on average, (so that the dropping of grains is still a Poisson-random event) we encounter the possibility, for every *finite* driving rate, that a grain of sand is dropped *before* the last avalanche is completed. We see immediately that if we increase the driving rate in such a manner that we do *not* wait for the avalanche to finish most of the time, we reach a situation of *steady flow* of sand. Clearly such a state is neither self-organized nor critical. Indeed, we shall see later that SOC only results in the limit where the driving rate is infinitesimally small. In this limit, the system will always return to its critical state, which is so prone to disruption. It is this self-tuning feature that has attracted the most attention, as all standard statistical systems that possess a critical point (such as the freezing transition from water to ice, or the freezing transition in magnetic Ising systems) sport a parameter (the temperature in the latter examples) that has to be fine-tuned to obtain this state. In SOC, the system apparently self-tunes to this state. Before attempting to understand this feature, we first describe another model that displays SOC and then show how SOC may be an important ingredient in the behavior of populations of self-replicating entities, specifically artificial ones.

6.2 SOC in Forest Fires

Here we investigate another simple model that displays SOC but that is easier to analyze in a systematic manner. The analytic treatment yields some important insights into the limits of self-organized behavior and points to possible generalizations. The Forest Fire model was first introduced by Bak, Chen and Tang [Bak et al., 1990]. We will henceforth refer to their model as Model I. It was subsequently improved by Drossel and Schwabl [Drossel and Schwabl, 1992] (Model II). Let us first consider Model I.

Imagine a two-dimensional lattice where each site can be in any one of the following states:

- T [tree, susceptible to being burned]
- B [a burning tree]
- A [ashes: a tree that has burned down]

The dynamics of the model are determined by the following update rules. In one time-step, any B→A, i.e., a burning tree ceases to burn after being reduced to ashes. At the same time, any BT→AB, i.e., a burning tree will ignite an adjacent one, while leaving ashes only at the next update. Also, new trees can grow from ashes: A→T with a small probability $p < 1$. As we shall see, Model I turns out *not* to be critical. Rather, the dynamics are more similar to disease-spreading dynamics. The reason for this is that the system is *not* driven, so rather than returning to a critical state in a dissipative manner, we are witnessing waves of live and dead trees in the system. The crests of these waves are separated by a fixed distance that provides a *scale* in the system, as was reported in [Grassberger and Kantz, 1991]. We thus notice that the absence of a dissipative element renders the dynamics *periodic* rather than critical.

The required driving was added in the form of a small probability for trees to start burning spontaneously, i.e., a probability for lightning strike f, in Drossel and Schwabl, 1992. Thus, the added rule is that a tree will start burning: T→B with a small probability $f \ll 1$. Implementing this algorithm reveals that the dynamics of the forest is such that, after a transition period, the forest settles into a steady state with a constant mean forest (nonburning trees) density $\bar{\rho}$. Let us estimate the dynamics of the forest in this steady state.

Following Drossel and Schwabl, let us calculate the time between two lightning strikes hitting a tree. This can be done in a general manner for forests of all dimensions, even though it becomes a stretch to imagine any forest with dimension larger than $d = 2$. If the forest is d-dimensional with linear dimension L (such that the total volume of forest is L^d), we obtain this time scale—the time between two strikes—by multiplying the probability for a tree to be hit, f, by the mean tree density $\bar{\rho}$ and the volume of trees, to obtain the rate at which trees are hit:

$$R_f = f\bar{\rho}L^d \; . \qquad (6.15)$$

Naturally, the time between hits is R_f^{-1}. We can then calculate the average number of trees *growing* between two lightning strikes, which is just the time between such strikes times the density of ashes $(1 - \bar{\rho})$ multiplied by the probability of tree growth p and again the volume of forest L^d. Thus, the average number of growing trees $\bar{s}$ is

$$\bar{s} = \frac{p(1-\bar{\rho})L^d}{f\bar{\rho}L^d} = \frac{p}{f}\frac{1-\bar{\rho}}{\bar{\rho}} . \tag{6.16}$$

Since we are in a steady state situation, this is also the average number of trees *burning*. Due to the nearest-neighbor interactions in this model, fires form clusters of burning trees surrounded by ashes. As a consequence, the average number of burning trees is the same as the average cluster size in this model. As we shall see, the average cluster size in a model in which there is no scale other than the size of the system diverges as the system size is made arbitrarily large. This is a key element in SOC systems, so let us study this in more detail.

Clearly, the distribution we are interested in is the size distribution of burning clusters, $N(s)$. Armed with this distribution, we can write down the total number of burning trees as

$$N_b = \sum_i^{s_{\max}} s\,N(s), \tag{6.17}$$

and thus the probability for any site to be burning is

$$P(s) = \frac{s\,N(s)}{N_b} , \tag{6.18}$$

where $s_{\max}$ is the maximum size of a cluster. Then, the *mean* number of burning trees is

$$\bar{s} = \sum_i^{s_{\max}} s\,P(s) = \frac{\sum_i^{s_{\max}} s^2\,N(s)}{\sum_i^{s_{\max}} s\,N(s)} . \tag{6.19}$$

Let us determine what happens if we *assume* a power law form for the distribution of clusters $N(s)$:

$$N(s) \propto \frac{1}{s^\tau} \tag{6.20}$$

with a critical exponent τ. We find, not surprisingly, that the average size of the cluster is determined entirely by the only scale in the problem, $s_{\max}$, as

$$\bar{s} \propto \begin{cases} s_{\max}^{3-\tau} & \text{if} \quad 2 < \tau < 3 , \\ s_{\max}/\log(s_{\max}) & \text{if} \quad \tau = 2 . \end{cases} \tag{6.21}$$

Consequently, as $s_{\max}$ diverges when the system size tends to infinity ($L \to \infty$), the average size of clusters also must diverge. Let us go back to the estimate we derived previously, Eq. (6.16). It becomes clear then that the average cluster size diverges only in the limit

$$f/p \to 0 , \tag{6.22}$$

which we recognize as the condition for an infinitesimally small driving rate! Thus, the system approaches the critical point in this limit. Note that condition (6.22) more precisely is a condition for the *separation of time scales*. According to this, we should only observe SOC behavior in the limit where the time scale for growing trees is much smaller than the time scale to ignite them.

6.3 SOC in the Living State

Let us now make the concepts introduced in these abstract models more palpable as far as living systems are concerned. If self-organization to criticality is a feature of living systems, what distribution is showing power laws? What is the agent of self-organization? How are fluctuations transported throughout the system? What is the critical threshold parameter? We shall attempt to answer all of these questions for a simple system of self-replicating strings. First, we shall try to argue for the existence of a self-organized critical state in populations of self-replicating strings and then carry out experiments on the computer to support the claim.

There is no shortcoming of distributions in natural living systems that show power law behavior, i.e., are of the same type as Eqs (6.1–6.3). We must be careful, however, not to attempt to find an SOC explanation for all of them. Most intriguing is perhaps the distribution of extinction events throughout the fossil record depicted in Fig. 6.6 [Raup, 1986], and the episodic nature of evolution associated with it: (the punctuated equilibrium scenario advocated by Gould and Eldredge [Gould and Eldredge, 1977; Gould and Eldredge, 1993]. The general idea is that some sort of ecosystem forms between all the competing species that makes them intimately connected. A small perturbation of this equilibrium would then result in extinction events of all sizes, which will leave a system that slowly creeps back to the self-organized state. Here, we shall investigate

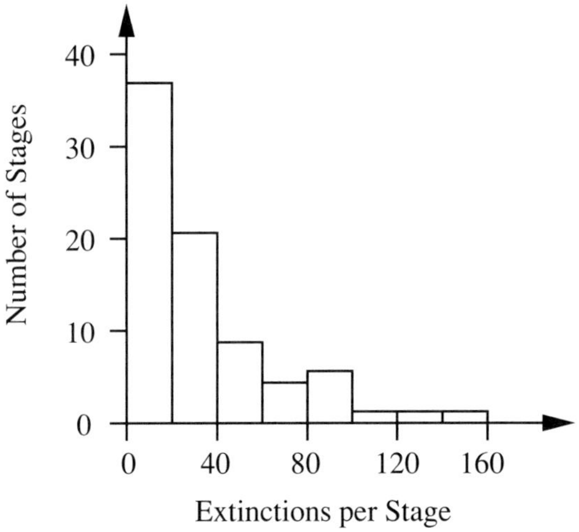

FIGURE 6.6 Distribution of extinction intensities based on recorded times of extinction of 2316 marine animal families (adapted from Raup, 1986).

a simpler model that does not assume such a strong interconnection between the species. Rather, the self-organizing agent we consider is *information*. We shall assume a population of strings of code whose only distinguishing mark is its genotype, reflecting (in the cold, i.e., non-volatile, spots) the information it has stored about its environment. The idea will be that if the population is sufficiently equilibrated, almost all of the strings (in a certain confined region) exhibit the same information. Consequently, they are susceptible to *avalanches of all sizes* instigated by rare mutation events that introduce new, powerful information into the population, akin to the seeds of first-order phase transitions spawned by nucleation events.

Newman et al., 1997 challenged the view that criticality is apparent in the tierra experiments that we shall discuss here, on the grounds that SOC is usually associated with *second-order*, rather than first-order, critical phenomena. While indeed one expects that SOC based on the connected ecosystem mechanism should display second-order dynamics (and we shall investigate the picture of SOC in terms of second-order phenomena later in this chapter), there is good reason to believe that SOC can also be described in terms of first-order transitions (see Gil and Sornette, 1996), i.e., the waves of invention occurring in tierra as well as avida.

Let us set up our model (theoretically at first) such that we can identify the critical threshold parameter (the analogue of the sandpile's critical slope). Imagine a population of N strings self-replicating in an environment with limited resources (so that the number of strings is constant). Each string is composed of ℓ instructions taken from an alphabet of size D. The total number of different strings is the maximum number of genotypes $N_{\rm max} = D^\ell$, usually a very large number. In the current population at time t, let there be N_g different such genotypes in existence. Remember from Chapter 4 that the maximum number of genotypes represents the total volume of phase space available to the population, while N_g is the volume currently occupied by the population. The population of N strings falls into subpopulations of n_i strings each of genotype i currently in existence, with each genotype i furthermore characterized by its replication rate ϵ_i. We then know that

$$N = \sum_i^{N_g} n_i \,, \tag{6.23}$$

and we can define the average replication rate of the population:

$$\langle\epsilon\rangle = \sum_i^{N_g} \frac{n_i}{N}\epsilon_i \,. \tag{6.24}$$

Note that the average replication rate is intrinsically time-dependent, while for Eq. (6.23) we have the additional constraint $\dot{N} = 0$. We can write down a simplified equation that determines the time dependence of the genotype occupation-number n_i, if we introduce the mutation rate R as the probability for mutating a site per unit time,[2] and the corresponding probability for a single string of length ℓ to be hit by such a mutation, $p \approx R\ell$. Then, in an approximation where the average number of strings *removed* per unit time is proportional to the average replication rate (this is the *chemostat* approximation, where all strings exceeding the total number of allowed strings N are simply removed), we can convince ourselves that [Adami, 1994]

$$\dot{n}_i(t) \approx (\epsilon_i - \langle\epsilon\rangle - R\ell)\, n_i(t) \,. \tag{6.25}$$

[2]This is a cosmic-ray mutation rate that affects each string in the population in the same manner. As $R\ell \ll 1$ always, we can approximate $1-(1-R)^\ell \approx R\ell$, unlike in copy mutations where $R\ell$ can be of the order 1 or larger.

Several assumptions go into the writing of Eq. (6.25), which here serves only to identify a critical parameter. On the one hand, we assumed that mutations are Poisson-random events that simply remove the genotype under consideration. In more realistic systems we need to include errors arising from an erroneous copy operation in the replication process. Such mutations behave differently than the external mutations we are considering here. An explicit model for populations subject to copy mutations will be presented in Chapter 11. Also, we ignore the fact that a mutation might *produce* a genotype of the type i by a mutation hitting some genotype j. In the limit where $N_g \ll N_{\max}$ this is a safe assumption, but it does imply that Eq. (6.25) violates the number conservation equation $\dot{N} = 0$. If we ignore all these problems for the moment and introduce the *growth-factor*

$$\gamma_i = \epsilon_i - \langle\epsilon\rangle - R\ell \,, \tag{6.26}$$

we can immediately solve Eq. (6.25) in the *static* limit, i.e., in the limit where the average replication rate is approximately independent of time, as

$$n_i(t) = N_0 e^{\gamma_i t} \,, \tag{6.27}$$

where N_0 represents the population of genotypes of type i at time $t = 0$. This equation implies that each genotype must either grow exponentially if its growth factor is larger than zero, or shrink in the same manner if it has a replication rate that renders the growth factor negative. Such a behavior is, however, incompatible with our assumption of equilibrium that guaranteed that the average replication rate is approximately time-independent. Thus, growth factors unequal to zero, while entirely possible, cannot possibly be a fixed point of the population, i.e., a point to which the population always returns. Rather, this point must be characterized by

$$\gamma_i \approx 0 \tag{6.28}$$

for *almost all* genotypes. This must then be the critical threshold parameter that organizes the entire population. Let us imagine a population poised at this state, where almost all γ_i vanish (except for those few negative γ_i that are results of mutations and are quickly purged from the population). Since no genotype dominates another, the situation is quasistable, even though genotypes are still being created and go extinct at a small rate that is determined by the number of hot instructions of the main genotype. Indeed, it is in general fair to assume that the population is dominated by a *quasispecies* (a term introduced by Eigen, 1971), which we can imagine as the genotype specified by its cold instructions only. While there can in principal be several quasispecies in the population, they must all be "degenerate" in the replication rate (must all have the same replicate rate), as otherwise one would drive the other to extinction. If enough time has elapsed since the establishment of the quasispecies, we realize that the information that characterizes it has spread throughout the population. The population then is very much connected, held together by the invisible thread that is their common cold sites: their complexity. Every mutation that changes the replication rate of a string to one higher than the previous $\epsilon_{\rm best}$, the replication rate of the quasispecies, constitutes a *ripple*: a fluctuation in γ_i. Because the population has organized itself to a state where $\gamma_i \approx 0$ almost everywhere, such a fluctuation can indeed travel throughout the entire population, a criterion that we remember as being on our shopping list for self-organized criticality.

Thus, we find that in this simple model information indeed organizes the population, and since the information is part of the population, we can safely say that the population *self*-organizes. But is it a critical state, and which parameter determines the infinitesimal driving rate? The answer to both of these questions lies in the magnitude of the mutation probability per string. First, we easily recognize the small driving of the sandpile (initiated by the dropping of the grains) in the driving which results from mutations creating ever-fitter genotypes. In the same manner we see that a condition for the emergence of a self-organized critical state is that avalanches due to new inventions (i.e., events that created a substantially better replication rate) must be rare, so rare, in fact, that most avalanches are long over (i.e., have eradicated and driven to extinction each and every inferior genotype) before a new invention sweeps through the population. Otherwise, we would witness overlapping avalanches just as in the continuously driven sandpile, and no critical behavior could emerge. On the other hand, the probability of a mutation leading to a fitter genotype must not be too small, as otherwise we would not find any avalanches on the time scale of our observation, and again no SOC would result. Thus, SOC in living systems is predicated upon a mutation rate *just* right so that the critical state is formed. While this sounds much like the tuning of parameters in standard statistical phase transitions, we shall argue that this critical mutation rate is actually achieved for most reasonable parameters of the living state, and furthermore there is evidence, both experimental and theoretical, that living systems actually arrange for this mutation probability to be just in the critical range [Domingo et al., 1980; Adami and Schuster, 1997]. How this is achieved is part of a different discussion, which we will enter when we describe the "race" to the error threshold (Chapter 11).

Let us now turn to observable consequences of self-organized criticality in such simple systems of self-replicating strings. First, we expect to see a punctuated picture of evolution, where fitness does not change gradually, but in bursts that are initiated by a genotype that found a better way to exploit the environment. The time between such events must be large compared to the time it takes to communicate this new information to the rest of the population. If you remember the discussion in Chapter 4, you will easily identify the latter time as the relaxation time of the system. Thus, we note again a separation of time scales, a recurring theme. Let us look at Fig. 6.7, a typical time-evolution of fitness (i.e., replication rate) of a population of self-replicating strings, here computer code from the tierra world, the system introduced by Ray, 1992 (the direct ancestor to our avida system).

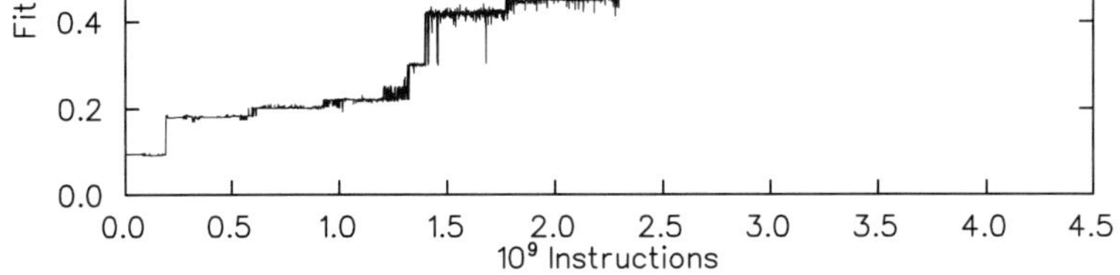

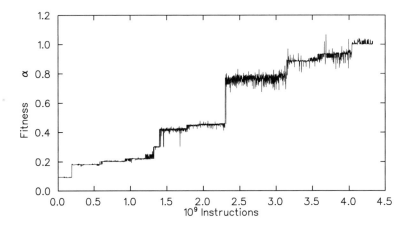

FIGURE 6.7 Normalized replication of best of population measured every million executed instructions for a population of tierran strings, with $D = 32$, $\ell \approx 100$, and $N \approx 1000$, (from Adami, 1995b).

The staircase structure of the fitness curve is very evident, underlining that evolution in that system indeed proceeds in such a way that periods of long stasis (the plateaus in Fig. 6.7) are interrupted by very brief periods of invention that move the population to a new plateau rather quickly, compared to the length of the plateaus. Time is measured in total number of instructions executed by the population, which is a convenient measure even though it does not scale well with population size.

Let us first look for any signs of periodicity in the dynamics. This would be revealed by a preferred frequency in the power spectrum of a time-series such as in Fig. 6.7. This spectrum (obtained by the so-called maximum entropy method) is shown in Fig. 6.8, and reveals no bumps of any kind: the spectrum is, except for the finite size effects at small and large frequencies, a pure power law:

$$P(f) \sim \frac{1}{f^2} \,. \tag{6.29}$$

Note that power law behavior for the spectrum is a necessary, but *not* a sufficient, condition for self-organized criticality. It reveals that there are no periodic dynamics in the population, but does not rule out certain

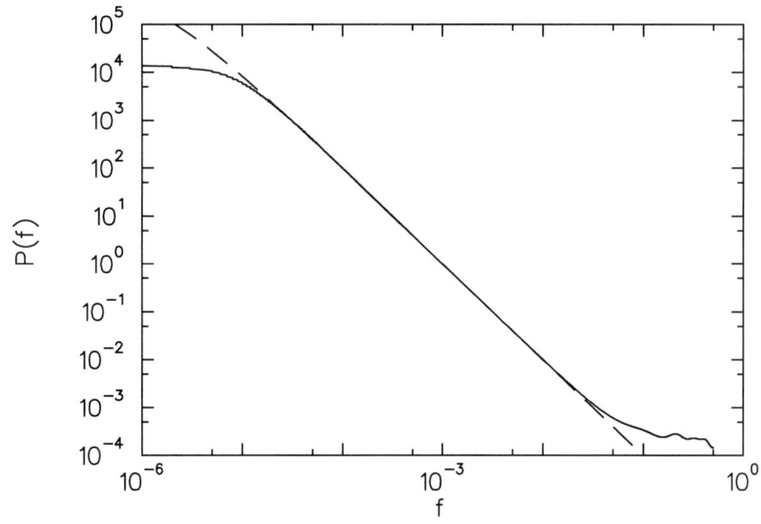

FIGURE 6.8 Square of Fourier transform (power spectrum) of the time-series shown in Fig. 6.7. The dashed line is a fit to $P(f) \sim 1/f^2$ (from Adami, 1995b).

random processes that have a power law frequency spectrum but show no signs of self-organization or critical behavior (see, e.g., Problem 6.2). To extricate the latter, we must look at temporal correlations and thus ask about the distribution of plateau lengths (in units "millions of executed instructions"). This seems to be a natural question, since the beginning of a plateau certainly signals the birth of a new quasispecies, while the end of a plateau heralds its demise. Thus, the length of a plateau reflects the time of domination of a quasispecies. Since the total number of representatives of that quasispecies that ever lived must be proportional to the length of time it dominated the population, the length of the plateaus may also tell us something about the absolute size (total number that ever lived) of a species. Clearly, however, a single run such as depicted in Fig. 6.7 will not be able to reveal this size distribution to us.

Let us couch the question about the distribution into a more technical language. We are here interested in the inter-event-interval distribution, where an event is defined as the emergence of a genotype noticeably fitter than those present before that event. The restriction to noticeably larger fitness improvements is a rather practical one, enabling us to pick out visually the beginning of a new epoch. In each run, we witness between five and ten such events on average, but sometimes less and sometimes more. In the particular set of runs that are reported here, the strings were adapting to a specific landscape created by the user: one in which the strings were guided to learn how to add integer numbers (see Chapter 11 for more details). The specific kind of landscape is not our concern here, even though this is a subject of fundamental importance that we will look into in Chapter 8. We shall only demand that the landscape contains enough information (to be discovered by the strings) that no population will ever exhaust it during the time that we spend in observing the population. This, of course, is just the requirement that there always be a very small driving rate for the population: if the landscape is exhausted, or in other words, if there can be no more fitness improvement, we expect to lose the SOC behavior.

For such effectively infinite landscapes, we observe the adaptive behavior of the population at a fixed mutation rate and count the number and lengths of epochs. The distribution of number of epochs as a function of the length of the epoch is our inter-event-interval distribution. Let us look at the results, obtained in 50 runs under identical conditions, at a moderate mutation probability of the order 10^{-6}. The sizes were obtained by measuring (painstakingly) the lengths of every epoch by hand (visual inspection first, then extracting the beginning and end of the epoch through the identification of the jumps in fitness), so as not to bias the analysis. As the data obtained with tierra are somewhat noisy, a program designed to find the beginning and end of an epoch might easily be fooled. Still, the 512 epoch sizes that were extracted from these runs need careful statistical interpretation. For example, let us try first to bin the data into bins of length 100 (millions of executed instructions). For each data point that we obtain (the first one covering epochs of lengths between 1 and 100, centered at 50), we estimate the error in that number to be due entirely to statistics, i.e., we assume a $\sqrt{N}$ error, which is certainly appropriate for those bins with very few entries (at large epoch sizes), but could underestimate the error for small epochs, which may be affected by systematic trends. The result of such an analysis is shown in Fig. 6.9. We attempt to fit the data with the smallest number of parameters that seems reasonable to us, in the form of a power law

$$N(\tau) = A\frac{1}{\tau^{\beta}}e^{-\tau/T}\,, \tag{6.30}$$

where T is our cutoff parameter. The cutoff simply reflects the fact that our measurements are finite in time: most runs were terminated after

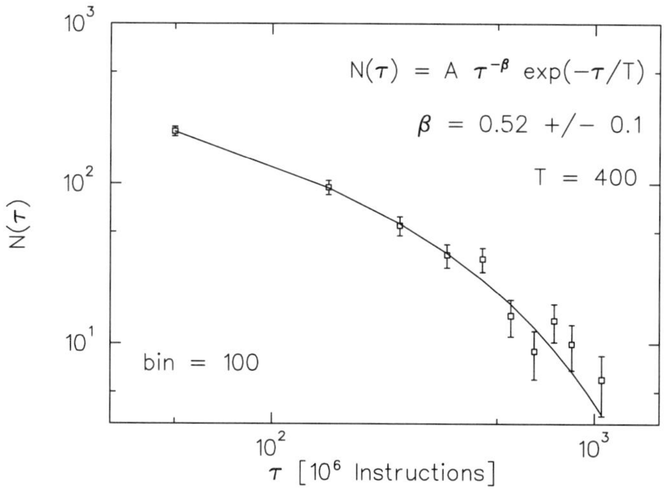

FIGURE 6.9 Fit of the binned abundance distribution with bin size 100.

between 500 and 2000 million instructions had passed (while a few were allowed to exceed the 4000 million mark, such as the one in Fig. 6.7). Nevertheless, we will certainly *undersample* lifetimes larger than 500 million. This is taken into account here by an exponential decay. There is a certain amount of freedom in the choice of a cutoff function, which can affect the result of the fit. Consequently, care must be taken in that choice also. In general, a full finite-size scaling analysis, which would involve repeating the experiment and cutting off the experiment at different times, should substitute for choosing a cutoff function.

As the resulting fit with $\beta = 0.52$ and $T = 400$ has a $\chi^2 = 1.1$ per point, we can at least be confident that we are not over-fitting the data. Yet we should not be overconfident, as we find that β depends strongly on the size of the bins chosen (a fact that can without effort be shown analytically, see Problem 6.3). Choosing a bin size of 200 results in data with smaller error bars, but the fit changes accordingly, as we can see in Fig. 6.10. At the same time, our measure of confidence, the χ^2 coefficient, drops to $\chi^2 = 0.44$, suggesting that we overfit the data.

In the face of such adversity, we can choose another path to extract meaningful results. Consider a different distribution $M(\tau)$, obtained from our distribution $N(\tau)$, by asking about the distribution of events with a size *larger* than τ. As this distribution, which has the same power law

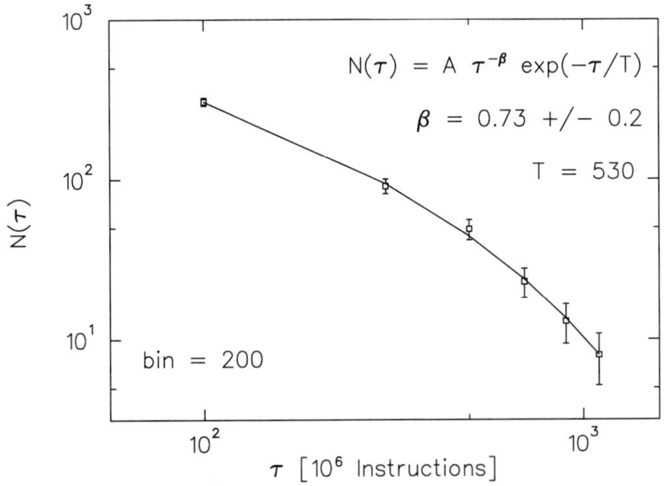

FIGURE 6.10 Fit of the binned abundance distribution with bin size 200.

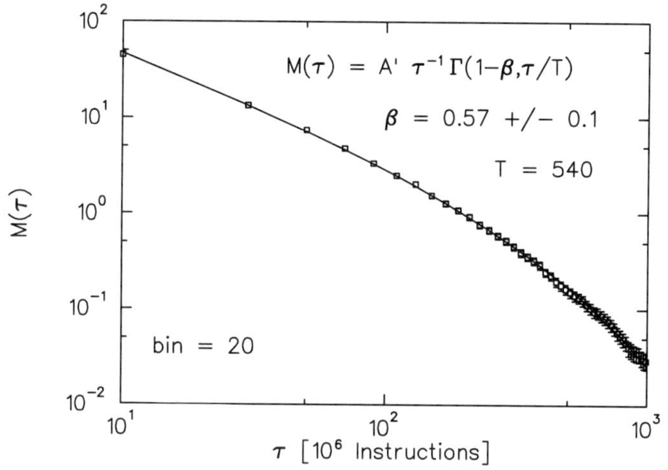

FIGURE 6.11 Fit of the integrated abundance distribution.

behavior, can be obtained from the first:

$$M(\tau) = \frac{1}{\tau}\int_{\tau}^{\infty} N(t)\,dt\ , \tag{6.31}$$

but has much better statistics (due to the effect of summing the distribution at each τ), we may have better luck fitting that distribution *without* the need of large bins. Fortunately, the functional form for the fit of $M(\tau)$ is dictated to us, as the integral appearing in Eq. (6.31) can be done analytically, resulting in an incomplete Γ-function:

$$M(\tau) = A'\frac{1}{\tau}\Gamma(1-\beta, \tau/T)\ . \tag{6.32}$$

Note that fitting $M(\tau)$ should yield the same result as fitting $N(\tau)$ (as we use the same parameters and the same functional form for N), except for the statistics and a much smaller bin size. The result is shown in Fig. 6.11, and suggests $\beta = 0.57$. The analysis presented here leaves us fairly confident that in the populations we studied, the time between avalanches is indeed distributed as a power law. Still, we cannot be very confident in the *numbers* that resulted from the fits, and we must wait for a much more extensive investigation with at least an order of magnitude more epochs, and an analysis of the influence of system size and mutation rate on the critical parameter β, to be convinced that such systems self-organize to a critical state. Fortunately, with the avida system, such an investigation is within the realm of possiblility.

Even though the results we have in hand now are not fully conclusive, let us speculate about their significance. It appears that there is no temporal scale of the order of the observational time in the system that would determine an average plateau length. In other words, a numerical estimate of the average plateau length $\bar{\tau}$ will always reflect the length of time the system is observed, instead:

$$\bar{\tau} \propto \tau_{\max} . \tag{6.33}$$

This implies that $\bar{\tau}$ is, in self-organized critical systems, a meaningless quantity that is never independent of the way we observe the system. Therefore, it can also not be used to make any *predictions* about the temporal behavior of same. In other words, it is impossible, armed just with statistics gathered from the past, to infer how much longer, on average, the current epoch will last. This is a somewhat satisfying side effect of criticality and the absence of scales.

If the time between events has a scale-free distribution (or, more precisely, is governed only by the size of the system), what about the distribution of *sizes* of events? The latter can be defined in two ways. If we consider as the size of an event the total number of members of the new dominating species that will be produced (or equivalently the total number of subspecies it will spawn) before it is driven to extinction by a new species, we have very good evidence (both from natural and artificial systems) that this distribution is also scale free. Indeed, this has been investigated in detail using the avida system in Adami et al., 1995 (see also Chapter 9).

As far as the distribution of sizes of *jumps* (i.e., the vertical axis in Fig. 6.7) is concerned, the situation is somewhat more tricky, as this may depend on the kind of landscape on which the population is evolving. The landscape used in the analysis above is probably too poor, and the runs not long enough, to take a stab at estimating this distribution. Still, the results are compatible with the assumption that there is also no scale in the distribution of fitness increases, at least in an infinite landscape. This would indicate that any time the population climbs a new peak in the landscape, the number of possible increases in fitness, and the size of such possible increases, is *unchanged*. This is in stark contrast to finite landscapes, where each improvement reduces the number of possible future improvements. We will discuss fitness landscapes in much more detail in Chapter 8.

In the meantime, let us speculate about the nature of fitness curves in landscapes where neither the time nor the size of a fitness jump is dictated by any scale. Such curves are *fractals*, and look similar at *any* scale of observation. So, if, for example, Fig. 6.7 were a true fractal, then what appears as a plateau to the naked eye would, under the magnifying glass, be resolved into many very small jumps (as is almost discernible at the beginning of the fitness history in Fig. 6.7). Because of the finiteness of the actual landscape in the experiment, and the noisiness of the data, this aspect is difficult to discern for the later part of the history. Fractals of the kind we have in mind are termed *Devil's Staircases* in the literature (see, e.g., Mandelbrot, 1977). If evolutionary histories are generally Devil's staircases, a number of interesting consequences can be obtained for evolutionary systems in general. First, it would imply that no fitness jump is too big to be explained by common mutational events such as the ones that drive the populations in tierra. As the evolutionary system self-organizes to the critical state, it poises itself for such grandiose events to occur, albeit in an unpredictable manner. Thus, there is no need for more than one theory to explain all sizes of evolutionary advances. Second, it would imply that, because the fitness histories are self-similar, certain global aspects of evolution occurring on very large time scales may already be present in the microscopic histories, and can be inferred from them.

Before treading on arguably more solid ground in the next section, let us issue a few *caveats* about the preceding discussion. The arguments presented there applied to finite populations of self-replicating strings, in which the only means of competition was acquiring a higher replication rate. While this may have been a scenario present on the very early earth, such populations are unrealistic in the present world. Specifically, we must keep in mind that if speciation occurs on a higher level of taxonomy (such as to a genus), we must remember that the population may then segment into parts that do *not* compete directly anymore. Also, the evolution of cells, or more generally *hosts* that carry the information present in the genome, may affect the dynamics of populations. While as a consequence of these limitations the lessons learned from simple systems of self-replicating strings must not strictly be applicable to all living systems, it is conceivable that the central trait, self-organization to criticality, is a universal characteristic of all evolving systems.

6.4 Theories of SOC

From speculations about SOC in the living state, let us return to the mathematics of SOC. While we have stressed many times how the avalanches, the trademark signal of SOC, are indicative of critical phenomena, we have not attempted to connect the transitions occurring spontaneously and apparently without the tuning of parameters, to the ordinary phase transitions of equilibrium statistical physics. Most researchers approach SOC as a phenomenon intrinsically *different* from that kind of physics. On the other hand, there seems to be some evidence that SOC might just be an equilibrium phase transition in disguise, indicating that there may be a unified theory of critical phenomena that describes both tuned and self-tuned transitions at the same time.

The views presented in this section are due to two groups of researchers, one in France led by Didier Sornette [Sornette et al., 1995] and a group led by J.M. Carlson of UC Santa Barbara [Carlson et al., 1990]. Because both views are, in fact, raising the same point (albeit in a different language), we present them together. In essence, they claim that SOC is just the manifestation of an *unstable* critical point that is reached in certain dynamical systems by tuning the control variable to a small but positive number. The unstable critical point, in turn, is described by the usual physics of equilibrium phase transitions. If such a description of SOC within conventional statistical theory succeeds, we might be one step closer to a consistent formulation of the physics of the living state.

In this attempt to understand SOC, we witness again the importance of the concept of slow driving. In essence, the contention is that the slow driving itself constitutes the tuning of an external parameter. To understand this in more detail, we must consider the physics of *transport*. For sandpile models, this is the transport of grains of sand; in the artificial living systems we are concerned with, it is the transport of information. (The transport of information in systems of self-replicating code is taken up in much more detail in Chapter 10). In any case, the spatial and temporal aspects of transport need to be studied in order to pin down the essence of the mysterious self-tuning.

We start by writing down the reaction-diffusion equation that underlies dissipative transport. If $\rho(x,t)$ is the density distribution (concentration) at point x at time t, its time evolution is given by

$$\frac{d}{dt}\rho(x,t) = \frac{\partial}{\partial x} D(\rho) \frac{\partial}{\partial x}\rho(x,t) + f\,\rho(x,t) \ . \tag{6.34}$$

The first term on the righthand side of Eq. (6.34) is the diffusion term, while the second is the *reaction* term. The latter adds or subtracts concentration, i.e., it describes sources and sinks. The diffusion coefficient $D(\rho)$ depends in general on the concentration ρ, but often is independent of it. The simplest case, with $D(\rho) = D$ and $f = 0$, results in the *diffusion equation*, a partial differential equation that describes the spatial and temporal distribution of substances diffusing in space and time:

$$\frac{\partial}{\partial t}\rho(x,t) = D\frac{\partial^2}{\partial x^2}\rho(x,t) \ . \tag{6.35}$$

For example, if we consider the diffusion of *particles*, $\rho(x,t)$ might describe the probability that a particle is found at x at time t. Particles moving solely according to (6.35) are said to perform a *random walk* (also known as *Brownian motion*). The equation can (for usual boundary conditions) be solved exactly with

$$\rho(x,t) = \frac{1}{\sqrt{4\pi D t}} \exp\left(-\frac{x^2}{4Dt}\right) \ . \tag{6.36}$$

In Fig. 6.12, we show this function at different times as a function of the distance x. This gives an intuitive feel about how a substance, from a peaked initial distribution, diffuses out to larger distances with time (see West, 1995, for an introduction to random walks in biological systems). We may ask about the mean distance that a particle diffuses out to, starting from the center $x = 0$. This is obtained by integrating (6.35) over the square of the distance from minus to plus infinity, and yields

$$\langle x^2 \rangle = 2Dt \,, \tag{6.37}$$

the well-known relation for random walks (in one dimension). In Box 6.1 we discuss walks with different statistics that are relevant in natural systems. In the one-dimensional sandpile, we witness the action of the

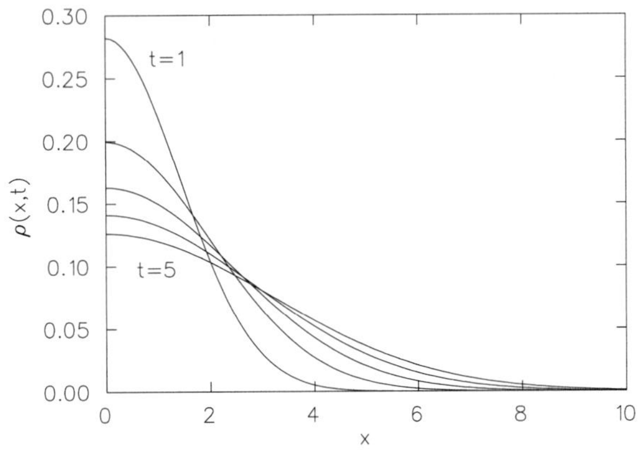

FIGURE 6.12 Solution of the diffusion equation for times $t = 1$ to $t = 5$ (with diffusion coefficient $D = 1$).

transport equation when a grain of sand that is added, say, at the origin, ends up at the edge of the pile only to flow off the table. The function $\rho(x, t)$ in this case is just the *slope* of the pile at x and t. The idea of SOC is then that $\rho(x, t)$ will always move to a critical value ρ_c. The continual addition of sand is not described by just the diffusion equation, but is specified by a *boundary condition* for the diffusion equation with $f = 0$:

$$\frac{\partial}{\partial t}\rho(x,t) = \frac{\partial}{\partial x}D(\rho)\frac{\partial}{\partial x}\rho(x,t) , \tag{6.38}$$

$$D[\rho(0)]\frac{\partial \rho}{\partial x}\bigg|_{x=0} = J(t) , \tag{6.39}$$

where $J(t)$ represents the *flow* off of the sand pile. Eq. (6.39) is just a flux balance condition relating the inflow of sand (via dropping of grains) to the flow off the table:

$$J(t) = D[\rho(L,t)]\frac{\partial \rho}{\partial x}\bigg|_{x=L} . \tag{6.40}$$

This conservation law can be obtained by integrating the condition that the density distribution $\rho(x, t)$ is normalized. The second boundary condition (at the right boundary) is just the condition

$$\rho(L) = 0 , \tag{6.41}$$

which marks the end of the table (open boundary condition).

The Gaussian probability distribution for random steps Eq. (6.36) has the very special property of *scaling*. If $P(x, t)$ is the probability that a random variable $X(t)$, which is just the sum of many individual steps X_j,

$$X(t) = \sum_{j=1}^{n} X_j \tag{6.42}$$

is between x and $x + dx$ at time t, we can easily show that

$$P(\lambda^{1/2}x, \lambda t) = \lambda^{-1/2} P(x, t) \ , \tag{6.43}$$

i.e., the distribution of steps for the random variable $X(t)$ is the same as for the variable $\lambda^{-1/2}X(\lambda^{1/2}t)$. It turns out that there are other processes that have self-similar walks, but the Gaussian walk is the *only* one that results in a finite diffusion coefficient. Indeed, consider the walk obtained when the individual steps X_j are drawn from a power-law distribution

$$p(X_j) \sim X_j^{-1-\beta} \ , \tag{6.44}$$

normalized such that $Y_N(t) = \frac{1}{N^{1/\beta}} \sum_{j=1}^{N} X_j$. One can show that the probability distribution P_L of walk lengths Y_N is *still* scale-invariant [West, 1995]:

$$P_L(\lambda^{1/\beta}y, \lambda t) = \lambda^{-1/\beta} P_L(y, t) \ , \tag{6.45}$$

and asymptotically scales just like the probability distribution for the individual steps (6.44). Such a walk is termed a *Lévy flight*. Yet, the second moment of the distribution that defines the flight's diffusion coefficient *diverges* (anomalous diffusion):

$$\langle Y^2(t) \rangle \to \infty \tag{6.46}$$

for $\beta < 2$. For $\beta = 2$ we reach the only point where $\langle Y^2 \rangle$ is finite: this is just the Gaussian random walk.

BOX 6.1 Lévy Flights

It was noted in [Carlson et al., 1990; Carlson et al., 1993] that Eq. (6.38), together with the boundary condition (6.39) is solved with a diffusion coefficient that depends on ρ as

$$D(\rho) \sim \frac{1}{(\rho - \rho_c)^{\phi}} \tag{6.47}$$

with an integer exponent ϕ. Thus, as $\rho \to \rho_c$ because of the finite driving with $J(t)$, the diffusion coefficient would grow and finally diverge, a phenomenon known as singular, or anomalous, diffusion. We can understand the phenomenon of singular diffusion in sandpiles more intuitively if we rewrite the flux J as the average number of grains per unit time that flow off the table, or equivalently (due to grain conservation) as the number of grains dropped on the pile per unit time

$$J = \frac{n}{\Delta t} . \tag{6.48}$$

Obviously, in the limit where $J \to 0$ (infinitesimal driving rate), the unit of time defined by the rate of dropping of grains must go to infinity:

$$\Delta t \to \infty . \tag{6.49}$$

At the same time, the diffusion coefficient measures how fast particles diffuse out to infinity, so if the time scale Δt diverges, this implies that the diffusion coefficient must diverge. In essence, dropping the grains infinitely slowly, i.e., letting all avalanches finish before a new one is started, is equivalent to letting the sand diffuse to infinity $\langle x^2 \rangle \to \infty$ infinitely fast, or *instantaneously*.

The previous discussion suggests that the BTW model, with the protocol for dropping of grains as it is, is not a dynamical model, and the power laws in avalanche size that we measure appear to be forced. However, this does not reduce the importance of the model as a paradigm. Indeed, if the experiments are performed with a finite (but small) rate J, we can *still* observe the scale-free dynamics. Of course, we realize that this is going to be the case as long as the temporal scale introduced, J^{-1}, is sufficiently different from the time it takes an avalanche to finish on average. Of course, the latter should only depend on the system size. The mystery of the ubiquity of power laws in Nature, then, is just shifted to a different question: Is there a mechanism that *ensures* a separation of time scales in certain dynamical systems occurring naturally? This is a question that is still under debate. In order to understand this phenomenon of self-tuning, we should have a handle on describing what is tuned, and to what. This is best achieved by borrowing a page from the theory of critical phenomena: second-order phase transitions in statistical physics.

The following description of the critical behavior of sandpiles is due to Sornette et al., 1995. It is a complementary description to the singular

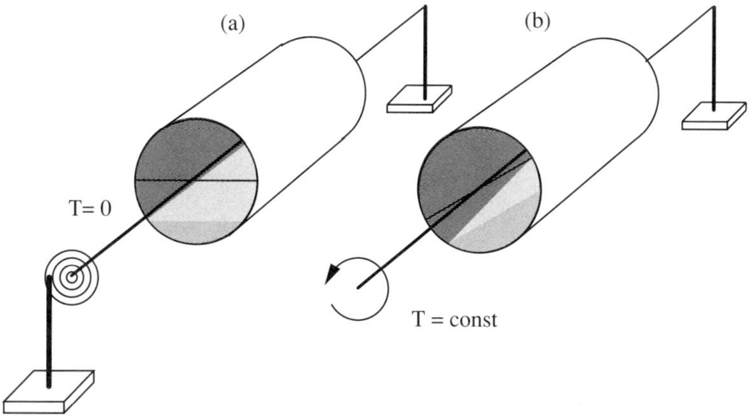

FIGURE 6.13 (a) Sandpile in a cylinder with zero applied torque, (b) finite torque.

diffusion picture outlined above, and is wonderfully intuitive. Imagine our pile of sand, instead of leaning against a wall as in Fig. 6.1 or being stationary on a table, to be enclosed in a cylinder that has the ability to rotate around its axis (see Fig. 6.13). We also imagine that there is a spring attached to the axis so that we can apply a finite *torque* T to the cylinder. Clearly, if the torque is zero, the sandpile is completely stationary. If we increase the torque, the angle that the sandpile makes with the horizontal increases linearly with the torque. If T is below a critical value, the slope of the sandpile will be below its critical value, and no grains will be dislodged, no sand will flow. If T is larger than the critical value, we note that the cylinder will un-stick, and the drum will start rolling with a constant angular velocity. At the same time, sand will flow off the pile (but onto itself) at a constant rate, and again we are in a stationary state. If the torque is just *at* its critical value, we will notice intermittent flows of sand of all sizes: we have reached the critical state.

In Fig. 6.14, we plot the flow of sand J against the applied torque T. This is very reminiscent of second-order phase transitions in statistical physics. Indeed, such a curve is found if the spontaneous magnetization of a ferromagnet is plotted against temperature. There is a critical temperature (the Curie temperature) at which the magnetization disappears, just as the flow of sand suddenly disappears if the torque drops below a critical value. In analogy to these second-order phase transitions, we

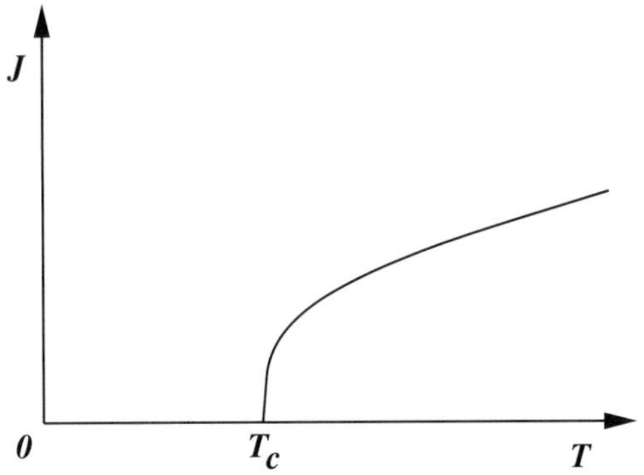

FIGURE 6.14 Sand flow versus applied torque for rotating sandpile.

assume a relation between the flow and the torque of the form

$$J \sim |T - T_c|^{\beta} \tag{6.50}$$

with an exponent β to be determined. At the same time, we can define a *correlation length* in such systems. It is the maximum distance of correlation between fluctuations, or in the sandpile case the maximum size avalanche that occurs if T is infinitesimally increased above T_c. More practically, it is the maximum size of an avalanche if the pile is disturbed by the addition of a single grain/hole pair. According to the theory of critical phenomena, we should then expect

$$\xi \sim |T - T_c|^{-\nu} , \tag{6.51}$$

where ν is yet another (positive) exponent that can be related to β. We will encounter relations of this sort in more detail in our discussion of percolation in Chapter 7. At this juncture our main point is to make a connection between the theory of critical phenomena and the SOC behavior of some dissipative dynamical systems.

In the conventional theory of critical phenomena, we distinguish between an *order parameter*, which generally will reflect the phase that we are currently in, and a control parameter. In magnetic systems, e.g., the order parameter is just the magnetization. If it is zero, we are in the disordered phase; if it is nonvanishing, we are in the ordered phase. The control parameter that takes us from one phase to the other is in that case the temperature. By direct analogy we can then identify the order parameter and the control parameter for the critical transition in sandpiles. The order parameter is the flux of sand J, and the critical parameter is the applied torque T. Note that the torque is directly related to the angle of repose of the pile, as noted earlier. Having identified the control parameter, let us repeat the question that we asked at the beginning of this chapter. What forces a self-organized critical system to the critical point? In this analysis, we see that the decision to wait before an avalanche has finished results in an infinitesimally small flow. If we therefore *force* the flow to vanish, we commensurately force $T \to T_c$ via Eq. (6.50). Furthermore, let us consider the diffusion coefficient in this scheme. In the theory of critical phenomena, the diffusion coefficient is directly related to the square of the correlation length

$$D \sim \xi^2 , \tag{6.52}$$

which, as we force $T \to T_c$, will indeed diverge owing to relation (6.51). Thus, we return to the conclusion that in systems that display SOC, an infinitesimal driving rate selects the order parameter to be critical and causes a divergence in the diffusion coefficient.

Even though it appears as if we have removed some of the layers that were hiding the true nature of SOC, there are still many unanswered questions. The preceding still does not constitute a *bona fide* theory of SOC, because we can still not be sure when and under what circumstance a system will display the traits we usually associate with SOC. It is clear, for example, that there has to be a feedback mechanism between the order parameter and the control parameter. The nature of this feedback, and its consequences for the dynamical systems that are subject to it, are still unclear. Another point that has yet to be clarified is the importance of *metastable* states in self-organized criticality. It appears that all SOC systems seem to jump from one metastable state to the other, and that the transitions between the states are akin to *first-order* phase transitions, initiated by nucleation. Yet, the phenomenology is adequately described with formulas borrowed from second-order critical phenomena. Also, the divergence of the diffusion coefficient reminds us that the dynamics in SOC systems may not be Gaussian, but rather of the Lévy type (see Box 6.1). Considering the pace at which research on SOC is advancing, however, we expect these issues to be resolved in the not-so-distant future.

6.5 Overview

The concept of self-organized criticality seems to capture the essence of the evolutionary transitions characterizing living and evolving populations. Self-organization to criticality appears to be the mode of choice of systems with dissipative transport properties, and which possess many degenerate, and therefore metastable, fixed points. In such systems, slow driving will move the dynamics towards a fixed point that is *not* stable, and the system responds to the driving with violent avalanches that restore a fleeting equlibrium. Self-organized criticality in living systems leaves telltale signs in the inter-event-interval distribution of transitions: the time between evolutionary transitions seems to be distributed in a scale-free manner, as a power law. All the necessary ingrediends of SOC can be easily identified: mutations take care of the slow driving, the threshold parameter is the growth factor of individual genotypes or species, and the self-organized critical state is characterized by a population of degenerate genotypes, all of which (or almost all) have a vanishing growth rate. Avalanches are then produced by mutations that create genotypes with positive growth rates which, due to the self-organization to vanishing growth rate of the population, can have catastrophic repercussions reaching *all* genotypes in the population. These unstable fixed points have been described by more conventional methods of statistical physics, in terms of second-order transitions tuned to the critical variable by an infinitesimal driving rate, as well as by the formalism of anomalous diffusion.

Problems

NOTE: *Problems indicated by an asterisk are of higher difficulty.*

6.1 A two-dimensional sandpile can be constructed that follows very simple cellular automata rules. The state of a cell is characterized by the number of grains of sand in it. The update rule states that if a cell has four or more grains of sand in it, it loses four, and from each of its four immediate neighbor cells with four or more grains in it, it gains one. A progression might therefore look as follows (cells affected by the avalanche are shown shaded):

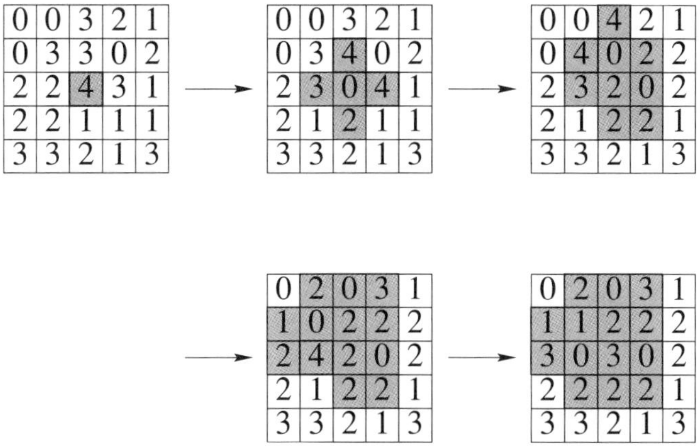

We can drive this lattice to a critical state by rapidly adding sand to the table until the rate of dropping sand is approximately equal to the rate at which the sand is dropping off the sides of the table. An equivalent way of doing this is to overload the table with sand (putting a random amount of five to eight grains in each square) and then updating (without adding any additional sand) until the table has reached a steady state.

(a) What is the longest lasting avalanche that you can construct on a 5 x 5 table? The initial sequence may only have up to three grains of sand per location, and a single fourth may be added to a cell to start the avalanche. Is it possible to construct an endless cycle? Why or why not?

(b) Using numerical simulations, determine the critical exponent for the *size* distribution of avalanches on a regular lattice, but for a size 64×64.

6.2 (a) Program a simple one-dimensional random walk, where the probability to take a step in either direction is $p = 0.5$. Obtain the time-series of such a walk $R(t)$ for a finite number of iterations, and take the *power spectrum* $P(\omega)$ of the walk, i.e., the absolute value-squared of the Fourier transform

$$\tilde{R}(\omega) = \int_0^\infty e^{i\omega t} R(t) dt \,. \tag{6.53}$$

and

$$P(\omega) = \tilde{R}^*(\omega) \cdot \tilde{R}(\omega) = |\tilde{R}(\omega)|^2. \tag{6.54}$$

What is the frequency dependence of the power?

*(b) Calculate the probability for a random walk to return to its starting point (the return-to-zero probability) as a function of the length of walk it took to return to zero. Compare this prediction with a numerical simulation of this probability using the program written for (a) [Adami et al., 1995].

6.3 For a distribution $N(s) = As^{-\alpha}$, show that the coefficient $\tilde{\alpha}$ obtained from the slope of a log-log plot of this distribution depends on the bin size Δ as

$$\tilde{\alpha} = \alpha - \frac{\log \Delta}{\log s} , \tag{6.55}$$

where s is the point chosen to extract the slope of the log-log plot.

CHAPTER SEVEN

Percolation

Πάντα ῥεῖ.[1]

Heracleitus

In this chapter we consider evolution as a dynamical process and focus on the geometrical and global properties of the underlying landscape without regard to the actual individuals evolving on it. Evolution appears, from this point of view, as a complex optimization process on a complicated landscape. This landscape appears to have very many peaks, leading to a rugged structure, while the process taking the population from one peak to the next seems to be nonergodic on the time scales of interest. In order to gain some insight into the ramifications of such landscapes for observable statistical measures, we consider an example process from the study of disordered systems: percolation. As we shall see, there is good reason to believe that certain aspects of the evolutionary process can be mapped without difficulty to percolation-type processes.

Percolation is a geometric process, the description of which does not require the notion of fitness or energy, or more generally, the existence of a *Hamiltonian*. Percolation is a *static* problem, i.e., we are not interested

[1]Everything flows.

in the temporal development of percolation, but only in the properties of the landscape itself and the distributions it generates. Our main concern, as far as genetic systems are concerned, will be: "How difficult is it to traverse genetic space via mutations?" To answer this question, we need to know how close (or how far) in genetic space, regions are that give rise to roughly the same phenotype, in other words, we would like to study the degeneracy of genetic space. This in turn leads us to investigate the *connectedness* of this space. For example, we may ask whether there can be an evolutionary path that spans the entire diameter of genetic space, and what the probability is of ever encountering such a path. Alternatively, we may ask what the probability is that a random genome (a random point in genetic space) supports self-replication at all, or is part of a specific species, or part of any other taxonomic group. Such questions are answered qualitatively by assuming that genomes that pertain to a specific taxonomic group form *clusters* in genetic space. Consequently, we may investigate the distribution of cluster sizes as a function of the fraction of occupied sites (fraction of genomes sharing a specific trait) for landscapes of different types. This is the object of percolation theory.

7.1 Site Percolation

Imagine a regular, i.e., *Euclidean*, lattice of points, each of which can be occupied or unoccupied. By a random process, let unoccupied sites be filled with a probability p between 0 and 1. For a sufficiently large lattice, the ratio between the number of occupied and the total number of lattice sites will then approach p. After this process, connect all those occupied sites that are nearest neighbors (only the cardinal directions are counted here; there can be no diagonal nearest neighbors). This process of connecting adjacent sites is called *site percolation* (see Fig. 7.1). There is an equivalent process, *bond percolation*, where instead of filling sites, we distribute bonds between sites with probability p in the lattice. All of the results that we obtain for site percolation can also be obtained for bond percolation, and so we shall not consider the latter.

One of the most important concepts of percolation theory is that of the *infinite cluster*. If p is sufficiently large, the possibility arises that a cluster forms that spans the entire lattice, i.e., a cluster that connects the left edge of the lattice with the right one, or the top edge with the bottom one.

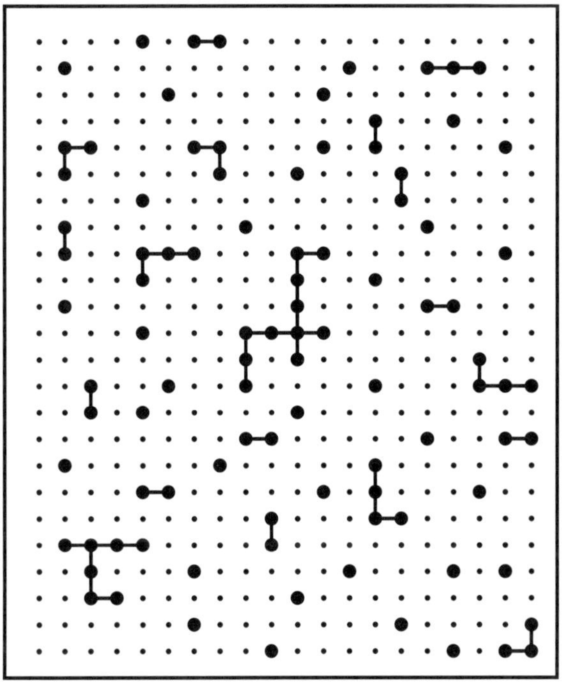

FIGURE 7.1 Two-dimensional lattice of sites filled to fraction $p = 0.2$. Occupied sites are indicated by dark dots and connected if they are adjacent.

If such a cluster exists, the system is said to *percolate*. The appearance of an infinite cluster changes the properties of the system drastically. If, for example, empty sites represent insulators and occupied sites represent electrical conductors, current can start to flow through the lattice only after the appearance of an infinite cluster. If occupied sites represent pores in an amorphous medium, fluid can flow through the medium when the critical occupation probability is reached. At the percolation threshold (which is just the critical occupation probability), the system undergoes a *geometrical* phase transition, known as the percolation transition. The question we would like to ask is: "At what filling fraction p does the probability for appearance of an infinite cluster become appreciable?" In general, this question is hard to answer. This critical probability for percolation p_c can be obtained analytically for only a few geometries, but can be estimated numerically by Monte Carlo simulations with good accuracy.

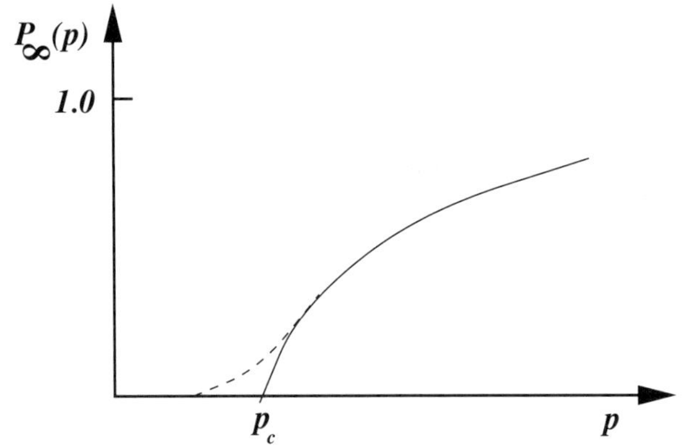

FIGURE 7.2 Probability to obtain an infinite cluster as a function of the filling fraction p. Solid line: infinite lattice. Dashed line: finite lattice.

As mentioned above, the lattice with $p < p_c$ shows qualitatively different characteristics than with $p > p_c$: the critical probability p_c is a genuine critical point, and the transition from a subcritical to a supercritical lattice shows the signs of a second-order phase transition. Fig. 7.2 shows the typical dependence of $P_\infty(p)$, of the probability to find an infinite cluster, on p. We remember this behavior from the dependence of sand flow on the torque (Chapter 6), or the magnetization on temperature. Note that the smooth rise of the probability in Fig. 7.2 is due to the finiteness of the lattice. For an infinite lattice, the probability to obtain an infinite cluster rises sharply at p_c, as indicated by the solid line in Fig. 7.2.

As is the case for thermodynamic systems undergoing second-order phase transitions, a percolation system at the critical point (i.e., for density p_c) is characterized by an *absence of scales*. In statistical systems, this usually means that fluctuations of all sizes at all scales can occur, just like in the self-organized critical systems. In the latter, it is observed that the system always returns to such a point. Here, we would like to investigate percolation systems *at or near* the critical point, because of the evidence gathered from SOC systems that this is an interesting fixed point, reflecting the idea that nature somehow has managed to adjust the evolutionary landscape in such a way that it is *critical*: poised at the critical point. When examining percolation at the critical density, we will also investigate what happens if the system is somewhat away from the critical point.

7.2 Cluster Size Distribution

One of the first aspects of percolation systems we turn to is the distribution of cluster sizes. Let us start by defining the cluster size distribution at density p, $n_s(p)$. In Fig. 7.1, for example, there are 34 clusters of size 1 (i.e., sites that do not strictly form clusters), 10 clusters of size 2, 4 clusters of size 3, 3 clusters of size 4, 1 cluster of size 7, and 1 cluster of size 11. This distribution is shown in Fig. 7.3. For convenience, the cluster size distribution n_s is defined as the number of clusters of size s *divided* by the total number of sites N (this is why we are multiplying n_s by N in Fig. 7.3). Then, $sn_s(p)$ is the probability for a site to belong to a cluster of size s (it is larger than n_s by a factor s, as any of the s sites are cluster sites), and the sum over those gives the ratio of occupied to total sites,

$$\sum_s sn_s = N_{\text{occ}}/N \,, \tag{7.1}$$

which is 96/480 in Fig. 7.1, and happens to be exactly 20 percent of the total number of sites: the probability with which we filled the lattice. In general, i.e., for filling fractions closer to the critical one, this relationship between p and $\sum sn_s$ is more complicated, as we shall see presently. Armed with these definitions, we can write the probability to observe an infinite cluster in terms of cluster size distributions. Indeed, for each site we can determine that it belongs to exactly one of three types: either it is unoccupied (with probability $1-p$), or, if occupied, it either belongs to

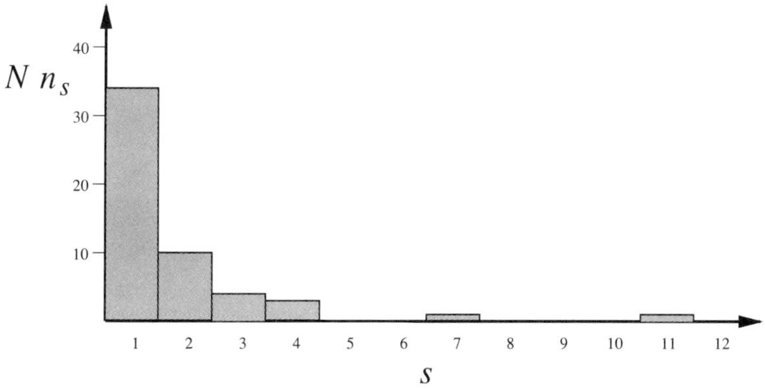

FIGURE 7.3 Cluster size distribution for the lattice in Fig. 7.1.

a cluster of size s (with probability $\sum_s s\,n_s$) or to the infinite cluster. The latter possibility has probability pP_∞, i.e., it is the probability that a site is occupied multiplied by the probability that an infinite cluster exists. Then, we can write down the *sum rule*

$$1 - p + \sum_s sn_s + pP_\infty = 1\ , \tag{7.2}$$

which, in turn, implies the strength of the infinite network:

$$P_\infty = 1 - \frac{\sum_s sn_s}{p}\ . \tag{7.3}$$

Let us continue by defining a few more useful quantities. For example, we can define the *mean cluster size*

$$\langle s \rangle = \frac{\sum_s s^2 n_s(p)}{\sum_s sn_s(p)}\ , \tag{7.4}$$

which, in turn, can be related to the *pair connectedness* $g_p(r)$. The pair connectedness is the probability that any two occupied sites that are a distance r away are in fact part of the same cluster if the lattice is filled with probability p. Clearly, this is a very important measure. For genetic landscapes, it would tell us something about the connectedness of the landscape, i.e., how distant (in genetic space) two genotypes can be while still belonging to the same (taxonomic) cluster. Intuitively, we expect a smooth landscape to be *infinitely connected*, i.e., two genotypes could be arbitrarily far away in genetic space and still belong to the same cluster. Put differently, we can expect to find a member of any cluster in any area of genetic space. On rugged, multi-peaked landscapes, on the other hand, we expect the opposite. There, the connectivity would be small, and members of the same cluster would necessarily have to be close genetically in order to belong to the same taxonomic group. In other words, on rugged landscapes the chance to find a representative of a specific cluster in an arbitrary region of genetic space would become vanishingly small. The idea of SOC is that living systems have arranged some sort of middle ground between these extremes. In percolation systems, we can dial the connectedness by choosing percolation lattices with different p. Of course, the self-organized landscapes would then correspond to those at the critical density p_c.

Mathematically, $g_p(r)$ is related to the average cluster size $\langle s \rangle$ via

$$\sum_{r=0} g_p(r) = \langle s \rangle (p) \,, \tag{7.5}$$

i.e., if we pick out an arbitrary site and sum over all the sites that it is connected to, we obtain the average cluster size. It is easy to see that $g_p(0) = 1$ and $g_p(1) = p$. More precisely, $g_p(r)$ is a *correlation function*. As such, we can express it in terms of a correlation *length* $\xi(p)$. This quantity defines the distance by which the correlation between two sites has dropped to approximately $\frac{1}{e}$ (from 1). Then, our representation of the functional form of the pair connectedness $g_p(r)$ is

$$g_p(r) \approx e^{-r/\xi} \,. \tag{7.6}$$

Note that the correlation length is implicitly defined as

$$\xi^2(p) = \frac{\sum_r r^2 g_p(r)}{\sum_r g_p(r)} \,. \tag{7.7}$$

Eqs. (7.5), (7.6), and (7.7) then allow us to relate the critical exponents γ and ν. Let us illustrate some of these concepts and definitions by example.

7.3 Percolation in 1D

The simplest percolating system is that in one dimension. It is simple because it is, as one can easily convince oneself, trivial. The critical density, for example, is $p_c = 1$. Let us calculate the number of clusters of size s. The probability that s arbitrary sites are occupied is p^s. The probability that a chain ends is $1-p$, and as a consequence the probability that a site is the end site of a chain of length s is $p^s(1-p)^2$. Thus, in one dimension,

$$n_s(p) = (1-p)^2 p^s \,. \tag{7.8}$$

Note that because the only possibility to have an infinite cluster is $p = 1$, and because $p > 1$ is inaccessible, there is *no* phase transition in the one-dimensional percolation problem. Let us nevertheless investigate this critical point [Stauffer, 1979]. We do this by expressing all quantities in terms of the difference $p - p_c$, and examine the limit $p \to p_c$. For example, in the limit $p \to 1$, we can approximate $\log p$ by $p-1$ by using the expansion of the logarithm (valid for $p \leq 1$)

$$\log p = (p-1) - \frac{(p-1)^2}{2} + \frac{(p-1)^3}{3} + \cdots . \tag{7.9}$$

and write

$$p = e^{\log p} = \lim_{p \to 1} e^{p-1} = \lim_{p \to p_c} e^{p-p_c} . \tag{7.10}$$

As a consequence, we write in the vicinity of p_c [from Eq. (7.8)]

$$n_s(p) \approx s^{-2}\left[(p-p_c)s\right]^2 e^{(p-p_c)s} \tag{7.11}$$

$$= s^{-\tau} f(z) , \tag{7.12}$$

where we introduced the critical parameter $\tau = 2$, the scaling function $f(z) = z^2 e^z$, and the variable $z = (p-p_c)s^\sigma$ with $\sigma = 1$. Thus, n_s for $p \ll p_c$ decays exponentially with s, and approaches a power law with $\tau = 2$ in the vicinity of p_c. The functional form (7.12) was chosen to illustrate this competition between power and exponential decay, even though it is trivial in the case treated here. However, this functional form turns out to be quite general, and for fixed dimension d the exponents τ and σ turn out to be *universal* across all percolation models.

In this one-dimensional theory, we can also calculate the average cluster size

$$S = \langle s \rangle = \frac{\sum_s s^2 n_s(p)}{\sum_s s n_s(p)} = \frac{1+p}{1-p}, \tag{7.13}$$

which clearly diverges as $p \to 1$. Again in order to make contact with the universal exponents we introduce later, we write this as

$$\langle s \rangle = (1+p)(1-p)^{-\gamma}, \tag{7.14}$$

with $\gamma = 1$ in one dimension. To conclude the one-dimensional example, we write the correlation length ξ in terms of a critical exponent ν. The correlation function $g_p(r)$ is

$$g_p(r) = e^{-r/\xi} , \tag{7.15}$$

whereas in the 1D model, the probability that an occupied site a distance r sites away belongs to the same cluster is

$$g_p(r) = p^r . \tag{7.16}$$

Then, using the expansion (7.9) of the logarithm again, we obtain

$$\xi = -\frac{r}{\log g_p(r)} = -\frac{1}{\log p} \approx (p_c - p)^{-\nu} \qquad (p \to p_c), \tag{7.17}$$

with $\nu = 1$ in this example. Note that we could have obtained this as well by calculating

$$\xi^2 = \frac{\sum_{r=1} r^2 g(r)}{\sum_{r=1} g(r)} = \frac{1+p}{(1-p)^2}, \tag{7.18}$$

which is commensurate with Eqs. (7.5) and (7.13).

Thus we see that, even though there is no percolation phase transition in one dimension, the average cluster size as well as the correlation length diverges in the limit $p \to p_c$. This divergent behavior fixes the values of the critical exponents γ and ν in one dimension. Furthermore, the cluster size distribution can be written in the form of a *scaling law* $n_s(p) \sim s^{-\tau} f(z)$, with $z = s^{\sigma}(p - p_c)$ defining two further critical exponents τ and σ. These exponents can be obtained for arbitrary percolating systems, and we shall treat the two-dimensional case next.

7.4 Higher-Dimensional Euclidean Lattices

As we shall see, things become much more complicated in two dimensions. Indeed, we cannot even calculate $n_s(p)$ analytically from first principles. However, the critical exponents do not depend on the geometry of the lattice at all, but only on its dimensionality. This is the concept of *universality*.

For a square lattice, we can only obtain the distributions for *small s* analytically. Indeed, it is clear that the distribution of clusters of size 1 is related to the probability p via

$$n_1(p) = p(1-p)^4, \tag{7.19}$$

i.e., the probability that a site is occupied multiplied by the probability that no adjacent site is occupied. In the same manner, we can obtain the cluster size distribution for $s = 2$,

$$n_2(p) = 2p^2(1-p)^6. \tag{7.20}$$

In general, we can write

$$n_s(p) = \sum_t g_{st} p^s (1-p)^t \,, \tag{7.21}$$

where g_{st} is the number of geometrically different cluster configurations of s sites with *perimeter* t. While expressions for distributions with higher s can easily be generated ($s = 3$ is left as Problem 7.1), no general expression for the infinite sum is known. However, the previous analysis shows that the average cluster size $\langle s \rangle$ admits a series expansion

$$\langle s \rangle = \frac{\sum s^2 n_s}{\sum s n_s} \approx \frac{\sum s^2 n_s}{p} = 1 + 4p + 12p^2 + 24p^3 + \cdots \,, \tag{7.22}$$

where the first approximation in Eq. (7.22) obtains from Eq. (7.2), and assuming that P_∞ is small for small p. Such series expansions can be compared to computer simulations of $\langle s \rangle$. Not surprisingly, however, the expansion (7.22) breaks down near p_c, and the computer simulations fail to converge. On the other hand, we have in the 1D example alluded to the fact that such sums become *universal* near p_c, and therefore their behavior can be predicted after all.

Let us take a look at this universality, and consider the quantity

$$F(p) = \sum_s n_s(p) \,. \tag{7.23}$$

Universality dictates that this zeroth moment of the cluster size distribution, which is the total number of clusters, displays power law dependence near p_c:

$$F(p) \sim |p - p_c|^{2-\alpha} \,, \tag{7.24}$$

where we defined *yet* another critical exponent α. While there seems to be an explosion of such exponents, we shall see shortly that there are relations between them, and that only a few are independent.

Via a numerical experiment, we can plot the logarithm of $F(p)$ versus the logarithm of $p - p_c$ to test universality, and extract the slope $(2 - \alpha)$ for small values of $|p - p_c|$. In thermodynamical systems, the exponent $2-\alpha$ is associated to the critical behavior of the *free energy*. This exponent is not independent of the ones previously introduced, and we note here without proof that $\alpha = 2\beta + 2\gamma - 2$.

The first moment of the cluster size distribution is related to the probability to find an infinite cluster, as can be seen from Eq. (7.3). This quantity also serves as an *order parameter*, i.e., a macroscopic observable that allows us to determine the phase the system is in. In magnetic systems, for example, the order parameter is the bulk magnetization, and the critical parameter the temperature. The disordered phase is recognized by a vanishing magnetization (at temperatures larger than the critical temperature T_c). Below T_c, the magnetization is non-zero due to the spontaneous alignment of the individual spins that make up the magnet, leading to an *ordered* state [see also the discussion following Eq. (6.50)]. The critical behavior of the order parameter near p_c is described by the critical exponent β

$$P_\infty(p) \sim (p - p_c)^\beta \,. \tag{7.25}$$

Of course the second moment of the size distribution is related to the average size of the clusters, and scales with critical exponent γ:

$$S = \langle s \rangle \sim |p - p_c|^{-\gamma} \,. \tag{7.26}$$

Again, there is an analogous quantity in thermodynamic phase transitions called the *susceptibility*. To complete the exposition of critical exponents, we mention the behavior of the correlation length ξ, given by

$$\xi \sim |p - p_c|^{-\nu} \,, \tag{7.27}$$

as anticipated in the 1D example.

An important aspect of percolation theory is the property of *scaling*, which relates properties of the system at the critical point to properties away from it. Before we enter this discussion, we analyze an important geometry of percolation systems, which just like the one-dimensional system, allows an analytic approach.

7.5 Percolation on the Bethe Lattice

The Bethe lattice (also called the Cayley tree) represents a percolation problem that can be solved rigorously. Furthermore, it undergoes a true phase transition (i.e., $p_c < 1$) and thus the regime above p_c can be examined as well.

The Bethe lattice is obtained by constructing a tree from a central site, by connecting z branches to the site, and by ending each branch

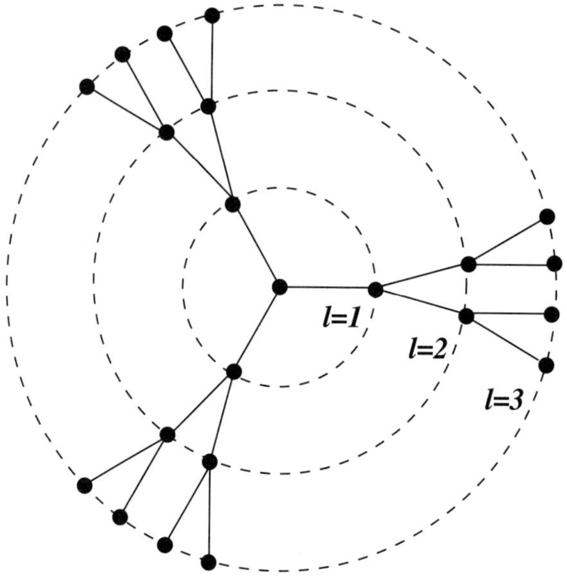

FIGURE 7.4 Bethe lattice with $z = 3$.

with a site. From each site thus formed, $z - 1$ branches emanate, such that each site in the lattice is connected to z other ones (see Fig. 7.4). There are no loops in the system, as any two sites are connected by only *one path*. As the *Euclidean* distance r is meaningless in this geometry, we introduce the *chemical* distance ℓ between two sites instead. For example, the chemical distance between the central site and a site on the ℓth shell is exactly ℓ. The shells are indicated by dashed circles in Fig. 7.4. Let us examine the dimensionality of this lattice. In a d-dimensional Euclidean lattice, the number of points inside a volume of radius R increases as $V \sim R^d$, whereas the number of sites on the surface grows like $S \sim R^{d-1}$. Consequently, for d-dimensional Euclidean lattices we find the surface to volume relation

$$S \sim V^{1-\frac{1}{d}} . \tag{7.28}$$

If we write down the corresponding relation for the Bethe lattice, we find, in the limit of an infinite number of shells,

$$S \sim V, \tag{7.29}$$

due to the exponential increase of sites with ℓ. This allows us to conclude that the Bethe lattice in effect represents the case of an *infinite-dimensional* Euclidean lattice ($d \to \infty$). As we shall see later, this makes the Bethe lattice the ideal candidate to model the genetic space spanned by very long genomes.

Let us calculate the critical exponents for the Bethe lattice. We start by calculating the correlation function $g_p(\ell)$ as a function of the distance between sites ℓ. We remind the reader that this function represents the probability that two sites that are a distance ℓ apart in the lattice actually belong to the same cluster. In order for this to happen here, *all* $\ell - 1$ sites that separate the two sites have to be occupied, as there is only one path between any two sites. Let us count the number of sites contained inside the ℓth shell. Going from shell to shell, the number of sites is multiplied by $z - 1$; thus, counting the initial site, we obtain

$$N(\ell) = z(z-1)^{\ell-1}\,. \tag{7.30}$$

Because the sites are occupied with probability p, the probability that all sites are occupied in between two sites separated by $\ell - 1$ shells is

$$g_p(\ell) = z(z-1)^{\ell-1}\,p^{\ell}\,. \tag{7.31}$$

Note that this reduces to the correlation function of the linear chain if $z = 2$, except for a factor 2 due to ℓ playing the role of a radius rather than a diameter. Eq. (7.31) immediately now delivers to us the critical probability! Indeed, if there is an infinite cluster, this correlation function must diverge with ℓ, whereas it should decay exponentially if $p < p_c$:

$$g_p(\ell) \approx e^{\ell \log p(z-1)}\,. \tag{7.32}$$

Thus, the critical point must occur where $p(z-1) = 1$, as this is where the logarithm changes sign. We therefore deduce that

$$p_c = \frac{1}{z-1} \tag{7.33}$$

for the Bethe lattice of connectivity z.

Armed with $g_p(r)$ and p_c, we can immediately calculate a number of interesting quantities. For example, the correlation length is obtained via

$$\xi^2 = \frac{\sum_{\ell=1} \ell^2 g_p(\ell)}{\sum_{\ell=1} g_p(\ell)} = p_c \frac{p_c + p}{(p_c - p)^2}\,. \tag{7.34}$$

Note that this does not allow us to deduce that $\nu = 1$, as the correlation function is expressed in terms of ℓ rather than the Euclidean distance r. However, a simple argument allows us to relate the distance r for an infinite dimensional lattice to ℓ in the Bethe lattice. Indeed, as correlations are extremely weak in the very high-dimensional lattices, the average distance (chemically) behaves much like a random walk, and thus

$$r^2 \sim \ell \, . \tag{7.35}$$

Consequently, as the correlation function expressed in terms of the Euclidean distance is just the square root of the correlation length expressed in chemical distance, the critical exponent of the correlation length in the Bethe lattice is $\nu = \frac{1}{2}$, rather than $\nu = 1$. The average size of clusters follows using previous results as

$$S = 1 + \sum_{\ell=1} g_p(\ell) = p_c \frac{1+p}{p_c - p} \, , \tag{7.36}$$

which yields $\gamma = 1$.

Let us now consider $n_s(p)$, the probability that any site on the Bethe lattice belongs to a cluster of size s. We take our hint from Eq. (7.21), but realize that unlike for d-dimensional Euclidean lattices, we can derive a concise expression for the perimeter of a cluster of s-sites on the Bethe lattice. Clearly, a one-cluster is always surrounded by z sites, whereas a two-cluster is surrounded by $z + (z-2)$ perimeter sites. In general, a cluster of s sites always has $z-2$ more perimeter sites than a cluster of $s-1$ sites. If we write $t(s)$ for the number of perimeter sites surrounding a cluster of s sites, we have

$$t(s) = z + (z-2)(s-1) \, , \tag{7.37}$$

and we find for n_s on the Bethe lattice

$$n_s(p) = g_s p^s (1-p)^{2+(z-2)s} \, , \tag{7.38}$$

where g_s is the number of configurations for an s-site cluster. We can now take this expression, which holds for all p, and investigate its behavior close to the critical probability p_c. Doing this (we skip the details of this expansion), we find that

$$n_s(p) \sim n_s(p_c) f_s(p) \qquad (p \to p_c), \tag{7.39}$$

where

$$f_s(p) \sim e^{-s(p-p_c)^2} \, . \tag{7.40}$$

The function $n(p_c)$ is, just as we derived in one-dimensional percolation, a power law, and we can assume

$$n_s(p_c) \sim s^{-\tau} . \tag{7.41}$$

As we know the average size exactly (Eq. 7.36), we can use this to fix the exponent τ. Using (7.39) to calculate the average size and comparing to (7.36) yields

$$S \sim |p - p_c|^{(\tau-3)/\sigma} , \tag{7.42}$$

or consequently

$$\gamma = (3 - \tau)/\sigma . \tag{7.43}$$

With $\gamma = 1$ and $\sigma = \frac{1}{2}$, we find ultimately $\tau = \frac{5}{2}$. Finally, we would like to obtain β, which is related to the critical behavior of the order parameter (here the probability to find an infinite cluster). As we have an exact expression relating P_∞ and $n_s(p)$, Eq. (7.3), we can obtain another relation between the scaling exponents, this time yielding

$$\beta = \frac{(\tau - 2)}{\sigma} \tag{7.44}$$

and $\beta = 1$ for the Bethe lattice. It is worth reiterating that the critical exponents that we obtained for the Bethe lattice are *independent* of z, the connectivity of the lattice, even though the critical probability p_c depends on z. This independence is a reflection of the universality of critical phenomena.

7.6 Scaling Theory

In previous sections we witnessed the paramount importance of the cluster size distribution $n_s(p)$, which would allow us to calculate not only the probability for an infinite cluster P_∞, but also the mean cluster size S as well as the correlation length ξ. Only in the one-dimensional and the infinite-dimensional (the Bethe lattice) problem could we calculate this quantity exactly. Here, we are going to see how much information can be extracted about critical indices *without* knowing n_s exactly.

The main idea of the scaling theory is that the functional form for n_s, obtained, for example, for the Bethe lattice [see Eq. (7.39)], in fact holds not only close to the critical point p_c but also away from it. Such an assumption should of course be verified directly. Before we do this, let us reformulate the assumption of scaling in a different manner. We have seen that the different moments of n_s scale in a universal manner near p_c. This can be explained if the sums that enter in these moments are dominated by a particular *characteristic* cluster size s_ξ. The characteristic cluster size should not be mistaken for the *average* cluster size. Rather, the idea of the characteristic cluster size is that for any sum

$$I_k = \sum_s s^k n_s(p) \approx s_\xi^k n_{s_\xi}(p) , \tag{7.45}$$

s_ξ gives the *main* contribution to it, and clusters with $s < s_\xi$ effectively do not contribute. Clusters with $s \gg s_\xi$ are exponentially rare, and therefore also do not contribute significantly. In this case, it is easy to see that the moment I_k should only depend on the *ratio* s/s_ξ. At the same time, the characteristic size s_ξ should also exhibit scaling behavior, of the form

$$s_\xi \sim |p - p_c|^{-1/\sigma} . \tag{7.46}$$

Note that the definition of the scaling exponent σ in this way (we have encountered it briefly when considering the 1D problem) is mainly historical. The important point is that for positive σ, the characteristic cluster size diverges at the critical point, such that the ratio s/s_ξ vanishes. We can then posit that the cluster-size distribution is given by its behavior at the critical point, $n_s(p_c)$, multiplied by a function that only depends on the ratio s/s_ξ, $f(s/s_\xi)$, provided $f(0) = 1$:

$$n_s(p) = n_s(p_c) f(s/s_\xi) . \tag{7.47}$$

This assumption can be tested, as armed with Eq. (7.46) we can deduce the function f by numerical experiments. Indeed, if Eq. (7.47) holds, we ought to find that

$$\frac{n_s(p)}{n_s(p_c)} = f\left(s(p - p_c)^{1/\sigma}\right) , \tag{7.48}$$

i.e., that if the ratio $n_s(p)/n_s(p_c)$ is plotted for different values of s and different values of $p - p_c$, the points should *all* lie on the same curve that defines the function f, if plotted against the product $s(p - p_c)^{1/\sigma}$. In other words, the function f would, if the scaling hypothesis holds, depend only on the product of s and $p-p_c$, rather than on each of them independently, as would be the general case. Such a numerical analysis is presented in Fig. 7.5 for a regular two-dimensional lattice. There, $(p-p_c)s^\sigma$ is plotted *vs.* $n_s(p)/n_s(p_c)$, for different s and $p-p_c$, and the points all lie reasonably close to a universal function, corroborating the scaling assumption. The power law scaling at the critical point

$$n_s(p_c) \sim \frac{1}{n_s(p)^\tau} \tag{7.49}$$

can also be verified numerically. In Fig. 7.6 we show the result of such a numerical experiment for a regular Euclidean lattice of 1000 × 1000 sites, where the power law is obeyed over several orders of magnitude, and the critical exponent can therefore be extracted with high accuracy.

The scaling law (7.48) can in principle not be valid far away from the critical point, as is obvious from remembering our derivation of the scaling function in one dimension, Eq. (7.12). For regimes further away from the critical point, it has been suggested that a universal law of the

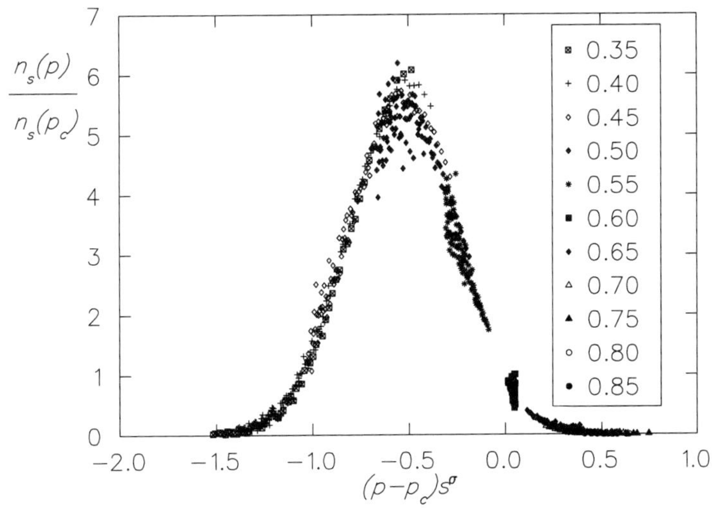

FIGURE 7.5 Number of clusters of size s (normalized to the value at the critical point), as a function of the scaling variable $(p-p_c)s^\sigma$ on a 1000 × 1000 square lattice. Included are clusters of size $5 < s < 100$ for filling fractions from $p = 0.35$ to $p = 0.85$, and with $\sigma = 36/91$.

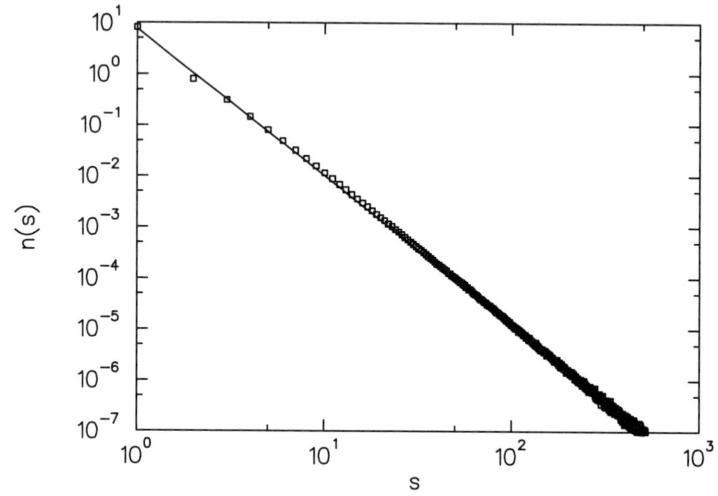

FIGURE 7.6 Number of clusters of size s at $p = p_c = 0.5927$, in two dimensions on a 1000×1000 lattice. The power law exponent is fitted with $\tau = 1.94$, not far from the theoretical value $\tau = 187/91$.

form

$$n_s(p) \sim s^{-\theta}\text{const}^s \qquad (p < p_c, s \to \infty) \tag{7.50}$$

might hold, with $\theta = 1$ in one dimension, and $\theta = \frac{3}{2}$ for $d = 3$.

In Table 7.1, we list the currently known values for the critical exponents for different dimensions, which are obtained analytically wherever possible or otherwise by numerical experiments.

TABLE 7.1 Universal critical exponents for lattices of different dimensions (from Campi, 1987).

Exponent	$d = 1$	$d = 2$	$d = 3$	$d = \infty$
β	—	5/36	0.45	1
γ	1	43/18	1.74	1
ν	1	4/3	0.88	1/2
σ	1	36/91	0.46	1/2
τ	2	187/91	2.20	5/2

7.7 Percolation and Evolution

In this section we speculate on the application of percolation theory to extract some information about the underlying fitness landscape of living and evolving systems from measurable distributions.

As mentioned at the beginning of this chapter, it is one of the more intriguing speculations that life self-organizes toward a critical state. One possible way to achieve this is to arrange for genotypic space to be occupied at a critical level, i.e., for the ratio of the number of (local) fitness maxima to total possible genomes to be given by the critical density for such a lattice. In this case, clusters of similar genomes would not be controlled by any scale, and evolution can proceed on such a lattice in an optimal way. There is some evidence [Burlando, 1990; Burlando, 1993] that taxonomic abundance distributions (which can be interpreted as cluster size distributions) show power-law behavior over many scales, but conclusive data is difficult to obtain. Here we would like to investigate the predictions of a theory that assumes that evolution is much like percolation on a Bethe lattice, where the connectivity of the lattice is controlled by the length of the genome.

Ordinary genetic space can be viewed as a Euclidean hypercube (or Hamming space: a space where the distance between two points is given by the number of mutations that takes one into the other), and where therefore the number of nearest neighbors of each genome is given by the number of possible one-step mutations. Clearly, the connectivity of such a space depends crucially on what is considered a one-step mutation. Here, we shall make the (unrealistic) simplification of bit-flip mutations only, such that all mutational neighbors are also neighbors in Hamming space. In truth, many mutations are of the crossover type, where entire sections of code are interchanged, and insertions and deletions occur as frequently. Such mutations cover vast distances in Hamming space, and as such give rise to new clusters in this model. In principle, it is possible to construct a space where all one-step mutations are also genetic neighbors, but we do not deem it necessary for outlining the main geometric aspects of the model.

At first sight there appears to be a fundamental difference between the occupation of sites in a percolation model and genetic evolution. In percolation, the occupation of a site is an event that is entirely *proba-*

bilistic, while in evolution *two* principles are at work. On the one hand, the occupation of a site is due to random mutation of an existing genotype, and therefore is probabilistic. However, sites near to an occupied site have an exponentially larger chance of being occupied than sites that are far from occupied sites. Thus, the distribution of clusters throughout genetic space will be maximally uneven. The reason for this is, as you will probably recognize, the lack of ergodicity of the evolutionary process in genetic space. As usual, we sidestep this problem by considering only those pieces of genetic space that have any appreciable concentration of occupied sites, i.e., we only consider regions where there are clusters, and ignore the (vast majority) of space where no clusters can be found at all. In this reduced space, the process of evolution can be considered as quasiergodic, and we may move on with our analysis of evolution from the point of view of percolation theory.

Let us first determine the geometry of the lattice we are dealing with. For purpose of definiteness, let us consider the space spanned by self-replicating genomes of length ℓ, made from an alphabet of size D. Then, every genotype has exactly $(D-1)\ell$ mutational neighbors, and the Hamming space then is a Euclidean lattice of dimension $(D-1)\ell$. For any reasonable genome length, this dimension is high enough that there is an exponentially small chance for any evolutionary path to return to a site, i.e., we can ignore *loops* in such a space. Thus, the space is effectively infinite-dimensional, and we may use a Bethe lattice of connectivity $z = (D-1)\ell$ as a replacement. On this lattice, we know the analytic form of the cluster size distribution and its moments, and we can therefore test the hypothesis that evolution is akin to a percolation process on a Bethe lattice *directly*. However, there is one major unknown: what is the occupation probability? We know from first principles [see Eq. (7.33)] that in this model the critical occupation probability is

$$p_c = \frac{1}{z-1} = \frac{1}{(D-1)\ell - 1} \sim \frac{1}{\ell}\,. \tag{7.51}$$

But is the occupation probability in a typical evolving and adapting system above, below, or exactly at the critical threshold? Is there a mechanism that selects the occupation probability in a universal manner, or is it different in different regions of genetic space? In order to answer some of these questions, we must turn to Artificial Life models such as avida to obtain hints, as it is difficult to obtain data with adequate precision from real systems. But first let us examine some of the predictions. Scaling theory on the Bethe lattice provides us with an expression of the cluster size distribution at and slightly away from the critical threshold. Indeed, for $p_c \sim \ell^{-1}$, we expect

$$n_s(p) \sim s^{-5/2} \exp\left(-(p - \frac{1}{(D-1)\ell})^2 s\right) \tag{7.52}$$

and an average size of the clusters

$$S = \langle s \rangle \sim \left(\frac{1}{(D-1)\ell} - p\right)^{-1} . \tag{7.53}$$

The latter equation is of little use to determine p, as it is determined by the difference of two very small quantities:

$$p = \frac{1}{(D-1)\ell} - \frac{1}{S} . \tag{7.54}$$

Similarly, we cannot use the relation to check the critical exponent ($\gamma = 1$), as we can only dial ℓ, but not p. Eq. (7.52) is more promising. For one thing, we can measure the cluster size distribution and check whether it is a power law, and what is the power law exponent. Such an investigation could reveal, at least approximately, whether percolation on a Bethe lattice is an analogy worth pursuing or whether the model fails completely. At the next step, we would look into *violations* of power law scaling that are introduced by an exponential factor in (7.52). Of course, we first need to find a criterion by which to identify clusters in real or artificially living systems in the first place. This was alluded to in the introduction to this chapter, and we figured there that we should look at *taxonomic* hierarchies to define clusters.

The lowest taxonomic cluster in the usual hierarchy is either the *species* or the *subspecies*. Thus, a cluster could be formed by counting the number of different *genotypes* (each genotype corresponding to a site in genetic space) a species has given rise to, from the time that the species was formed up to its extinction. Similarly, one may go up higher in the taxonomic hierarchy and consider a cluster as a *genus*, and determine the size of the cluster as the number of species that the genus has given rise to. Such data was collected early on by Willis (1922), and more recently by Burlando (cited above, see also the discussion in [Adami et al., 1995]). In the statistical analysis of cluster sizes taken throughout the taxonomic hierarchy, throughout the cataloged *flora* and *fauna*, as well as from the paleontological record, Burlando finds exponents ranging from 1.5 to 2.5 with error bars large enough such that no firm determination can be made from this data set alone. However, he also saw no trend that would indicate a difference in power law exponents throughout the taxonomic system, which led him to suggest that a fractal geometry was dominating the process of evolution. We take from this analysis the suggestion that the choice of taxonomic level may not affect the analysis, and first check the simplest taxonomic abundance distribution that can be obtained in avida, that of species. Abundance distributions of *genotypes* are measured in Chapter 9 later, but such distributions are not strictly cluster-size distributions in the sense of percolation theory, as the members of a cluster do not occupy distinct sites. The exponent of the *species* abundance distribution, which measures the number of genotypes contained in each species, can be taken as a first hint at critical exponents. The concept of species in avida is explained in more detail in Section 9.4. A straightforward experiment running avida for a reasonable amount of time and counting the total number of genotypes produced by each species produces Fig. 7.7. The distribution shows what appears to be a power law,

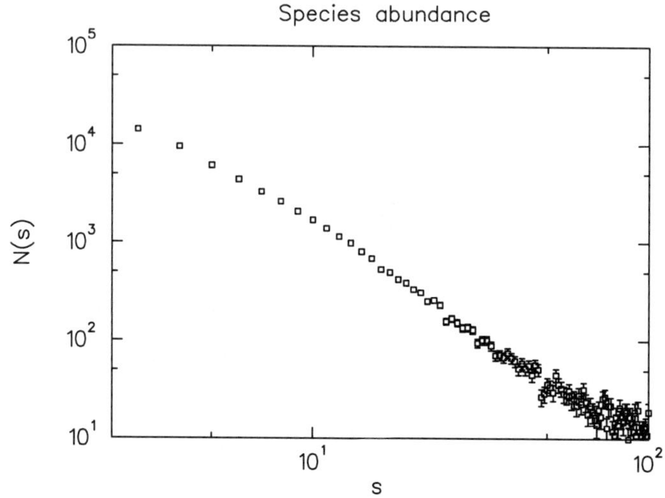

FIGURE 7.7 Number of clusters of species with s genotypes, obtained in a run lasting $T = 50{,}000$ updates at low mutation rate ($R = 0.03$).

but with an exponential component that cannot be fitted with a single exponential.[2] Also, the distribution seems to fall off with an exponent smaller than the one predicted from percolation on the Bethe lattice, raising the possibility that the current species definition is not very close to a cluster in percolating systems. For the future, it is important to examine whether this exponential component at large s is due to an intrinsic scale of the system (such as being away from the critical point), or whether it is due solely to finite-size effects: here the finite time that we ran the simulation, or the finite size of the population. This can only be ascertained by making runs with different population sizes, and which are run for different lengths of time. It is clear from Fig. 7.7, however, that a simple power law with exponential cutoff does *not* fit the observed cluster size distribution, warranting a more thorough investigation of this problem.

7.8 Overview

Percolation is the simplest system that displays geometrical phase transitions of second order. The concepts of cluster-size and critical parameters are defined with ease, even though the system is not simple analytically. The chapter on SOC hinted at the possibility that living systems self-organize to a critical state that displays scale-free distributions in physical observables. In percolation theory, we can investigate this state by tuning to the critical occupation probability p_c, and calculating universal exponents that occur in such systems. As a bonus, it is conceivable that evolutionary clusters can be viewed as percolation clusters in a very high-dimensional genotype space. A space with similar geometry is that of the Bethe lattice, for which percolation can be solved exactly. Thus, it appears possible that the subject of percolation on the Bethe lattice may provide insight about the distribution of taxonomic clusters and the approach to criticality.

[2]Quite possibly this is a *multi-fractal* distribution (several overlapping power laws) which points to much more complicated dynamics, presumably due to the fact that species of different ℓ contribute to the distribution.

Problems

7.1 For a regular square lattice, derive the expected number of clusters of size three as a function of the occupation probability p.

7.2 For a 64×64 regular (square) lattice, determine (computationally) the critical occupation probability p_c for the emergence of an infinite cluster. This is accomplished by choosing random configurations at fixed p often enough to obtain the fraction of cases in which an infinite cluster emerged, and repeating for different p. In order to get a good estimate for the critical filling fraction, it is advised to check p's close to the expected one. Note that the estimate only becomes precise in the limit of very large lattices, which is not the case here. Instead, find the p at which the probability to find an infinite cluster rises to 0.5.

7.3 Derive Eq. (7.39) from Eq. (7.38) in the limit $p \to p_c$.

CHAPTER EIGHT

Fitness Landscapes

For the things of this world cannot be made known without a knowledge of mathematics.

Roger Bacon, 1267

In this chapter we intend to look at the process of evolution from yet another angle, using the concept of the *fitness landscape*. Such an analysis is not without problems: there are a number of researchers who object to the notion that a fitness landscape for real evolving populations exists at all, never mind whether the process of evolution can be understood as adaptation to such an underlying landscape. Nevertheless, and keeping a number of caveats in mind, we introduce the concept of fitness landscapes and methods and tools to study them, in order to obtain more insight into one of the questions we have been asking repeatedly, and that we have been addressing from different perspectives. In the language of this chapter the question is: Are natural fitness landscapes fractal?

The concept of the fitness landscape was introduced by the eminent mathematical biologist Sewall Wright [Wright, 1932] long before the deciphering of the genetic code. It has proved to be one of the most powerful concepts in evolutionary theory by virtue of its imagery (the idea of mountainous terrain, valleys, ridges, and peaks) and by its mathematical accessibility. A drawback of the formalism as presented here is that it does not provide for a mechanism by which the population that evolves on the landscape *feeds back* on it, i.e., helps *form* the landscape. This feature of *coevolution* appears to be crucial for understanding real evolving populations. Whether or not an analysis of evolution in terms of fitness landscapes is directly applicable to real living systems, it certainly provides a fertile mathematical framework in which to cast very simple living systems, and certainly simple artificial ones where coevolution can be neglected.

8.1 Theoretical Formulation

The idea of fitness landscapes turns out to be just an extension of the percolation approach outlined in the previous chapter. There, each site in the metric space (a space where a distance measure between sites is defined) is assigned either one of two possible values—occupied or unoccupied—in a probabilistic manner. As an obvious extension, we might attempt to assign a *real number* to each site representing some intrinsic property of it, and more generally speak of a real-valued *function* that returns the fitness, given the coordinates of the site. The attentive reader realizes that in doing this, we move from a purely geometric discussion of the underlying space to a dynamic one: the function defined can play the role of a Hamiltonian, or more generally a Lyapunov function. Such a function provides a ranking between sites that determines their occupation probability according to the functional value. A simple example of such a function is the potential energy of a spherical ball in a bowl. The lowest point in the bowl confers the lowest energy to the ball, and thus this lowest point represents its preferred coordinates. If the ball occupies a higher position it will, in time, approach the preferred one: the minimum. The temporal dynamics of the ball is then inexorably controlled by the approach to the minimum. While this seems very simplistic as opposed to the problem of evolution, we can imagine a bowl filled with syrup, and a labyrinth of mazes on the inner surface of the bowl, such that the mass point has to sometimes move upwards in order to reach the point where it takes on the lowest potential energy. We can also imagine that, if the bowl is not shaken often enough, the mass point could get stuck at an intermediate ridge on the surface of the bowl. The shaking, of course, is an agent of noise that can dislodge balls that get stuck. Furthermore, we can imagine that the surface of the bowl is not fixed, but actually moves and changes, so that a ball that is stuck can get loose because the ridge that was preventing its descent to lower energies disappears. Instead of this energy function that is minimized in the dynamics just described, we consider a fitness function that is *maximized*, and ask about what type of landscape gives rise to which dynamical behavior.

We begin by an elementary description of binary sequence space. The size of the space is obviously determined by the length of the sequence, and we depict the sequence space for binary strings of length 1, 2, 3, and 4 in Fig. 8.1. The vertices of these hypercubes are shaded in such a manner that sites with the largest mutational distance have the most contrast. Each vertex of the binary hypercube can be represented by a

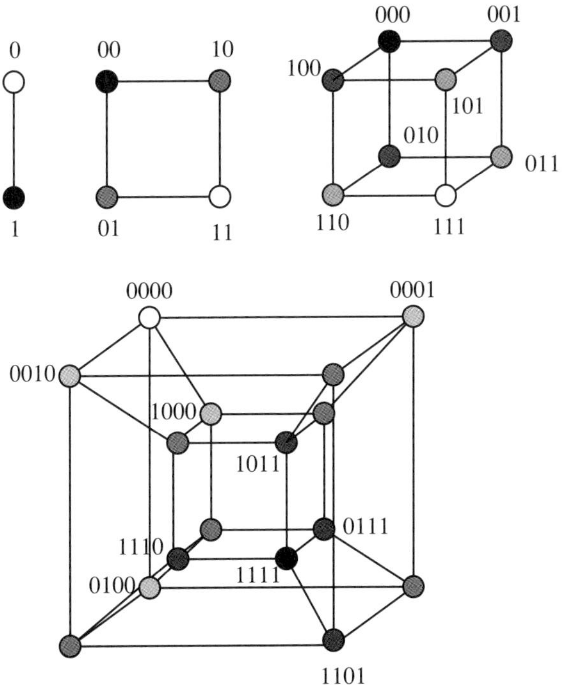

FIGURE 8.1 Hypercubes representing sequence space for binary strings of length 1 to 4. Sites that have the same distance from a reference point (say, the vertex 0000 in the four-dimensional hypercube), are shaded in the same manner.

vector

$$\boldsymbol{x} = (x_1, x_2, x_3, \ldots, x_\ell), \tag{8.1}$$

where x_i takes on the values 0 or 1 and ℓ is the length of the string. More generally, we can define hypercubes formed from any alphabet, so that $x_i \in \{A, B, C, \ldots\}$ with alphabet size $\mathcal{D}$. On this space, which is generally known as a *vector space*, we can define a real function $f(\boldsymbol{x})$, which is the fitness function:

$$\begin{aligned} f : \mathbb{V} &\to \mathbb{R} \\ \boldsymbol{x} &\to f(\boldsymbol{x}) \; . \end{aligned}$$

All of this chapter is concerned with properties of the function f, which defines the type of landscape in which the evolution on $\mathbb{V}$ takes place. For example, we may want to know what is the *average* value of the fitness. In principle, this can be obtained by summing the fitness function over all sequences in the space:

$$\langle f \rangle = \frac{1}{N_g} \sum_{i=1}^{N_g} f(\boldsymbol{x}_i), \tag{8.2}$$

where N_g is the number of sequences $N_g = \mathcal{D}^\ell$. For small sequences this is not difficult; we have seen, however, that for realistic sequence spaces it is impossible to obtain this average, simply because the number of sequences in the space is too large. The same is true for other global observables of this function, such as $\langle f^2 \rangle$ and the variance $\sigma^2 = \langle f^2 \rangle - \langle f \rangle^2$. We shall see that the only measure that can realistically be extracted from effectively infinite landscapes are *correlation functions* on the landscape. Before we define those, we need to introduce the concept of *stochastic* landscapes.

It is clear from a moment's reflection that the idea that a specific sequence can unambiguously be mapped to a fitness is overly simplistic. Rather, the sequence will result in a core fitness for the organism that harbors it, while the final survival of the particular member will be subject to a random component. This idea is central to the construction of stochastic landscapes, where a function $f(\boldsymbol{x}_i)$ is generated by drawing it from a *pool* of landscapes, which are distributed according to a probability distribution. Each *individual* landscape is fixed, however. Then, we can define averages over *ensembles* of landscapes rather than averaging over all sequences. For example, we can define the expectation value of f for sequence $\boldsymbol{x}_i$ (where i ranges from 1 to $\mathcal{D}^\ell$)

$$\mathbb{E}[f(\boldsymbol{x}_i)]$$

as the average over the value of $f(\boldsymbol{x}_i)$ over all the landscapes in the ensemble. Similarly, we can then define the variance of the fitness as

$$\mathrm{var}[f(\boldsymbol{x}_i)] = \mathbb{E}[f(\boldsymbol{x}_i)^2] - \mathbb{E}^2[f(\boldsymbol{x}_i)] \ . \tag{8.3}$$

With these tools, we can define the *correlation* function of the landscape defined by f by imagining a *random walk* Γ of length n in sequence space, that starts at sequence $\boldsymbol{x}_t$

$$\Gamma = \{\boldsymbol{x}_t, \boldsymbol{x}_{t+1}, \ldots, \boldsymbol{x}_{t+n}\} \ . \tag{8.4}$$

To this walk Γ corresponds a walk in fitness

$$f(\Gamma) = \{f(\boldsymbol{x}_t), f(\boldsymbol{x}_{t+1}), \ldots, f(\boldsymbol{x}_{t+n})\} \ . \tag{8.5}$$

Consider then the autocorrelation function $R(s)$ defined by

$$R(s) = \frac{\mathbb{E}[f(\boldsymbol{x}_{t+s})f(\boldsymbol{x}_t)] - \mathbb{E}[f(\boldsymbol{x}_{t+s})]\,\mathbb{E}[f(\boldsymbol{x}_t)]}{\sqrt{\mathrm{var}[f(\boldsymbol{x}_{t+s})]\,\mathrm{var}[f(\boldsymbol{x}_t)]}} \ . \tag{8.6}$$

It is easy to see that $R(s)$ is normalized $[R(s = 0) = 1]$, and decays for $s > 0$. It represents the correlation between the fitness of two sequences $\boldsymbol{x}_t$ and $\boldsymbol{x}_{t+s}$ separated by s mutational steps. Again, the general idea here is that if the fitness landscape is reasonably smooth, two sequences that are separated by only a few mutations should be reasonably close in fitness. A rugged landscape, on the other hand, will be characterized by a correlation function that drops precipitously after only a few steps in the walk. Then, sequences just a few mutations away may turn out to be very different in fitness. It is this characterization in terms of ruggedness in which we are interested.

A simplification of Eq. (8.6) occurs for *isotropic* landscapes. In such landscapes the expectation value of the fitness $\mathbb{E}[f]$ does *not* depend on the precise location in the landscape, i.e.,

$$\mathbb{E}[f(\boldsymbol{x}_t)] \approx \mathbb{E}[f(\boldsymbol{x}_s)] \approx \mathbb{E}[f(\boldsymbol{x}_0)] \qquad (\boldsymbol{x}_t \neq \boldsymbol{x}_s) \ . \tag{8.7}$$

With the isotropy condition (8.7), Eq. (8.6) simplifies to

$$R(s) = \frac{\mathbb{E}[f(\boldsymbol{x}_0)f(\boldsymbol{x}_s)] - \mathbb{E}[f(\boldsymbol{x}_o)]^2}{\mathrm{var}[f(\boldsymbol{x}_0)]} \ . \tag{8.8}$$

We should remember at this point that all averages are performed for specific sequences $\boldsymbol{x}_0, \ldots, \boldsymbol{x}_s$, with the averages taken over fitnesses obtained by polling from a pool of landscapes. Let us define another measure of correlation for a single fixed landscape, averaging over all possible *walks* that can connect sequences $\boldsymbol{x}_0$ and $\boldsymbol{x}_s$

$$\bar{R}(s) = \frac{\langle f(\boldsymbol{x}_0) f(\boldsymbol{x}_s)\rangle - \langle f(\boldsymbol{x}_0)\rangle^2}{\sigma^2[f(\boldsymbol{x}_0)]} \ . \tag{8.9}$$

There is a wide range of landscapes for which the correlation functions (8.8) and (8.9) agree. Such landscapes are called *self-averaging*, i.e.,

$$R(s) = \bar{R}(s) \ . \tag{8.10}$$

In the following, we shall focus only on self-averaging landscapes, where we also have

$$\mathbb{E}[f(\boldsymbol{x}_0)] = \langle f \rangle = \frac{1}{N_g} \sum_{i=1}^{N_g} f(\boldsymbol{x}_i) \ . \tag{8.11}$$

For self-averaging landscapes we can define yet another correlation function which, rather than depending on the length of a walk between two sequences, is defined in terms of the *Hamming* distance d, i.e., the *shortest* mutational walk, between the sequences. Thus, we define

$$\rho(d) = \frac{\langle f(\boldsymbol{x}) f(\boldsymbol{y})\rangle_\Delta - \langle f(\boldsymbol{x})\rangle^2}{\sigma^2[f]} \ , \tag{8.12}$$

where $\Delta = d(\boldsymbol{x}, \boldsymbol{y}) = d$ denotes the Hamming distance between $\boldsymbol{x}$ and $\boldsymbol{y}$. The notation $\langle \ldots \rangle_\Delta$ reminds us that we are averaging over *all* sequences in the space which are a Hamming distance d apart. Naturally, we expect a relation to hold between $\rho(d)$ and $\bar{R}(s)$. (The latter will be denoted just $R(s)$ in the future, as we assume that the landscapes we are dealing with are self-averaging.) Denote by ϕ_{sd} the probability that a walk of s steps ends at a sequence that is a (genetic) distance d away from the starting sequence. Then

$$R(s) = \sum_{d=1}^{s} \phi_{sd}\, \rho(d) \ , \tag{8.13}$$

that is, the correlation between sequences that are connected by s random steps is obtained by weighting the correlation function of two sequences that are a distance d apart with the probability that a walk of s steps takes you to a sequence a distance d away, and summing over all possible d. For a genetic space of very high dimension (hypercubes spanned by very long sequences), a walk of length s will almost always lead to a sequence approximately s *mutational* steps away, i.e., few walks backtrack genetically. In this case

$$\phi_{sd} \approx \delta_{sd} \,, \tag{8.14}$$

and the two measures of correlation approximately agree. Note that for a genetic space that has the topology of the Bethe lattice, Eq. (8.14) is exact. For some other landscapes, ϕ_{sd} can be obtained recursively (see, e.g., Weinberger and Stadler, 1993 and Problem 8.1).

The autocorrelation function $\rho(d)$ provides, if it can be measured, important information about the dynamics of the adaptation process on the landscape. Let us consider a number of examples in the next section.

8.2 Example Landscapes

One of the simplest landscapes to construct is that of independent Gaussian variables. For such a landscape, the fitness $f(\boldsymbol{x})$ is a random variable, independent of any other neighboring string $\boldsymbol{y}$. Such a landscape is unrealistically rugged, as we can convince ourselves easily. Indeed, if neighboring strings have independent fitnesses, the autocorrelation function $\rho(d)$ must decay abruptly from one to zero by just taking *one* step, i.e.,

$$\rho(0) = 1, \qquad \rho(1) = 0. \tag{8.15}$$

Such a model describes disordered media rather than evolution, and is known as the *random-energy model* [Derrida, 1981]. We call such landscapes "Derrida" landscapes.

From the most rugged of all landscapes, let us move to smoother ones. An important class of landscapes is represented by so-called AR(1) landscapes. They are characterized by a single scale: the correlation parameter ρ. For such landscapes one finds that the fitness of one site depends only on the fitness of a nearest neighbor. In other words, for a process that takes genotype $\boldsymbol{x}$ into genotype $\boldsymbol{y}$, the fitnesses are connected by

$$f(\boldsymbol{y}) = \rho f(\boldsymbol{x}) + \Delta f \ , \tag{8.16}$$

where Δf is a stochastic variable drawn independently from the same Gaussian distribution for $\boldsymbol{x}$ and $\boldsymbol{y}$. Stochastic processes for which the value of the function at $\boldsymbol{y}$ only depends on the value of the function at $\boldsymbol{x}$ are called *Markov* processes. A random walk is the simplest Markov process: the position of the walker at the next point in time only depends on the position at the previous time (and of course the random variable which determines the next move). Clearly, for a random walk on an AR(1) landscape, we find for the autocorrelation function

$$R(s) = \rho^s \equiv e^{-s/\xi} \ , \tag{8.17}$$

where we defined the correlation length

$$\xi = \frac{1}{|\log \rho|} \ . \tag{8.18}$$

Thus we see that indeed ρ (or ξ) is the single fundamental descriptive parameter of the AR(1) landscape. In such landscapes, sequences that are more than ξ steps apart are effectively *uncorrelated*. Thus, the correlation length ξ defines an effective radius of influence between sequences. This radius was zero for the Derrida landscape. We also note that the correlation function in this case is exponential and not a power law. Thus, we do not expect scale-free behavior unless ξ is proportional to the size (i.e., maximal diameter) of the landscape. We shall come back to this point below.

The next example we describe is Kauffman's $N - k$ model [Kauffman and Levin, 1987; Kauffman and Johnsen, 1991]. It is also defined on a Boolean hypercube (i.e., the genomes are binary sequences of length N), but their fitness function depends on a parameter k that allows us to tune the landscape from rugged to smooth. As such, it is a useful mathematical tool to investigate a range of landscapes.

Imagine a fitness function that assigns a fitness f_i to each bit x_i on a string of length N, and where the fitness of the string is obtained by just averaging over the contribution of each bit. If the f_i are obtained from a random probabilistic distribution (e.g., a uniform random number on the interval [0,1]), we obtain a relatively smooth dependence of the fitness on the genotype, as the flipping of one bit alters the fitness only by an amount of order $1/N$. Accordingly, the autocorrelation function for such a landscape is

$$\rho(d) = 1 - \frac{d}{N}\,. \tag{8.19}$$

This is the $N - k$ model for $k = 0$, as k denotes the number of *neighboring* bits that the fitness f_i depends on, and we just stipulated that the f_i are obtained randomly; we may draw for example, $f(x_5 = 0) = 0.45$ and $f(x_5 = 1) = 0.12$ independently.

For a $k = 2$ landscape, the fitness of each bit x_i depends on *two* other bits on the string, for example its nearest two neighbors. (Other models can be constructed where the fitness of each bit depends on k *random* neighbors, but we shall not consider these here.) If the fitness of a bit depends on its two nearest neighbors, we need to draw a random number for *eight* different possible neighborhoods just to define the fitness contribution of one bit. An example is given in Table 8.1 below, where we consider the case $N = 8$ and $k = 2$.

TABLE 8.1 Example fitness contributions for an $N - k$ model with $N = 8$ and $k = 2$ (from Weinberger and Stadler, 1993), showing the tables used for computing the fitness of the fifth and sixth sites. The numbers in the column labeled "Contrib." (below) are drawn randomly, whereas the $f(x_i)$ depend on the specific sequence.

Bit	1	2	3	4	5	6	7	8
Symbol	1	0	1	1	0	1	0	1
$f(x_i)$	0.39	0.46	0.91	0.18	0.73	0.29	0.84	0.70

Fitness of string: $f(10110101) = \frac{1}{8}\sum_i f(x_i) = 0.56$

Table for computing contribution of bit positions

Bit position 5				Bit position 6			
Bit 4	Bit 5	Bit 6	Contrib.	Bit 5	Bit 6	Bit 7	Contrib.
0	0	0	0.32	0	0	0	0.99
0	0	1	0.21	0	0	1	0.10
0	1	0	0.19	0	1	0	0.29
0	1	1	0.93	0	1	1	0.22
1	0	0	0.87	1	0	0	0.86
1	0	1	0.73	1	0	1	0.39
1	1	0	0.64	1	1	0	0.48
1	1	1	0.88	1	1	1	0.61

Note that to assign nearest neighbors to boundary sites, you can choose *periodic* boundary conditions, i.e., you can assume that the sites are arranged on a circle, such that the first and the eighth site are neighbors in this example.

On the other extreme, when choosing $k = N - 1$, we find the Derrida landscape. In that case, a single mutation changes the fitness of the string to a completely independent random number, and there is no correlation between adjacent genotypes. The correlation function can be obtained by considering the probability that a string survives d mutations (which act on different bits, as otherwise the d mutations would not result in a string at distance d). Here is the result (see Problem 8.2):

$$\rho(d) = \frac{(n-k-1)!\,(n-d)!}{(n-k-1-d)!\,n!}\,. \tag{8.20}$$

This function is shown in Fig. 8.2 for $N = 8$ and k-values ranging from 0 to 4. Note that for $k = 0$ we recover Eq. 8.19, while for $k = N - 1$ the function vanishes at $d = 1$ (not shown).

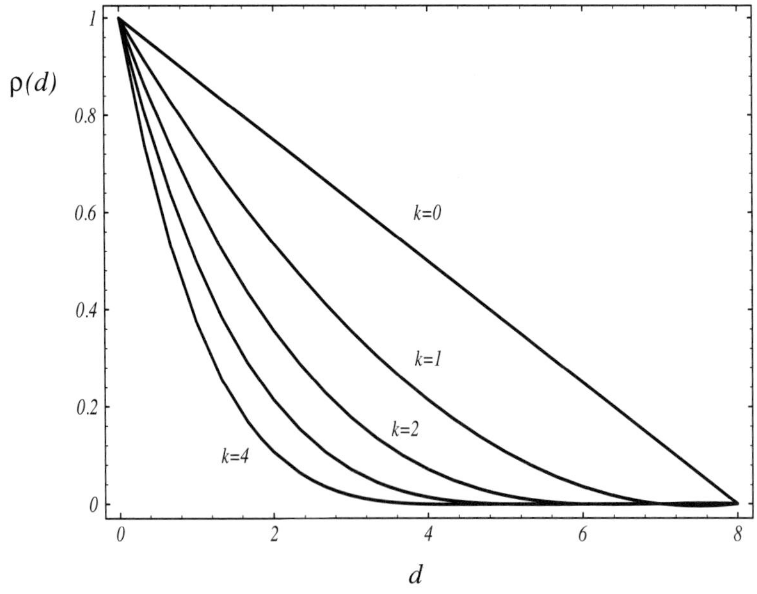

FIGURE 8.2 Autocorrelation function $\rho(d)$ for the N–k model (nearest neighbors or random neighbors) for $N = 8$ and $k = 0 - 4$.

8.3 Fractal Landscapes

Before formally discussing the concept of *fractality*, or self-similarity with respect to landscapes, let us take an intuitive look at the idea.

When imagining ourselves performing an adaptive walk in an infinite landscape, we can assume that we would only be able to see the local properties of the landscape, which may change as we climb any peak that we see. Of course this implies that we cannot, from our vantage point, see all the peaks of the landscape; neither can we see how high the peak is that we are about to climb. However, we do notice which way is uphill (and how steep the climb is), and we also see how *many* different paths are leading uphill. One way to classify the local properties of the landscape, then, would be to always keep track of the number of ways that lead uphill, as well as what the local slope is. Imagine a landscape that is of infinite extent *spatially*, or equivalently, one that does not have any boundaries. If there is one *global* maximum in this landscape, it is clear that, no matter where our journey starts, we are one day going to end up at this peak, and the local properties there will be very different from those observed during the ascent. Indeed, the number of ways to move uphill in that case is zero. Such landscapes, which we shall term *finite* landscapes (even though their spatial extent, or more precisely the number of different points $\boldsymbol{x}$ in the landscape, may be infinite) have the property that the number of ways uphill does *not* stay constant as we climb the peaks. Rather, the number of ways to improve oneself typically decreases by a constant factor each time a new (local) maximum is found. Thus, since the local properties change while we move around, such landscapes can hardly be called self-similar.

Typically, in a search process, the precision with which we examine a landscape's potential for improvement depends on the level of height (or fitness) already achieved. We may, for example, start to look for a maximum very broadly, and then when found, move towards it with more and more delicate and detailed moves. In a manner of speaking, we increase the *resolution* of our "radar" the closer we get to the goal. In finite landscapes, the landscape changes as the resolution is increased, until it looks (at the highest resolution) completely flat. Quite the contrary in self-similar landscapes: the higher the resolution we use to probe for further ways uphill, the more such ways we find! Such self-similar or fractal landscapes are thus *infinite*; there can never be a maximum, because the achievable height depends on the resolution with which we choose to examine our local landscape. In fractal landscapes, then, there is *no* scale that sets the average distance between peaks in the landscape (as expected), and in fact the only scales that could be present in realistic fractal landscapes should be those related to the size of the system and the distance between two *adjacent* vectors $\boldsymbol{x}$ and $\boldsymbol{y}$, sometimes called the *lattice constant*.

At this point it appears opportune to emphasize that, unlike in the percolation process outlined in the previous chapter, in the discussion of fitness landscapes the geometry of the space is as important as the *process* that takes place on it. Indeed, when we define the landscape to be a *metric* space, we assume that there is a means to measure the distance between any two points on the landscape. Apart from the usual Hamming distance, the process itself might impose a distance measure by declaring two points as neighbors if they can be reached by a single mutational event. As mutational events in genetic landscapes include cross-overs, such a measure can in principle make the landscape quite complicated, and *diffusion* on them may take on an *anomalous* character, i.e., be different from the usual Gaussian diffusion process.

Let us attempt to capture the idea of fractal landscapes mathematically. The concept of fractal landscapes was introduced almost a decade ago by Sorkin (1988). He defined a landscape to be a fractal of type h if

$$\langle (f(\boldsymbol{x}) - f(\boldsymbol{y}))^2 \rangle_\Delta \sim d^{2h} , \tag{8.21}$$

i.e., if the variance of the difference between the fitnesses of two configurations scales with their genetic distance. Here, we define $\Delta = d(\boldsymbol{x}, \boldsymbol{y})$ to be the *metric* distance between two strings. This could be the usual Hamming distance, or, in genetic landscapes, the smallest number of *operations* (including maybe insertions and deletions) necessary to obtain $\boldsymbol{y}$ from $\boldsymbol{x}$. In the following, we shall refer to $d(\boldsymbol{x}, \boldsymbol{y})$ simply as the genetic distance without specifying its construction. We will discuss different such measures in Chapter 9.

Sorkin's definition seems to connect with our intuitive discussion earlier: in such a situation there is no scale that sets the difference in fitness between two sequences, except for their genetic distance (and the size of the landscape as we shall see later). Let us rewrite this criterion in terms of the autocorrelation function. We find then that, since

$$\frac{\langle (f(\boldsymbol{x}) - f(\boldsymbol{y}))^2 \rangle_\Delta}{\sigma^2} = 2(1 - \rho(d)), \tag{8.22}$$

we can rewrite Sorkin's criterion as

$$1 - \rho(d) \sim d^{2h}. \tag{8.23}$$

Let us examine a few examples and discover what the fractal *type* h determines.

First, we examine the AR(1) landscape introduced earlier. There, we found that approximately

$$\rho(d) \approx e^{-d/\xi}, \tag{8.24}$$

where ξ is the correlation length of the landscape. For distances much smaller than this length,

$$\rho(d) \sim 1 - \frac{d}{\xi} + \mathcal{O}\left((d/\xi)^2\right), \tag{8.25}$$

i.e., for such small distances the term of the order (d^2/ξ^2) can be ignored, and *all* AR(1) landscapes then look like fractals of type $h = 1/2$. Such landscapes are also sometimes called *trivial* fractals, as they are the kind that arise in a random walk problem (see the discussion of scaling in random walks in Box 6.1 and below). In order to discuss what happens at larger distances, it is convenient to rescale the genetic distance by the *maximal genetic distance* in the landscape, i.e., the size, or *diameter* of the landscape L. Thus, we introduce the new distance $\tilde{d} = d/L$, and define the correlation function in terms of this distance $r(\tilde{d})$, which can then be written as

$$r(\tilde{d}) = 1 - \frac{L}{\xi}\tilde{d} + \mathcal{O}(\tilde{d}^2) \ . \tag{8.26}$$

Now we can see that an AR(1) landscape is a fractal of type $h = 1/2$ if and only if its correlation length is *proportional* to the diameter of the landscape, as only then can the higher order terms in the sum be ignored. In a fractal of type $h = 1$, for example, the autocorrelation function does *not* have a term linear in d, i.e.,

$$\rho(d) = 1 - \beta d^2 + \mathcal{O}\left(d^4\right) \ . \tag{8.27}$$

Clearly, we are more interested in nontrivial fractals ($h \neq 1/2$), and we may ask what are the conditions to achieve such a landscape.

Let us now examine the Kauffman landscape. From the autocorrelation function (8.20) we find, to lowest order in d,

$$\rho_{Nk}(d) = 1 - \frac{d}{N} - \sum_{m=1}^{k} \frac{d}{N-m} + \mathcal{O}\left(d^2\right) , \tag{8.28}$$

so that in the limit of large N

$$\rho_{Nk}(d) \to 1 - \frac{d(k+1)}{N} + \mathcal{O}\left(d^2\right) . \tag{8.29}$$

Note that for the Kauffman landscape, the sum is already given in terms of the genetic distance divided by the diameter of the landscape, as no strings can have a higher genetic distance than N. The $k = 0$ landscape is a trivial fractal of type $h = 1/2$, as all higher-order terms in the sum (8.29) vanish. Even though for higher k there is always a linear term in the sum (8.29), it is misleading to conclude that the landscape remains $h = 1/2$. Indeed, as the sum converges less and less well for higher k, the fractality of the landscape is less easy to determine. Yet, as it is clear that the Kauffman landscape is *finite* (one string is always the best in the landscape), the Kauffman landscape can only be fractal locally, i.e., for small d. Here we have put our finger on the single most important property a fractal landscape must have. Anytime the string length is *fixed*, or cannot exceed a certain value, the landscape will necessarily become simple once the highest level of fitness is reached. Indeed, in this case the landscape will appear flat: as no improvement can be achieved, the only nondeleterious mutations will be *neutral* ones. We shall be able to witness this aspect in a discussion of RNA landscapes in Section 8.5. Still, as long as the system is in a stage where the current string length is much smaller than the maximal one, the landscape may *appear* fractal, as the defining scale (the maximal genetic distance L—(or N in the $N-k$ model) has not yet been reached. We will return to this particular point when we discuss the fitness landscapes of the artificial chemistry in avida in Section 8.6. Before this, as a last *theoretical* example, let us construct a fitness landscape based on the *Lévy-flight* paradigm: self-similar jumps that occur with a probability that is inversely proportional to the size of the jump, and where the average squared fitness *diverges* in the limit of infinite landscapes.

8.4 Diffusive and Nondiffusive Processes

As discussed earlier, in a conventional random landscape we can imagine the fitness $f(\boldsymbol{x})$ to be drawn from a probability distribution that is Gaussian, whereas the fitness of a *walk* is

$$F(t) = \sum_{j=1}^{n} \Delta f(\boldsymbol{x}_j). \tag{8.30}$$

For such a Gaussian random walk, the distribution of the single steps Δf scales in the same manner as the distribution of the whole walk F, i.e., if $P(F, n)$ is the probability that in n steps we have achieved fitness F, then

$$P(\lambda^{1/2}F, \lambda n) = \lambda^{-1/2}P(F, n), \tag{8.31}$$

with a scaling parameter λ. In this case, the second moment of the distribution of F's, which is directly related to the correlation function (if we assume that $\langle F\rangle = 0$) scales like the number of steps taken,

$$\langle F^2\rangle \sim Dn\,, \tag{8.32}$$

where D is a constant that sets the scale of the process. In earlier chapters we saw that it plays the role of a diffusion coefficient. Since, as we have seen, in high-dimensional landscapes the number of steps taken is just the genetic distance [cf. Eq. (8.14)], we note once again that we are dealing with a random fractal of type $h = 1/2$. This is the case of *normal* diffusion. *Anomalous* diffusion (which we have encountered earlier in Section 6.4) is said to occur if the second moment of the walk either does not scale linearly with the number of steps taken (or, if a step is taken in unit time, does not scale linearly with time)

$$\langle F(t)F(0)\rangle = 2D\,t^{2h} \tag{8.33}$$

with $h \neq 1/2$, or if the second moment simply does not exist. The latter can occur if the probability to take a single step is *unnormalizable*, which can occur if we draw fitnesses (or step-sizes) from a distribution of the power-law type:

$$p(\Delta f) \sim (\Delta f)^{-1-\beta}\,, \tag{8.34}$$

where $\beta = 2$ corresponds to the Gaussian ($h = 1/2$) fractal (see, e.g., Klafter et al., 1996). For general β, this implies that the second moment of the fitness scales as

$$\langle F^2 \rangle \sim n^{2h}, \tag{8.35}$$

where now

$$h = \frac{3-\beta}{2}. \tag{8.36}$$

In such a manner, thus, we can construct landscapes with arbitrary h. Note that for infinite landscapes (no maximal string length), we cannot distinguish whether anomalous diffusion is due to a power law distribution in jump sizes or due to a power law distribution in waiting-times between jumps.

Intuitively, landscapes with higher h can be thought of as leading to *persistent* random walks. As compared to the random walk with $h = 1/2$, in a walk with a higher h the walker has a higher tendency to persist in the direction taken, whereas walks with $h < 1/2$ are *anti-persistent*. In evolution, we expect a walker to *locally* perform an $h = 1/2$ walk. The crossover operation, however, can throw the walker far away in genetic space, leading to anomalous diffusion and the fractal character of the landscape. Fig. 8.3 shows an example of such a walk, where the walker is trapped locally for some time and performs a Gaussian random walk

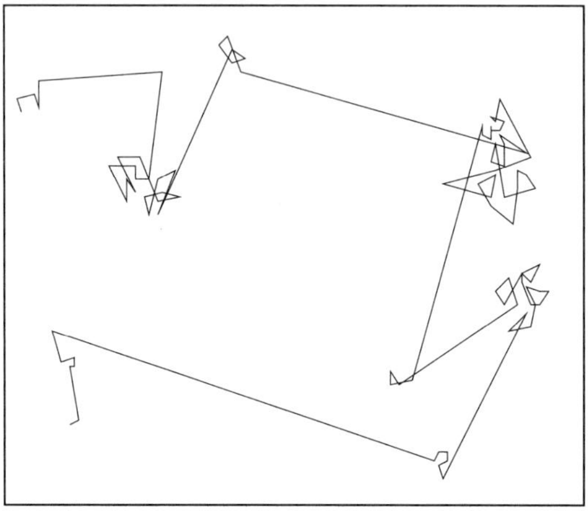

FIGURE 8.3 Lévy flight in two dimensions. The turning points of the walk tend to cluster in self-similar patterns.

there, but intermittently performs large jumps, with probability inversely proportional to the length of the jump. As a consequence, the mean-square displacement of the walker diverges. This is a familiar concept that we encountered in the chapter on self-organized criticality. Applied to the discussion of landscapes, fractality implies that inside *any* radius around a particular sequence we must be able to find maxima of *any* height if we are willing to adjust the scale finely enough (in the language of evolution: wait long enough). There is no doubt that this is an idealized concept, but one that is familiar from the physics of disordered media, especially the so-called *spin-glasses*. We shall return to discuss landscapes when we investigate learning in avida and the approach to the error-threshold.

8.5 RNA Landscapes

In this section we discuss an example of a realistic landscape: the mapping between nucleic acid chains and folded proteins. Such a mapping can be seen as occurring between symbolic sequences, such as ACCUCGCCCUUU... , and secondary structures, such as the tRNA depicted in Fig. 8.4. In the folding algorithm of the Vienna group [Hofacker et al.], the secondary structure is computed from the sequence using procedures based on thermodynamics. As the secondary structure covers most of the *free energy* of the tertiary structure (the energy required to *unfold* it), the folding algorithm can be used as a predictor of the sequence's fitness. In the absence of a cellular metabolism to determine fitness (as the RNA strings do not self-replicate), in the experiments described below [Huynen et al., 1996] *two* different measures were tested to determine the value of a particular structure: one in which the structure's *stability* alone is the determining factor (i.e., the structure with the highest free energy is declared the "winner"), and one where a particular *type* of secondary structure (here, the cloverleaf structure of tRNA) is declared optimal, and any candidate's structure is compared to the target to determine how close it is. Two different questions were addressed in the context of fitness landscapes, namely "How frequent is an 'optimal' structure in this landscape?" and "What is the dynamics of the adaptive process that has the optimal structure as its endpoint?" Naturally, these questions reveal instantly that the landscape considered cannot be globally fractal, as there is a global optimum. This can be traced back to the requirement of keeping the sequence length *constant* in these experiments (here $\nu = 76$), which makes the analysis much easier. As remarked earlier, no landscape with a fixed length requirement can ever be globally fractal.

In this landscape defined on symbolic strings of length 76, and an alphabet {G,C,A,U}, the first experiment concerned *neutral diffusion* on a *flat* landscape. This was achieved by taking the *target* sequence (which folds into the above-mentioned cloverleaf depicted in Fig. 8.4) and creating a population of 1000 copies of it. Note that the mapping of sequence into structure is many-to-one, i.e., many sequences fold into the same structure. Then, as the population is subjected to mutations, it diversifies while still keeping the target structure. The fitness of the strings is measured by a *tree-editing* procedure, where the folded structure of a candidate string is compared to the target structure, and the number of editing steps necessary to obtain the target from the candidate is termed the distance *d* between the two. Each candidate string is then replicated according to a fitness function. In the case at hand, the replication rate of a string a

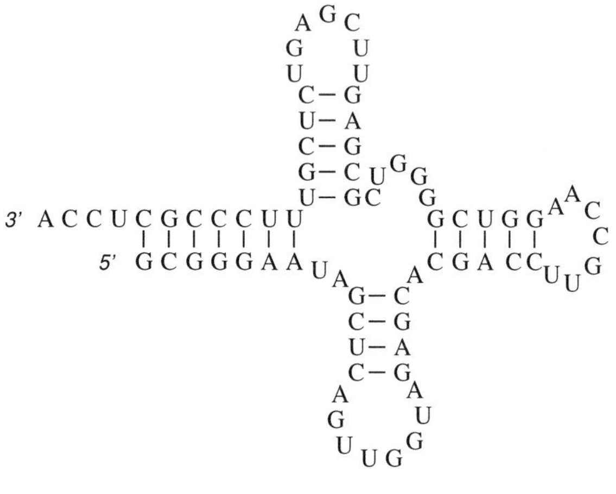

FIGURE 8.4 Secondary structure of the tRNA molecule, obtained from folding the symbolic sequence. The secondary structure can subsequently be folded into a tertiary structure.

distance d from the target is (arbitrarily) taken to be

$$A(d) = 1.06^{146-d}. \quad (8.37)$$

Clearly, any substitutions in the genetic string that leave the folded structure zero steps away from the target one will be replicated at the same rate as the original string. During the neutral walk, the original genotype of the ancestral string is soon lost, and the initially homogeneous population of strings diffuses in genetic space (see Fig. 8.5). Soon, indeed, the original genotype disappears from the population, while the *phenotype*, i.e., the target structure, is unchanged. For this population, then, the landscape appears *flat*: there is no possibility for improvement, but the *entire* landscape can still be traversed by the genotypes, i.e., the genetic distance of the final genotype can be maximally far away from the ancestral one. This seems to suggest that in such smooth landscapes, *any* structure can be found arbitrarily close to a starting one: an important ingredient in fractal

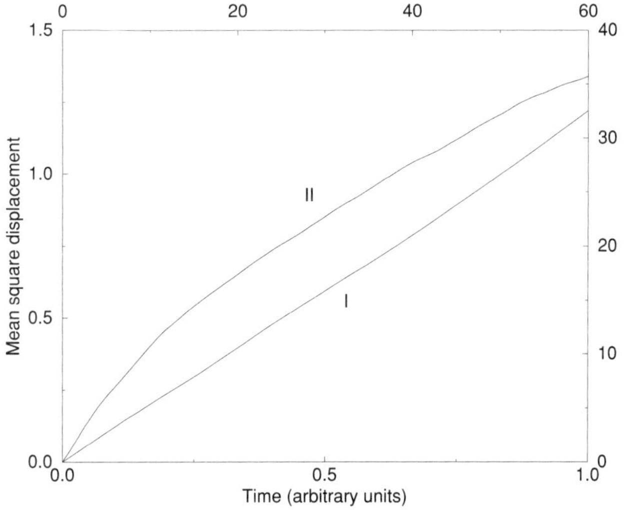

FIGURE 8.5 Mean square displacement of sequences, as measured by calculating the barycentric average (see Huynen et al., 1996) of the population, from the ancestral sequence. The curve is roughly linear in time (at least at small times, curve I, lower scale) which confirms that Gaussian diffusion is taking place in genetic space.

landscapes. In this particular experiment, this property of fractal landscapes was tested in a situation that was manifestly nonfractal (by keeping the string length constant). Nevertheless, this observation may point to the possibility that, if the string length is unrestricted, realistic RNA landscapes in realistic environments (a functioning cell-cycle) may indeed be fractal.

In another experiment, Huynen et al. started with a population of *random* sequences, and watched how the fitness of the population improved as the structure came closer and closer to the target cloverleaf (Fig. 8.6). Like in avida, the adaptation process is characterized by long periods of stasis, in which the population performs an approximately Gaussian random walk, punctuated by intermittent adaptation events, that can be likened to the Lévy flights of Fig. 8.3. Because of the finite population, equilibration occurs rapidly, and the population can freeze into a metastable state that is close to the target. Still, for large genetic distances (structures far away from the optimum), the adaptation curves seem to display a *Devil's staircase* character, that we identified with a power law distribution of waiting times in Chapter 6. Such a distribution may

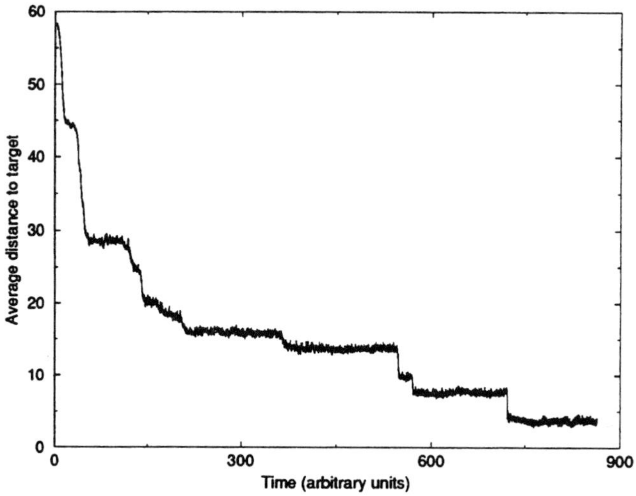

FIGURE 8.6 Evolutionary optimization of a population of random RNA strings toward the target structure with a mutation probability of $p = 0.001$ and 1000 strings in the population. Due to structural constraints, the exact target structure is not usually reached in these experiments.

signal a fractal landscape, as it leads to anomalous diffusion. Because of the finiteness of the landscape, however, this hypothesis cannot be tested here for the same reason that the experiments covered in Chapter 6 were inconclusive.

In the next section, we describe qualitatively the fitness landscape of the avida system, in order to gauge whether the inherent dynamics may give rise to a fractal fitness landscape. Because the length of strings in avida is not fixed, a major necessary condition for a fractal landscape is met.

8.6 Fitness Landscape in avida

Finally, at the end of this chapter we are prepared to discuss the fitness landscape of the avida world as a paradigm for landscapes created by the artificial chemistry of populations of self-replicating code. We shall see that what type of landscape emerges is to a large extent in the hands of the user: we can construct flat landscapes and we can construct complicated ones. In the end, it is the *amount of information* that is present in the environment in which the population evolves that decides whether the landscape is asymptotically flat or fractal.

Quite generally, we remark that in avida there is effectively no constraint on the length of the code, except for a *minimal* length requirement (which is trivial, as the smallest length for self-replicating programs with the default instruction set is eleven), and a maximal length requirement that simply assures that no string can attempt to allocate all of the computer's memory for itself. For this very reason then, evolution is *open-ended* in avida: anytime a fit string is found, there is no guarantee that there is not a fitter string that just happens to be much longer and therefore can encode a smarter algorithm. In practice, if there is no information in the landscape (we shall make this statement more precise below), this open endedness is merely academic. For example, it appears possible to design the best self-replicator (the one with the shortest gestation time) for a given instruction set within avida, and the chance that a much longer creature could beat such a designed program appears remote. So, in this simplified situation the avida landscape appears definitely finite

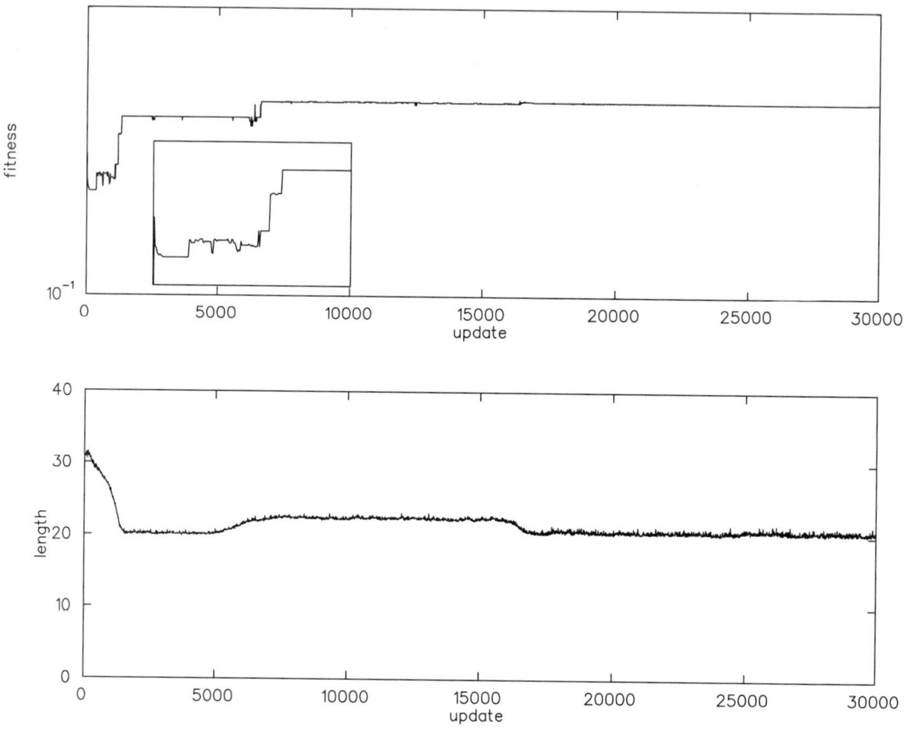

FIGURE 8.7 Fitness (upper panel) and length (lower panel) of the most abundant (the dominating) genotype in a typical avida population of 3600 strings at a mutation rate $R = 0.02$ in a landscape where *only* replication is of benefit. The inset shows the fitness for the first 2000 updates.

and far from fractal. An example of adaptation on this flat landscape is shown in Fig. 8.7, where we show the fitness of the most abundant genotype (the quasispecies in the language of Eigen), and the sequence length of this dominating genotype. After an initial period of adaptation where the gestation time is decreased (and consequently the fitness is increased), all adaptive activity has stopped after about 20,000 updates, and the population has settled into what appears to be a relative optimum. Note that during the adaptive phase, the length of the dominant replicator usually shrinks to about 15–20 (20 in the run showed in Fig. 8.7). Still, in this landscape fitness does *not* explicitly depend on length, in other words, while the gestation time of a string is roughly linear in length, the *allocated* time is also, such that evolution is size-neutral. This point will be discussed in more detail in the next chapters. The reason that the length of the strings consistently shrinks in this scenario is probably to evade the relatively high copy mutation rate of 2 percent. The optimal length appears to be around fourteen, and the typical length achieved in a flat landscape (15–18) is not far from that. The important point here is that the population explored the local confines of the landscape relatively quickly (note that during that time, the adaptive process looks vaguely fractal; see the inset in Fig. 8.7) and then hovers around the optimum: a fixed, flat landscape.

This situation is dramatically changed if the landscape offers something to be discovered. In avida, we can provide extra CPU time for strings with certain computational capabilities, which in our artificial chemistry paradigm can be likened to strings with certain (chemical) catalytic capabilities. Mimicking the speed-up of the metabolism of a cell that discovered the right genetic code to catalyze chemical reactions (so that it can extract more energy from its surroundings), we speed up the CPU of a program that discovered how to perform computations on numbers that are available to them. As we give out the rewards independently of the code used to accomplish the computation, there are *many* sequences that can trigger a specific reward, and the number of solutions can become *dense* in the number of possible strings, a key requirement, as we shall see soon. In a landscape where simple *logical* operations are rewarded, a typical run (started with the same initial conditions as Fig. 8.7) shows adaptive activity taking place for a long time, as new bonuses can be discovered building on the computational capabilities already achieved (Fig. 8.8). Note also that in such a run, the length of the strings increases dramatically, as this much memory is needed to store all the information acquired from the environment (cf. the discussion in Chapter 5). Still, the moment the population has discovered all the "reactions" that we give bonuses for, the landscape may become effectively flat, and little improvement is possible. Thus, the landscape in avida can only appear fractal for times shorter than the time it takes to discover all of the information that is coded in the environment. From Fig. 8.8 we may surmise that the landscape will *always* appear fractal as long as the population has not exhausted the information present in the environment. This is only partially correct, as it is possible to construct landscapes that are full of information, but where the maxima are so sparse (so rare) that

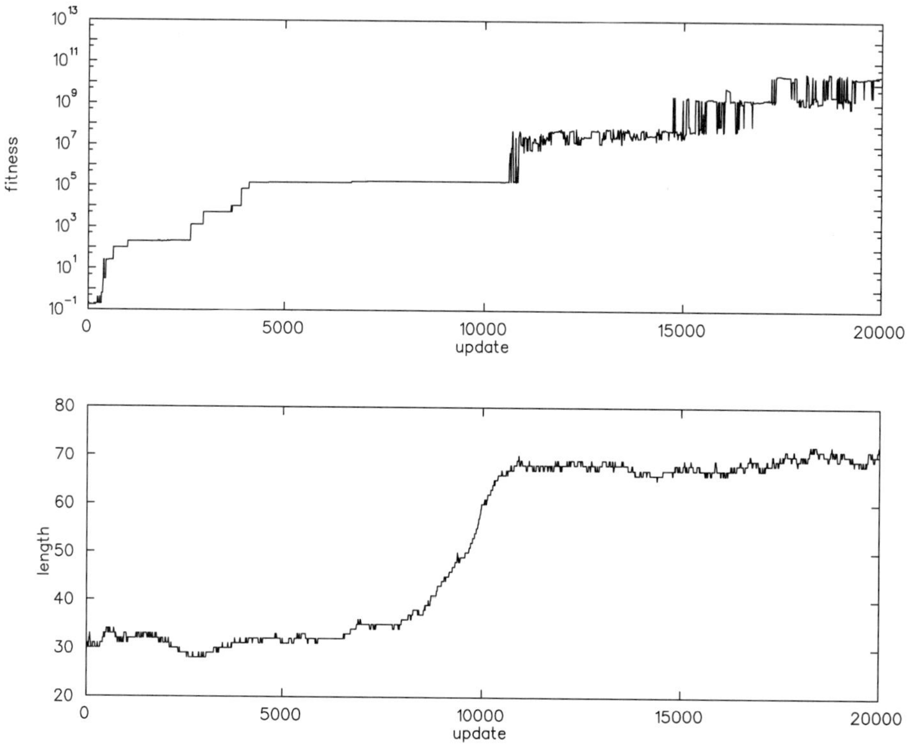

FIGURE 8.8 Fitness (upper panel) and length (lower panel) of the most abundant (the dominating) genotype in an information-rich landscape.

the population has no chance to ever find them. In this case, such a landscape will *appear* flat. We will be discussing the requirements for discoverable peaks (and consequently a fractal landscape) in much more detail in Chapter 11.

8.7 Overview

In this chapter we have looked at different ways to classify fitness landscapes. The examples we have looked at included *random* landscapes, where no two points have any correlation whatsoever as far as fitness is concerned, over smooth landscapes, to so-called AR(1) landscapes, which are characterized by a *correlation-length* that sets the scale for the average distance between maxima. Two explicit examples that we examined were the Kauffman landscape, which may have a fractal appearance for large k and small fitness, as well as RNA landscapes, which have similar characteristics.

Finally, we introduced the concept of fractal landscapes intuitively and mathematically. Such landscapes are not characterized by a scale, and therefore any fitness can be reached from any point within a radius of sufficient genetic distance. As a consequence, such landscapes must be *infinite* (for infinitely long strings). Conversely, any landscape that describes *fixed* length strings cannot be globally fractal, but it may be *locally* so, for strings that are small compared to the maximal length. Finally, we have introduced the landscape in avida and found that it has the potential for being a fractal landscape, but that this depends on the amount of information stored in the landscape (which the strings can discover) as well as *how* the information is stored. If exponentially few (isolated) points in the landscape have a high fitness, the population cannot discover them, and such a landscape may appear essentially flat rather than fractal.

Problems

NOTE: *Problems indicated by an asterisk are of higher difficulty.*

8.1 (a) Derive a recursion relation for the probability ϕ_{sd} that a random walk of length s on a Boolean hypercube (see Fig. 8.1) of dimension ℓ leads to a sequence separated by a Hamming distance d, by writing ϕ_{sd} in terms of ϕ at $s-1$.

*(b) What is this relation for arbitrary alphabet-size $\mathcal{D}$?

8.2 Prove Eq. (8.20). Note that the result quoted in the Appendix of Fontana *et al.*, 1993, is incorrect.

CHAPTER NINE

Experiments with avida

Measure what is measurable, and make measurable what is not so.
Galileo Galilei

After having treated adaptation and evolution of simple living systems from a more theoretical point of view in the previous chapters, we now turn to the kind of *experiments* that can be performed with avida to test some of the hypotheses advanced there.

Avida implements a simple artificial chemistry in which the molecules are computer programs, and the chemical action of these molecules is obtained by executing those programs. The geometry of the world that these programs live in is flexible (for details involving the operation of and options in avida, consult the *Avida User's Manual* in the Appendix), but for the sake of definiteness will be fixed here to be a two-dimensional regular lattice with wrapping boundary conditions. The size of the lattice itself (i.e., the size of the population) is arbitrary, but for practical reasons should be kept below 10^5. A version of avida that runs on parallel supercomputers is called sanda, and allows populations of up to 10^8 programs (see Chu and Adami, 1997). In the following, we shall learn about the main design parameters of avida and how to configure the software in order to perform dedicated experiments. Finally, we describe the concept of species in avida, and simple experiments related to speciation.

9.1 Choice of Chemistry

The dynamics of a typical run with avida depend strongly on a number of important settings that determine the character of the world. One of these is certainly the mutation *rate* and the *type* of mutations that the population is subject to; the other is the fitness landscape. Beyond those, the dynamics are controlled by the "physics" and the low-level "chemistry" that is being chosen. Some of the physics, for example, is determined by the mechanism by which cells are replaced in the population (the method of *selection*), while the chemistry is selected by choosing an *instruction set*. This choice may be likened to choosing a set of *amino acids* to be the basis of the biochemistry, and it is obvious that such a choice crucially determines the kind of chemistry. Ideally, we would like the choice of natural building blocks to be emergent itself, rather than being chosen in an arbitrary manner by the user. This would perhaps be accomplished by starting a simulation with a large number of very different, possibly redundant, instructions, and *no* self-replicating ancestors. Then, in the manner of the amoeba experiments reported on in Chapter 2, we might hope to see simple self-replicating programs emerge from nonreplicating ones, and witness the *origin of life* in the avida world. Such a natural ancestor would then be used to seed a population, and after a while we might examine this population to see which (and how many) of the initially available instructions are being used, and which are not.

While compelling, this is a difficult experiment to carry out. The first step, choosing a large number of *basic* instructions, may seem daunting already (as we should hope to have a universal set of instructions for the emergent populations to choose from). It pales in difficulty to step two, however. Indeed, if we restrict the number of possible instructions to 24 (the default instruction-set in avida), it is known that the smallest self-replicator (albeit not a very efficient one), is of length eleven. In this class, there are about 1.5×10^{15}, i.e., over one-and-a-half million billion different programs, while there are probably only a handful of self-replicators. Thus, these constitute the proverbial needle-in-the-haystack, and will be extremely rare to emerge in a random search. This difficulty seems to only be compounded by working with a starting set of, say, one hundred different instructions, and programs of arbitrary length. Still, such an experiment is not without hope. Indeed, as taught to us by the earlier attempts, notably by Rasmussen and by Pargellis (see Chapter 2), the spontaneous emergence of a short self-replicator need not be the only avenue to Artificial Life. Indeed, it has been argued in the framework of an artificial chemistry based on *logical expressions*, whose chemistry is based on *replacement* rules (the *AlChemy* project of Fontana, based on the λ-calculus, see Fontana and Buss, 1994) that the origin of life may have involved, at first, a *self-maintaining*, rather than self-replicating, population. Indeed, this was also Rasmussen's *ansatz*, and it is also explicit in the theory of the *Hypercycle* of Eigen and Schuster [Eigen and Schuster, 1979], as well as Kauffman's theory of autocatalytic networks (see Kauffman, 1993). Thus, we may hope that there is a *path* to self-replication that involves groups of programs that are self-maintaining (on the level of the group rather than the individual), and that will give rise to replicators and then self-replicators. Still, even if such a feat could be accomplished, it will be difficult to apply the lessons learned in the artificial medium to the natural world. Indeed, before such an investigation becomes credible, we need to prove the *universality* of life, i.e., we need to extricate those aspects of living systems that are *independent* of the substrate from those that clearly *are*, before rushing to apply the lessons learned in Artificial Life to the natural world. This is the goal of most of the initial experiments performed with avida.

Foregoing experiments to determine an emergent instruction set, we have to settle on one that appears promising enough to result in a rich chemistry with powerful instructions, while still maintaining computation universality and flexibility. The tradeoff between the power of single instructions and the universality of the set has been discussed in Chapter 2. Clearly, an instruction such as `replicate` would be very beneficial and powerful; it cannot, however, be used for any other purpose and universality might be jeopardized. The insistence on computation-universality of the instruction set is largely based on the intuition that the set of amino acids that the cellular metabolism is based upon is computation-universal. From an abstract point of view, this means that one should be able to build a universal computer out of the polypeptides that can be made from the 20 amino acids that form the instruction-set of natural life. From an intuitive point of view, this just means that almost anything can be made out of these building blocks. It can be argued that, because a select group of biological organisms can perform abstract calculations in their head, they can therefore simulate Turing machines without difficulty. This would then constitute a proof that DNA is Turing-complete. More seriously, there may be good reasons to believe that any universality in life is just a reflection of Turing universality, mirroring perhaps von Neumann's old idea that systems can evolve towards higher complexity only if the system is complex enough, just as the *Principia Mathematica* are complex enough. We shall leave such speculation behind and instead start with the *assumption* that a Turing-complete set of instructions is a necessary ingredient in a nontrivial implementation of life. The emphasis here is on non-trivial, as Pargellis's experiments have shown that there are *non-universal* implementations that can give rise to self-replicating programs. However, a proof that Artificial Life based on this chemistry shares features with the *universal* implementations (as far as dynamics and adaptability are concerned), does not as yet exist.

The universality of the default instruction-set in avida hinges on a number of crucial features of the CPU on which the instructions are executed. Largely, the CPU mimics the design of ordinary CPUs, albeit in extreme simplification. Consequently, the basic underlying computations are arithmetic on registers, shuffling of numbers between register and stacks, as well as conditional program flow. To maintain strict Turing universality, the avida CPU has the capability of using *two* stacks, which allows a simulation of the arbitrarily long tape of the universal Turing machine. Experience has shown, however, that the dual-stack capability is not crucial for the emergence of simple arithmetic and logical operations, but may become so if *sorting*, or any other *input-intensive* operation, is required to accomplish a task. While we defer a detailed exposition of the precise functionality of each instruction to the Appendix (Section A.5), we should briefly address their chemistry here. In Table 9.1 we list the instructions that constitute the default set for the avida software that accompanies this book. Most experiments are carried out with this particular set (any exceptions are explicitly noted).

TABLE 9.1 Mnemonic of the 24 instructions available in the avida distribution as default.

Inst.	Inst.	Inst.	Inst.
nop-A	call	pop	allocate
nop-B	return	push	divide
nop C	shift-r	add	get
if-n-eq	shift-l	sub	put
jump-f	inc	nand	search-f
jump-b	dec	copy	search-b

The no-operation (no-op) instructions `nop-A`, etc., play a central role in this set, similar but not identical to the no-ops in tierra. While the no-ops are used in pattern-based addressing, as was Ray's `nop0` and `nop1` (see Chapter 2), they serve an additional function in avida. No-ops can modify the chemistry of a *preceding* instruction in a definite way, and as a consequence give rise to a more flexible, and therefore redundant, instruction set. Consider for example the `pop` instruction, which pops the top of the stack (as the `switch-stack` command is disabled in the default set, only one stack is active here) into one of the three registers `AX`, `BX`, or `CX` that the CPU has at its disposition. If *no* no-op is following the `pop`, the contents of the stack will be popped into register `BX`. However, a `nop-A` following `pop` will result in the top of the stack being popped into register `AX`, and so on. The same is true for the `push`, as well as all operations that involve arithmetic or conditional decisions on registers. As a consequence, most operations can be written in *many* different ways: the chemistry is redundant and flexible.

Important for the self-replicative capability of the set are the instructions `copy`, `allocate`, and `divide`. The `copy` command copies an instruction from one memory location to another, more specifically from the location pointed to by the `BX` register to the location pointed to by the location `AX` instructions *away* from `BX`, i.e., in memory location `BX+AX`. This is a satisfactory procedure that allows a range of different modes of copying, even though such a scheme is neither general nor universal. A more general scheme involving `read` and `write` instructions only (rather than `copy`) should be investigated in the future. As we shall discuss below, the `copy` command places a *random* instruction at the destination with a given small probability to simulate copy errors. The `allocate` instruction is also crucial for reproduction, as it allocates a stretch of memory at the *end* of the code that issues the command, in preparation of filling this memory with its own genome. The amount of space to be allocated is read from the `BX` register if no no-op is following the command, or else the `AX` or `CX` register if the respective no-ops follow the `allocate` instruction immediately. The `divide` instruction, finally, splits off the newly grown code from the mother program, at an address specified by the `AX` register (by default). A no-op following the `divide` as usual changes the register used to store the address at which the splitting is to take place. After the daughter code is split off, the physics in avida takes over and places the new code into a nearest-neighbor spot, determined by an algorithm that can be chosen by the user (see Section A.4 in the Appendix).

A number of instructions in this set are not strictly necessary for replication, but are included in order for the programs to be able to perform arithmetic and/or logical operations on numbers provided from "outside". These are the `get` and `put` instructions, which read and write numbers from I/O buffers, as well as the `nand` instruction, which performs a simple (bitwise) logical operation on registers `BX` and `CX`, and puts the result into the register specified by the following no-op (`BX` if no no-op follows). As emphasized before, the successful performance of arithmetic or logical operations can be viewed as the successful catalysis of exothermic chemical reactions, leading to a *speed-up* of the metabolism, and consequently to a higher replication rate.

Finally, we should take a look at some of the other instructions that are available for use, but that do not form the default set. An example is the `switch-stack` instruction, that may be very useful for breeding more complicated multi-input tasks. Also, as programs in avida have a *facing*, i.e., they always face one of their eight nearest neighbors directly, there are commands that *change* the facing: the commands `rotate-r` and `rotate-l`. These commands are important in experiments involving *parasitism*. As the instruction pointer automatically wraps back to the beginning if it is not explicitly returned by a `jump` instruction at the end of the code, parasitism must be implemented in a different manner as in tierra. In avida, a special instruction allows for a *dedicated* jump into the neighboring creature that is currently faced: `jump-p`. Note that parasitism is disabled in the standard mode of operation, as the `jump-p` instruction is not part of the default instruction set.

9.2 A Simple Experiment

In this section we shall take the first steps towards extracting data from an avida experiment. We shall take a look at a few simple observables (while pointing out the salient steps in preparing avida to deliver this data), and discuss some of the results obtained.

The first experiment will involve taking the *genotype abundance* distribution. In avida, from the moment that the ancestral string is introduced into the memory, many millions of different genotypes are produced during a run due to the effect of mutations. Any genotype that arises via mutation and that has lost the ability to self-replicate will (almost always) be present only in an abundance of *one*, i.e., the mutation event was simultaneously the creation and the extinction event for the genotype. Other genotypes are much longer lived, and extinction can be many thousands of updates removed from creation. The *abundance*, or size, of a genotype is then the *total* number of members of that genotype that ever existed from creation to extinction. The question we ask is about the *distribution* of the frequency of these abundances.

Configuring avida

In this experiment, we shall confine ourselves to genotypes created by copy mutations, i.e., a copy command that should simply copy an instruction from one location in memory to another will, with a small probability, place a *different* instruction at this location instead, while the identity of this miscopied instruction is chosen randomly from the 24 of the set. To set the rate of mutations, and to ensure that only copy mutations happen, we need to adjust the parameters in the file that constructs the avida world according to the user's whim: the `genesis` file. This file is structured into sections that concern certain aspects of the world. Here, we are interested in the section entitled "Mutations", which is reproduced below (see the *User's Manual* for a description of the other mutation mechanisms):

```
### Mutations ###
POINT_MUT_RATE  0    # Mutation rate (per-location per update) (x10^-6)
COPY_MUT_RATE  30    # Mutation rate (per copy).  (x10^-4)
DIVIDE_MUT_RATE 0    # Mutation rate (per divide). (x10^-2)
DIVIDE_INS_RATE 0    # Insertion rate (per divide). (x10^-2)
DIVIDE_DEL_RATE 0    # Deletion rate (per divide). (x10^-2)
```

All the different mutation modes are turned off by writing a 0 next to the rate descriptor. The copy-mutation rate is set to 0.3 percent in the above example, which means that (on the average) one in 333 copy operations will result in a random instruction being placed at the destination. Note that the mutation process is *Poisson-random*. Thus, the time between miscopies is exponentially distributed, while the average time between miscopies is just the inverse of the copy-mutation rate, i.e., approximately 333 copy events in the case at hand. Note that the default `genesis` file specifies a low rate of insertion and deletion mutations. This rate ensures a diversity in lengths of programs, but can be turned off here.

Also, for this experiment we need to record more than the usual metabolic data for the population. As we are interested in the genotype's abundance from creation to extinction, we need to log the total number of members of any genotype that ever lived, at the moment of its *extinction*. This is provided for in the **genotype.log** file, and this logging is turned on by toggling the switch in the respective field, in the "Data and Log Files" section of the **genesis** file reproduced below.

```
### Data and Log Files ###
SAVE_AVERAGE_DATA   10    # Print these files every x updates. Enter 0 for
SAVE_DOMINANT_DATA  10    # those which should never be printed
SAVE_COUNT_DATA     10
SAVE_TOTALS_DATA     0
SAVE_TASKS_DATA     50
SAVE_STATS_DATA     10
SAVE_GENOTYPE_STATUS  0 # Print these files every x updates. Enter 0 for
SAVE_DIVERSITY_STATUS 0 # those which should never be printed
LOG_CREATURES 0         # 0/1 (off/on) toggle to print file.
LOG_GENOTYPES 2         # 0 = off, 1 = print ALL, 2 = print threshold ONLY.
LOG_THRESHOLD 0         # 0/1 (off/on) toggle to print file.
LOG_SPECIES 0           # 0/1 (off/on) toggle to print file.
LOG_BREED_COUNT 0       # 0/1 (off/on) toggle to print file.
LOG_PHYLOGENY 0         # 0/1 (off/on) toggle to print file.
```

There are three types of output files in avida (see Section A.10 in the Appendix for more details). Data files are the first type, which output metabolic data such as replication rate, gestation time, size, number of births, etc., of the most abundant genotype in the population (**dominant.dat**), as well as the averages across the population (**average.dat**) at a specified time interval. This interval can be chosen in the **genesis** file as shown above.

The **log** files keep track of events that do not occur at regular time intervals. An example is the creation, or the extinction, of a genotype or a species (the concept of species in avida will be discussed later in this chapter). These are the **genotype.log** and **species.log** files, which we will need for this particular experiment. They are not usually printed, so they need to be toggled to on in the **genesis** file as shown above.

A *1* will log *all* (i.e., threshold[1] and nonthreshold) genotypes, while a *2* will only record the threshold genotypes. The latter is preferred if disk

[1]Genotypes are divided into *threshold* genotypes (which have more than a minimum number of copies in the population) and non-threshold ones. The latter usually are incapable of self-replication.

space is at a premium, as the total number of genotypes can exceed two million easily in a run going to 50 thousand updates.

Similarly, the `creature.log` file logs the birth and death of *every* program in the population in such a manner that this file can in principle be used to recreate (playback) an avida run from this file alone. The status files in avida can be used to reconstruct certain aspects of the population after a run. For example, the `genotype.status` file records the distribution of abundances of threshold genotypes at each point in time (rather than historically, as is done in the `genotype.log` file), while the `diversity.status` keeps track of the genetic distance of any *pair* of threshold genotypes in the population.

Finally, we need to decide what kind of world we would like to run this experiment in and when it should end. These details are spelled out in the "Architecture Variables" section of the `genesis` file, which we can see below.

```
### Architecture Variables ###
MODE 2                     # 1 = Tierra, 2 = Avida
MAX_UPDATES 50000          # Maximum updates to run simulation.
WORLD_SIZE 3600            # Number of creatures in GA or Tierra mode.
WORLD-X 60                 # Width of the world in Avida mode.
WORLD-Y 60                 # Height of the world in Avida mode.
RANDOM_SEED 0              # Random number seed. (0 for based on time)
### Configuration Files ###
DEFAULT_DIR ../work/    # Directory in which config files can be found.
INST_SET inst_set.24.base # File containing instruction set.
TASK_SET task_set       # File containing task set.
EVENT_FILE event_list   # File containing list of events during run.
START_CREATURE genebank/creature.base # Creature to seed the soup.
```

The first line in the architecture section sets the mode under which avida should run. Beyond the `avida` mode that is the default, the software can emulate tierra in mode 1. Note that the latter mode is not a perfect emulation of tierra, as neither the tierran instruction set is used, nor does the instruction pointer roam the core if it runs off the end of a program. However, the global reaper queue is implemented there, and there is *no* spatial geometry in the population. Thus, this mode simulates a stirred-reactor type of environment. Of course, we should set `MODE` to 2 for the experiment we are about to perform. The influence of the spatial geometry on the abundance distribution can be tested by rerunning the same experiment in tierra mode. The next line sets the time (in updates) at which avida will automatically quit. Furthermore, we have control over the world size, the random number seed, and finally the *chemistry*. The latter is controlled by setting up the *configuration* files. The first line in this section determines the default directory where avida is going to look for the configuration files, whereas the following lines specify those files. Avida loads the instruction set proper by specification of the `inst_set` file. The default instruction set is contained in the `inst_set.24.base` file, and does not need to be changed here. The rest of the configuration variables can safely be left unchanged for this experiment, and we are finally ready to start. More details about the configuration of avida runs can be found in Section A.9 in the Appendix.

Running avida

An avida run is started by just typing `avida` on the command line in the `work` directory. The visual appearance of avida on startup depends on the architecture variable `VIEW_MODE`. By default we start in `MAP` mode, which is reproduced below after 71 updates. Note that the appearance of the viewer depends on the platform on which avida was compiled.

The population starts with a single program of our own design (in this case, the program contained in `creature.base`, a file that is loaded via the `genesis` file in the section entitled "Population"). This particular program is 31 instructions long, and carries a certain amount of redundancy in its code. It is the default startup program and is well-suited for this experiment. In fact, the data that we are recording should not depend on the structure of the ancestor that we are seeding the population with: this can be checked explicitly by starting another run with the much shorter ancestor `creature.small`, and comparing the results.

When the ancestor is placed in the avida world, it immediately begins to self-replicate, filling up the grid with copies of itself. Shown below is the lowest denominator `ASCII` interface, where different genotypes are represented by different capital letters, rather than colors. The ancestor, for example, is represented by `A`. As can be seen, even though most of the cells are of type `A`, mutations of the ancestral genotype have already arisen (denoted by capital letters further down the alphabet), and are competing with the ancestor for space. Otherwise, the population expands into the empty space in a vaguely circular manner.

```
+---------------+-------------------------------------------+-----------+
| Update: 60    | [M]ap  [S]tats  [O]ptions  [Z]oom  [Q]uit |  Avida    |
+---------------+-------------------------------------------+-----------+


                             . A . .
                             A A A . . G G
                         K A A A A A G A   G   . G
                        K K K A . A A A G . G G G G
                          K K D A . G A . G . G E G
                          . A A D A . . G G . E G . G
                          . A . D A A G C C C . E C C J J
                          L A . A . . A A . B G C C E E J
                        A A A A A A A A D B G E C C E E E .
                        . L A A . A . . B B E E E . E .
                          A . F . A . . A A B E E . . .
                          A A + . A . A C C E E C C C . .
                          A A A F . C C C C C E E C E E . .
                            . . C . . C C C C E E E . E
                              A C C C C C C C . E . E E
                            .   . C C C C C C C . C E   E
                                C C C . C . . C C
                                      C   .
* Clipping last 39 line(s) *                         [<]  Genotype View  [>]
```

Any time a replicating program spawns an offspring that bears a mutation, this daughter program will necessarily have a different genotype than its mother. By default, just after birth such programs are denoted by a *period* in the Genotype view. Indeed, as most mutated programs are not viable after a mutation, we do not bother to give them an identifying letter until they have proven themselves in the battle for survival. Such proof is accepted by the software if a minimum of *three* members of the new genotype have appeared in the population. In such a case, the genotype is considered to have crossed the *threshold*, and it is assumed that the reason why there are three identical copies of that program in the world is that this program knows how to self-replicate. Then, an identifying tag is bestowed on the genotype (see the discussion of the Zoom screen later), and is displayed via a capital letter in the Genotype view. Of course, there is a possibility that three identical programs have arisen from three identical mutations of one particular type of program. Such occurrences are usually very rare, but in case they might become important in the analysis of data, it is possible to raise the threshold from three to higher in the `genesis` file to thwart impostors from acquiring threshold status.

Another important view of the avida world is entered by pressing S at the keyboard (these commands are case-insensitive). This brings up the Stats screen, which displays current information about metabolic activity, as well as statistics and averages over the population. Here, we focus only on a selection of data that appears in that window; as always, a detailed explanation of the information shown in this and other windows appears in Section A.8 in the Appendix. In the first column of the Stats screen in the upper left corner, we find information about the current status of the population as a whole: the number of births this update, the total number of programs that breed true, the number of parasites, the current average inferiority (energy), as well as the highest fitness and merit currently found in the population (the concept of merit is introduced in Section 9.3). Below those are kept the number of programs,

```
+-----------------+-----------------------------------------------+----------+
| Update: 81      | [M]ap  [S]tats  [O]ptions  [Z]oom  [Q]uit      |  Avida   |
+-----------------+-----------------------------------------------+----------+

Tot Births.:      89   -- Dominant Genotype --               Dominant  Average
Breed True.:     293   Name........: 029-aaaab   Fitness..:    0.1997   0.1437
Parasites..:       0   ID..........:         4   Merit....:        26       26
Energy.....:    0.33   Species ID..:         2   Gestation:       132    136.8
Max Fitness:  0.2472   Age.........:        66   Size.....:        29     29.9
Max Merit..: 3.3e+01                             Copy Size:        29     29.9
                                                 Exec Size:        26     25.9
             Current     Total Ave Age  Entropy  Abundance:        87     3.17
Creatures:       460   1.3e+03     3.8     6.13  Births...:        25    0.614
Genotypes:       145   3.2e+02    43.0     0.71  BirthRate:     0.230    0.166
Threshold:        17   1.8e+01    59.2
Species..:        17   1.8e+01    51.4     0.53

+----------------------------------------------------------------------------+
| Input...:       9       Not.....:    0       Nor.....:    0                |
| Output..:      28       And.....:    0       Xor.....:    0                |
| I/O.....:       0       ~A Or B.:    0       Equals..:    0                |
| Echo....:       0       ~A And B:    0                                     |
| Nand....:       0       Or......:    0                                     |
+----------------------------------------------------------------------------+
```

genotypes, and species currently extant (not surprisingly, the number of programs equals the number of grid points in the lattice that we defined once the lattice has filled up).

To the right of the abundances are the *total* abundances, the respective number of cells, genotypes, or species that *ever* existed. The latter information is important to us here, as we would like to collect enough statistics for the genotype abundance distribution. To obtain reasonable statistics, we should collect around 500 thousand genotypes. For a run at mutation rate $R = 0.003$, this takes about 10 thousand updates. Note that most of those are not viable, and therefore would appear as size-one genotypes. The number of *viable* genotypes is listed under "Threshold." This count is usually much less than the number of genotypes listed above that number. On the right side of the Stats screen, we find a listing of the properties of the dominant (i.e., the most abundant) genotype, as well as the value of those properties averaged over the entire population. The box below this data informs the user about the tasks that are learned by the population: the number next to the description of the task is the number of programs that have accomplished this task during the last update. Before we leave this description of screens and files and concentrate on the data we obtained, we should take a quick look at one of the most useful views into the avida world that is offered to us: the Zoom screen.

Performing Micro-Analysis of the Population

The Zoom screen is entered by typing `Z` at the screen. What we are offered there is a partial view of the grid on the righthand upper corner, with the cell in the middle enclosed in brackets. This is the cell that we are currently zoomed-in on, and we can observe its execution and its metabolism in every little detail, and at our own pace!

In the view below, avida is paused (as we can see from the word `Un-[Pause]` appearing in the lower left corner). It can be restarted at any time by just pressing `P` again. While the upper righthand corner displays a piece of the grid as mentioned (it behaves otherwise just as the Map Screen, and allows different views by cycling through the maps with the > and < keys), the box titled "Memory" displays part of the program of the cell that occupies the center of the map in the inset.

```
+--------------+---------------------------------------------+------------+
| Update: 81   |  [M]ap  [S]tats  [O]ptions  [Z]oom  [Q]uit  |   Avida    |
+--------------+---------------------------------------------+------------+
Current CPU.: (0, 0)          +--------------+-------------+----------------+
Genotype....: 028-aaaaa       | Memory: 56   | Stack A     | A . A . G G B  |
Species.....: spec-7          +--------------+-------------+ . A . B . B B  |
                              | 12:   sub    |           0 | . . . . B B B  |
Gestation...:      127        | 13:  nop-B   |           0 | F A B[B]B . C  |
CurrentMerit:       31        | 14:  nop-A   |           0 | A . . A . E B  |
LastMerit...:       33        | 15:  nop-B   |           0 | . . A . B E E  |
Fitness.....:   0.2598        |>16:  nop-A   |           0 | . C A . E B B  |
Offspring...:        0        | 17:  nop-B   |           0 |[<] Genotypes [>]|
Errors......:        0        | 18:   copy   |           0 +----------------+
Age.........:        1        | 19:   inc    |           0 | AX:         28 |
Executed....:       10        | 20: if-n-equ |           0 | BX:          0 |
Last Divide.:       10        | 21:  jump-b  |           0 | CX:         28 |
Flags.......:   AT            +--------------+-------------+----------------+
Facing......: ( 1,  0)        | Inputs       | Get.: 0   Not.:  0   Nor.: 0 |
                              +--------------+ Put.: 1   And.:  0   Xor.: 0 |
Un-[P]ause                    |    932719887 | GGP.: 0   ~Or.:  0   Equ.: 0 |
[N]ext Update                 |     89019955 | Echo: 0   ~And:  0           |
[Space] Next Instruction      |    998324565 | Nand: 0   Or..:  0           |
[-] and [+] Scroll Memory     +--------------+------------------------------+
```

The currently active instruction is highlighted in boldface (indicated above by an arrow instead), and pressing the spacebar executes exactly *one* instruction of the program. In this manner, we can check the precise operation of the program as it moves information from buffers to registers or to the stack, their precise value at that time also being displayed in this window. Also, the number of tasks completed by this particular program (since birth) is displayed in the lower righthand corner. In the upper left-hand corner of the Zoom window, the genotype that the program that we are zoomed-in on belongs to is identified by a code that consists simply out of the length of the program, followed by an identifying five-letter code that is particular to this genotype, but that otherwise does not reflect the character of the type as it is chosen automatically in order of arrival of the genotype. Thus, the first genotype to arise of size 28 would be named `028-aaaaa`, the second one `028-aaaab`, and so forth. Note that only *threshold* genotypes are cataloged in such a manner; nonthreshold genotypes do not receive any identifying tag. The species that this genotype belongs to (see below for the definition of species) is indicated by a simple four-digit code. The coordinates of this particular program on the two-dimensional map can be found above the genotype tag, whereas below the species tag are listed the metabolic data of the program: its gestation time, merit, fitness, the number of offspring generated since birth, etc. Also displayed is the cell that the active program is currently facing, as well as flags indicating that the active program has allocated some memory (`A`) and whether it is capable of breeding true (`T`).

This view allows for the most detailed examination of a program's function and interaction with its neighbors, and is especially helpful when testing newly designed instructions, as well as handwritten creatures. For a description of the Histogram view, as well as the Options screen, we defer again to the Appendix. We shall now, finally, bow to the impatient reader and take a look at the data that avida has delivered.

Analyzing Data

The file `genotype.log` contains the abundance (fifth column) and the age (in number of updates, seventh column) of a genotype at the time of its extinction. In order to use this data, we first need to *bin* it, so as to obtain a *histogram* of the frequency of each abundance. For data with low statistics (few genotypes), the bin size may be chosen to be larger than one. However, since the bin size influences the power law exponent in the fit, it is preferable to obtain enough data that this is not necessary. The binned data can then be displayed graphically on a log-log plot. From the earlier discussions about self-organized criticality as well as percolation, we do expect the genotype-abundance distribution to display a power law, i.e., it should appear as a straight line on a log-log plot. The slope of the line then reveals the critical exponent. For this run at a mutation rate $R = 0.003$, we collect a million genotypes (including nonthreshold ones), and obtain a log-log plot as shown in Fig. 9.1. Note that for the fit, we *excluded* the nonthreshold genotypes, i.e., those that have an abundance below three.

Fig. 9.1 displays power-law behavior over almost two orders of magnitude, showing convincingly that the abundance distribution is scale free in this regime. Note, however, that this is *not* the case when the distribution is taken at *high* mutation rate. In this case, the distribution acquires a

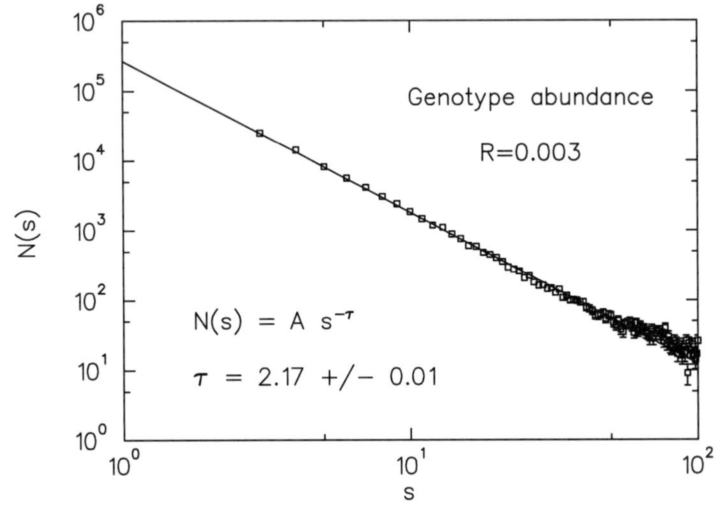

FIGURE 9.1 Abundance distribution of genotypes in a typical run at mutation rate $R = 0.003$. The distribution is fitted with a power law $N(s) = A\,s^{-\tau}$.

distinct exponential component that is not due to finite size. As we have learned earlier (Chapter 6), this is to be expected, as the criticality only manifests itself in the limit of *slow driving*, i.e., small mutation probability. Thus, even with a relatively small R, an exponential component may be observed in the abundance distribution if the *length* of the programs adjusts itself such that the mutation probability $p(\ell) = 1 - (1 - R)^{\ell}$ becomes large. A follow-up experiment, then, is to examine the power-law exponent (measured to be $\tau \approx 2.17$ in the above run) in the limit $p(\ell) \to 0$, by taking the distribution for smaller and smaller $p(\ell)$ and extrapolating to 0.

9.3 Experiments in Adaptation

Another simple experiment involves monitoring the *adaptation* of the population to a fitness landscape. As discussed earlier, and in more detail in Section A.3 of the Appendix, a fitness landscape rewarding logical and arithmetic operations on inputs is artificially constructed for the population to adapt to. The meaning of the logical operations is entirely arbitrary: what is important is that a genetic sequence (a code) has to be developed in order to trigger the bonus, while we do not specify exactly which sequence accomplishes this.

The basic metabolism of the programs is kept up by providing a slice of CPU time to each program according to a figure of *merit*. The *base* merit $\mathcal{M}_0$ of each cell is calculated according to a formula that is selected in the `genesis` file, and is usually proportional to the *length* of the program. The overall merit of the program is then obtained by multiplying $\mathcal{M}_0$ by a number $\nu_i > 1$ for each task acquired. These numbers are (except for $b = 0$, which implies $\nu = 1$) obtained from the *bonus* b in the `task_set` file by

$$\nu_i = 1 + 2^{b_i - 3} \; . \tag{9.1}$$

The `task_set` file and the bonuses b_i, reproduced below, allow the user to select the multipliers ν_i for each task that can be bred into the population (see Appendix for more details).

```
#Task      bonus(0=off)      Meaning          Difficulty

get              1          # I/O
put              1          # I/O
ggp              1          # I/O
echo             1          # I/O
not              2          # ~A              - 1 nand
nand             2          # ~(A and B)      - 1 nand
or_n             3          #  ~A or B        - 2 nands
and              3          #   A and B       - 2 nands
or               4          #   A or B        - 3 nands
and_n            4          #   A and ~B      - 3 nands
nor              5          # ~(A or B)       - 4 nands
xor              6          #   A xor B       - 5 nands
equ              6          # ~(A xor B)      - 5 nands
```

The first four tasks in this list simply breed input/output (I/O) capability into the programs. This happens, on the time scales considered, relatively fast. After the `echo` bonus is triggered, virtually all programs in the population read numbers from the input and write them into the output. This sets the stage for them to learn how to compute on these numbers.

In Fig. 9.2 we can see the time-evolution of the fitness α, defined as the merit

$$\mathcal{M} = \mathcal{M}_0 \prod_{i=1}^{n} \nu_i \,, \tag{9.2}$$

(where n is the number of tasks learned by the program) divided by the gestation time t_g,

$$\alpha = \frac{\mathcal{M}}{t_g}$$

for the dominant genotype in the population. The dominant genotype is simply *defined* to be that which has the highest *frequency*, or representation, in the population. For low mutation rates the dominant genotype usually stands out, whereas at high rates, where the population is very diverse and close to the melting point, the dominant genotype still may have only a handful of representatives. However, the dominant genotype is *always* selected from among the threshold genotypes. In case several genotypes have the same (highest) abundance, the genotype that was chosen as most abundant the update before is kept.

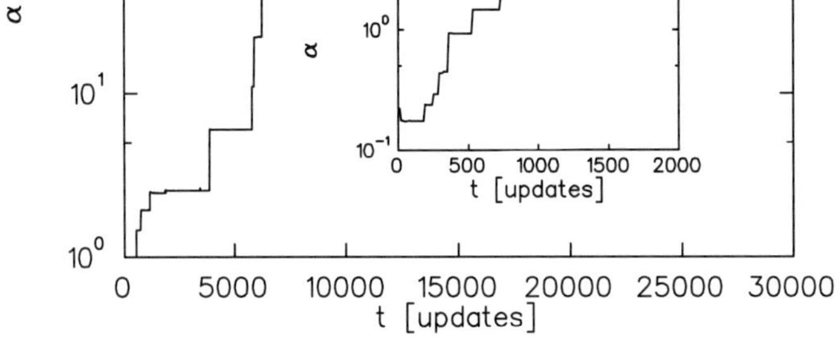

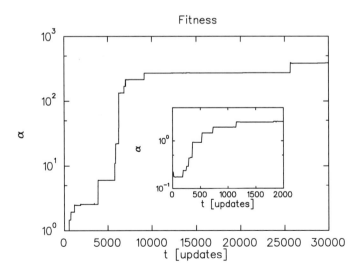

FIGURE 9.2 Fitness of the dominant genotype, for a run of 3,600 programs at mutation rate $R = 0.003$. Inset shows early adaptation to I/O in first 2,000 updates.

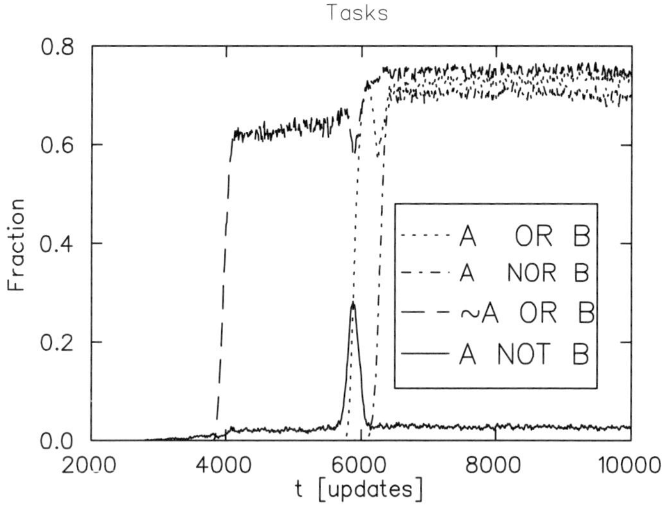

FIGURE 9.3 Fraction of programs that perform a specific task as a function of time, in the run depicted in Fig. 9.2. The legend identifies four of the taks learned in the first 10,000 updates.

The fitness curve in Fig. 9.2 shows the typical staircase structure, where the jumps indicate the adaptation events that are akin to first-order phase transitions (see Section 4.10). All of the I/O bonuses were acquired after about 2000 updates (see the inset). The next large transition involved the learning of the `~A or B` task, after around 4,000 updates (see Fig. 9.3). This transition and the concurrent drop in entropy is depicted in detail in Fig. 4.6. At that time, a small number of programs apparently had also figured out how to perform the `NOT` task, but did not do so consistently until just before update 6,000, when they started to take over the population. In the midst of this, this task seems to have been *converted* to the `OR` task, which carries a higher bonus. As a consequence, `NOT` was driven back into oblivion and `OR` prospered. The programs discovered how to trigger the bonus for `NOR` right on the heels of the discovery of `OR`, very likely using most of the same code. These adaptations, and corresponding optimizations of the code, account for most of the fitness jumps before 10,000 updates in Fig. 9.2. In general, the tradeoff and battle between programs having mastered different bonuses, and especially the temporal order in which they are discovered, is unpredictable. The phenomenon where one task is converted into another, and some discoveries following each other closely (because a small modification of an existing code will trigger additional bonuses), is reasonably common.

9.4 Experiments with Species and Genetic Distance

In this project, we are going to address the concept of *species* in avida, and the ways evolutionary changes affect the genome.

The lowest taxonomic level in avida is of course the genotype of a program. As the programs replicate asexually, the usual species definition of biology, as *groups of interbreeding natural populations that are reproductively isolated from other such groups* [Mayr, 1970], is inapplicable. However, from observations of the population of programs, it becomes clear that certain groups of programs, identified by their similarity in "genetic" code, do form. Therefore, let us consider a method to group such "species", and follow their development as a function of time, as well as geographical or ecological separation.

In close analogy to the usual biological species concept, we shall be determining the likeness of two replicating programs by lining up their genetic code (in such a way that they are identical in the maximum number of corresponding sites) and constructing crossover products. A crossover product consists of *part* of one of the organisms, taken from above a chosen crossover point, together with the part of the other, taken from below the crossover point. Such a "sexually" reproduced program is then *tested* with respect to its functionality. If this hybrid is capable of self-replication, we can assume that the parents were genetically similar. Note that the hybrid is not, in this case, introduced into the population. Different levels of taxonomy can now be constructed. For the lowest level of the taxonomic hierarchy, we require that a functional hybrid is formed if the crossover is performed at *all* possible crossover points. This definition of a species entails that the parents differ only at positions that are *neutral* as far as mutation is concerned, i.e., if they can be mutated with impunity without affecting the fitness of the organism. A higher taxonomic level, that contains as subtaxa all the species defined by the previous criterion, is obtained by only asking that genomes produce functional crossover products for one single crossover point chosen towards the middle of the genome. With such a definition we can obtain species-abundance distributions just like the one we obtained earlier for genotypes. This we did in Chapter 7 when we tested a simple theoretical model for cluster-size distributions in percolation.

In avida, we can also examine the *dynamical* aspects of species formation, as a function of geographical or ecological separation, and random genetic drift. For that purpose, we need to define a *genetic distance* to monitor the genetic kinship between programs, and observe taxonomic divergence as a function of either geographic separation, ecological separation, or random drift separately, and then combined.

A genetic distance between symbolic strings can be defined in different ways depending on the accuracy of the desired result, and the possible mechanisms that can lead to alterations of the string by single events. For strings whose length cannot change, and that are only affected by point mutations, the natural *Hamming* distance, which counts the number of positions at which the strings differ, is an adequate measure. For processes in which lines can be added, inserted, or deleted from the code, the simple Hamming distance is obviously inadequate. In such cases, the *Levenstein* distance [Levenstein, 1966] should be used, which compares two strings and returns the minimum number of mutations, insertions, and deletions it takes to obtain one string from the other. The Levenstein distance, on the other hand, returns inaccurate results if *transpositions* of code are common, i.e., if there is a finite probability that sections of code are swapped between two strings. Even though this is not a mechanism that is implemented *a priori* in avida, such alterations happen frequently in an *emergent* fashion. Thus, to keep an accurate count of the genetic divergence of the population, such processes must be taken into account. Such a *generalized* genetic distance measurement requires complicated, time-consuming algorithms. In avida, we have implemented an approximate algorithm that gives much more accurate results than Levenstein distance if transpositions are present, but is still manageable computationally. Armed with such measures of genetic distance and variability, we can explore genetic drift and speciation under controlled circumstances.

The first simple experiment concerns the amount of natural genetic drift occurring between organisms that are, artificially, separated by a geographic barrier. In preparatory experiments, we split a monotypic population (single species) after about thirty generations, and monitor the mating success between all pairs *across* the barrier (again, without

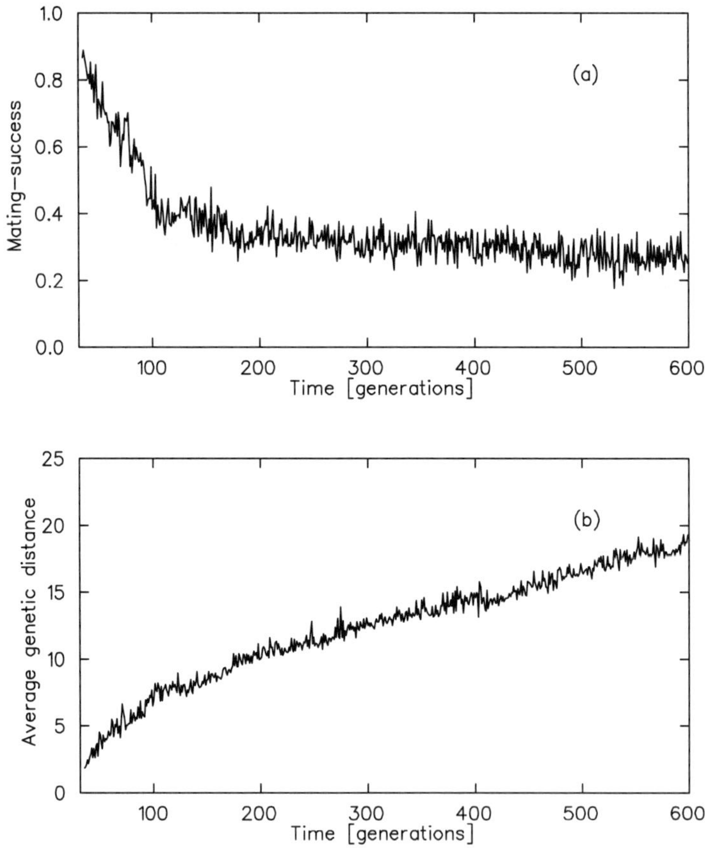

FIGURE 9.4 (a) Crossover mating success for two strings from separated populations as a function of time for 1000 strings separated after 30 generations; (b) Genetic distance between strings across artificial barrier.

reintroducing any of the hybrids). At the same time, we measure the genetic distance between all pairs to observe how mating success decreases as the genetic distance between the split populations increases. These results are shown in Fig. 9.4. From these experiments it is apparent that the mating-success between strings separated by the barrier drops rapidly at first, then levels out to an approximately constant value. In the meantime, the average genetic distance between the populations increases steadily. Note that these experiments require a modification of the avida software from the form included here to allow for the geographic separation.

9.5 Overview

Experiments with artificial chemistries that support self-replication allow insights into a number of real-life biological problems. First, experiments need to be designed that separate the *universal* features of the system from the ancillary ones, in order to gauge the system's predictive power with respect to natural living systems. Then, specific experimental conditions can be set to explore the dynamics and statistics of particular simple living systems. The flexibility of the configuration of avida allows experiments in such areas as taxonomic abundance distributions, evolutionary adaptation, speciation, and many more. As an example, an experiment to determine the genotype abundance distribution is carried out in detail, serving as a grand tour of the avida system. The result of this experiment suggests that genotype abundance distributions in avida are *scale free* in the limit of slow driving, i.e., small mutation probabilities. Experiments in speciation suggest that species formation can occur relatively fast given geographic separation coupled with random genetic drift.

Problems

9.1 Investigate the relationship between gestation time and program length from a theroretical and an experimental perspective. Can you make a statement about the smallest possible gestation time in avida? Compare your theoretical estimates and your attempts at breeding programs with the smallest gestation time. Can you "engineer" fitter programs?

9.2 Using the avida software, breed programs that perform the `NOT` and the `AND` operations. Find specific examples of creatures that perform these and extract them from the soup. Explain how these creatures work.

9.3 Obtain the frequency distribution of *ages* of genotypes and compare to the abundance distribution. What can you say about the connection between "size" (the abundance) and age of a genotype? (The famous Age–Area relation; see Willis, 1922.)

CHAPTER TEN

Propagation of Information

A theory has only the alternative of being right or wrong. A model has a third possibility: it may be right, but irrelevant.

Manfred Eigen

In the investigation of artificial systems with the goal of making predictions about real systems, it is important to check whether the artificial system at least shares some basic, universal features with the natural one that it is supposed to mimic. In this chapter, we shall investigate the basic modes of information propagation and diffusion across the lattice, and compare them to theoretical models that are known to describe the effect in natural systems. In such a manner, we can start to establish a *baseline* of characteristics that we know we can trust, and that are an accurate abstraction of the more complicated processes ocurring in real systems.

10.1 Information Transport and Equilibrium

As we discussed in Chapter 4, systems in thermodynamical equilibrium respond to perturbations with waves that reestablish equilibrium. This is a general feature of statistical systems, but it can also be observed in natural populations, where the disturbance of interest is a new species with either negligible or positive fitness advantage. The new species spreads through the population at a rate dependent on its relative fitness and some basic properties of the medium that can be summarized by the diffusion coefficient. This problem has been addressed theoretically [Fisher, 1937] and experimentally since early this century (see, e.g., Dobzhansky and Wright, 1943 and references therein). The application of the appropriate machinery (reaction-diffusion equations) to the spatial propagation of *information* rather than species is much more recent, and has been successful in the description of experiments with *in vitro* evolving RNA [Bauer et al., 1989; McCaskill and Bauer, 1993]. Systems of self-replicating information (cf. the replicating RNA system mentioned above) are often thought to represent the simplest living system, and it is for this reason that we would like to investigate information propagation in our artificial chemistry.

It has long been suspected that living systems operate, in a thermodynamical sense, far away from the equilibrium state. On the molecular scale, many of the chemical reactions occurring in a cell's metabolism require nonequilibrium conditions. On a larger scale, it appears that only a system far away from equilibrium can produce the required diversity (in genome) for evolution to proceed effectively. In the systems that we are interested in—systems of self-replicating information in a noisy and information-rich environment—the processes that work for and against equilibration of information are clearly replication and mutation. In the absence of mutation, replication leads to a uniform nonevolving state where every member of the population is identical. Mutation in the absence of replication, on the other hand, leads to maximal diversity of the population but no evolution either, as selection is absent. Thus, effective adaptation and evolution depend on a balance of these driving forces, as we have seen in the previous chapter and will show in more detail in Chapter 11. The relaxation time of such a system, however, just as in thermodynamical systems, is mainly dictated by the mutation rate, which plays the role of temperature in these systems. As such, it represents a crucial parameter that determines how close the system is to thermodynamical equilibrium. Clearly, a relaxation time larger than the average time between (advantageous) mutations will result in a nonequilibrium system, while a smaller relaxation time leads to fast equilibration. The relaxation time may be defined as the time it takes information to spread throughout the entire system (i.e., travel an average distance of half the diameter of the population). A nonequilibrium population therefore can always be obtained (at fixed mutation rate) by increasing the size of the system. At the same time, such a large system *segments* into areas that effectively cannot communicate with each other, but are close to equilibrium themselves. This may be the key to genomic diversity, and possibly to speciation in the absence of niches and explicit barriers.

In order to test equilibration through information transfer in an artificial chemistry, we shall investigate here the dynamics of information propagation in the Artificial Life system sanda,[1] a variant of the avida system designed to run on arbitrarily many parallel processors. This is a necessary capability for investigating arbitrarily large populations of strings of code. The purpose of these experiments is two-fold. On the one hand, we would like to validate our Artificial Life system by comparing the experimental results to theoretical predictions known to describe natural systems, such as waves of RNA strings replicating in $Q\beta$-replicase [Bauer et al., 1989; McCaskill and Bauer, 1993]. On the other hand, this benchmark should allow us to determine the diffusion coefficient and the speed of information propagation armed only with the relative fitness and mutation rate. Finally, we shall arrive at an estimate of the minimum system size which guarantees that the population will not, on average, equilibrate.

In the next section we briefly describe those parts of the sanda system that differ from avida.

10.2 The Artificial Life System sanda

Like avida, sanda works with a population of strings of code residing on an $N \times M$ grid with periodic boundary conditions. Each lattice point can hold at most one string. Each string consists of a sequence of instructions from a user-defined set. These instructions, which resemble modern assembly code and can be executed on a virtual CPU, are designed to allow self-replication. The set of instructions used is capable of universal computation.

[1] This software was designed and written by J. Chu [Chu and Adami, 1997].

Also like in avida, when a string replicates it places its child in one of the eight adjacent grid spots, replacing any string that may have been there. Which lattice point is chosen can be defined by the user. In the experiments that we are going to examine, both random selection and selection of the oldest string in the neighborhood are used, an option that can be tested in avida also (see Section A.6 in the Appendix). As we shall see, the selection mechanism has a significant effect on the spread of information. We note here that this birth process, and indeed all interactions between strings, are local processes in which only strings adjacent to each other on the grid may affect each other directly. This is important, as it both supplies the structure needed for studies of spatial characteristics of populations of self-replicating strings of code, and allows longer relaxation times—making possible studies of the equilibration processes of such systems and their nonequilibrium behavior.

What decides whether one particular sequence of instructions (or genotype) will increase or decrease in number is the rate at which it replicates, and the rate that it is replaced at. In the model that is described in Chapter 11, the latter is genotype independent (the chemostat regime). Accordingly, we define the former (i.e., its own replication rate) as the genotype's fitness. In other words, fitness is equal to the inverse of the time required to reproduce (gestation time):

$$n_i(t+1) \approx n_i(t) + \epsilon_i F n_i(t) - \langle\epsilon\rangle n_i(t), \tag{10.1}$$

where

$$\epsilon_i = \frac{1}{t_g^i} \tag{10.2}$$

and t_g^i is the gestation time of genotype i. In Eq. (10.1) above, we note the appearance of the *copy-fidelity* F, which takes values between 0 and 1 and which we define later. To consistently define a replication rate, it is necessary to define a unit of time. In avida, time is defined in terms of the *update*: the time it takes for every member of the population to execute the average time-slice. In sanda, physical time is defined by stipulating that it takes a certain finite time for a cell to execute an instruction. This base execution time may vary for different instructions (but is kept constant in all experiments presented here). The *actual* time a cell takes to execute a certain instruction is then increased or decreased by changing its *efficiency*. Initially, each cell is assigned an efficiency near unity,

$e = (1 + \eta)$, where η represents a small stochastic component. As a cell acquires a *merit*, its efficiency increases; this is translated into a speed-up of the cell's metabolism, i.e., its baseline time for executing an instruction drops. Self-replication consists of the execution of a certain series of instructions by the cell. Thus, the fitness of the cell (and its respective genotype) is just the rate at which this is accomplished and depends explicitly on the cell's efficiency. We can assign better (or worse) efficiency values to cells that contain certain instructions or that manage to carry out certain operations on their CPU register values. This allows us to influence the system's evolution (by effectively speeding up their CPU) so as to evolve strings that carry out predetermined tasks. A cell that manages to perform a user-defined task can be assigned a better efficiency for accomplishing it. Such cells, by virtue of their higher replication rate, then have an evolutionary advantage over other cells and force those into extinction. At the same time, the discovery that led to the better efficiency is propagated throughout the population and effectively frozen into the genome. In addition to the introduction of a real time, sanda differs from its predecessors in its parallel emulation algorithm. Instead of using a block time-slicing algorithm to simulate multiple virtual CPUs, sanda uses a localized queuing system that allows perfect simulation of parallelism. Finally, sanda was written to run on both parallel processors and single processor machines, its biggest advantage over avida in its current implementation. Therefore, it is possible, using parallel computers, to coevolve very large populations of strings. This permits studies of extended spatial properties of these systems of self-replicating strings and holds promise of allowing us to study them away from equilibrium.

10.3 Diffusion and Waves

Information in sanda is transported entirely by self-replication. When a string divides into an adjacent grid site, it is also transferring the information contained in its code (genome) to this site. Here, we shall investigate the mode and speed of this transfer in relation to the fitness of the genotype carrying the information, the fitness of the other genotypes near this carrier, and the mutation rate.

Consider what happens when one string of a new genotype appears in an area previously populated by other genotypes. We will make the assumption that the fitness of the other viable (self-replicating) genotypes near the carrier are approximately the same. This holds for cases where the carrier is moving into areas that are in local equilibrium. We will use f_c for the fitness of the newly introduced (carrier) genotype and f_b for the fitness of the background genotypes. If $f_c < f_b$, obviously the new genotype will not survive nor spread.

In the following, we study three different cases: diffusion, wave propagation, and wave propagation with mutation. The diffusion case represents the limit where the fitness of both genotypes is the same. It turns out that this can be modeled as a classical random walk. On average, if the carrier string replicates, it will be replaced before it can replicate again. This is effectively the same as the carrier string *moving* one lattice spacing in a random direction, chosen from the eight available to it (see Figure 10.1). The random walk is characterized by the disappearance of the mean displacement and the linear dependence on time of the mean squared displacement:

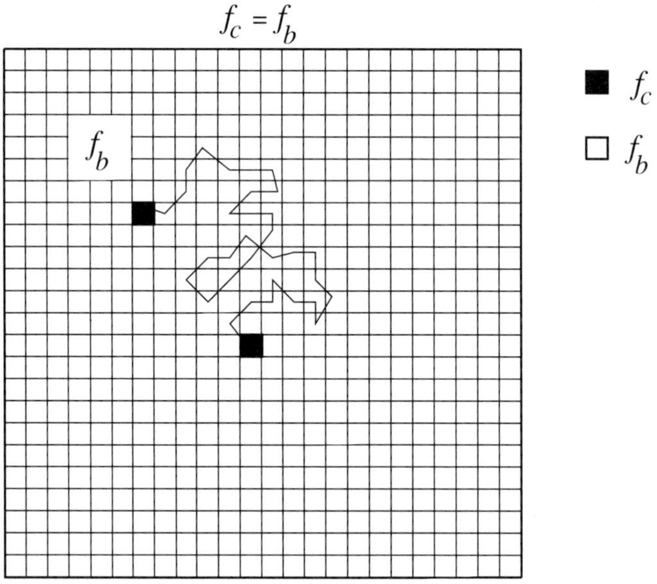

FIGURE 10.1 Random walk of the carrier genotype with fitness $f_c = f_b$ on the background of fitness f_b. On average, the carrier genotype is replaced as often as it replaces another cell, giving rise to the walk.

$$\langle r\rangle(t) = 0\ , \tag{10.3}$$

$$\langle r^2\rangle(t) = 4Dt\ , \tag{10.4}$$

where D is defined as the diffusion coefficient. Note that this equation differs by a factor of two from the diffusion equation that we encountered in Chapters 6 and 8, as diffusion takes place in *two* dimensions here. For our particular choice of grid and replication rules, it is easy to see that the diffusion coefficient of a genotype with fitness f is

$$D^{(b)} = \frac{3}{8}a^2 f, \tag{10.5}$$

where a is the lattice spacing. This holds for a *biased* selection scheme where we select the *oldest* cell in the neighborhood to be replaced (see below). This is indicated by the superscript (b) on the diffusion coefficient in Eq. (10.5). The factor 3/8 just reflects the number of choices that lead to a displacement in one direction, as indicated in the picture below.

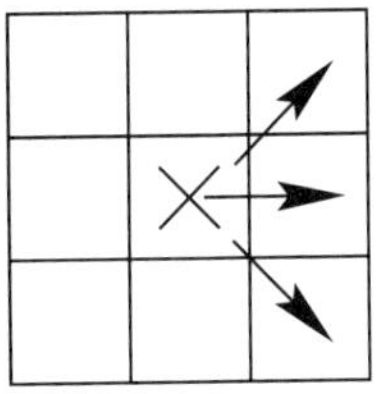

If $f_c > f_b$, we find that instead of diffusion we obtain a roughly circular population wave of the new genotype spreading outward, as depicted in Fig. 10.2. We are of course interested in the *speed* of this wavefront. Let us first treat the case without mutation. If the radius of this wavefront is not too small, we can treat the distance from the center of the circle r as a *linear* coordinate. In other words, even though propagation takes place on a two-dimensional lattice, we shall only monitor the progress in one direction, as indicated below.

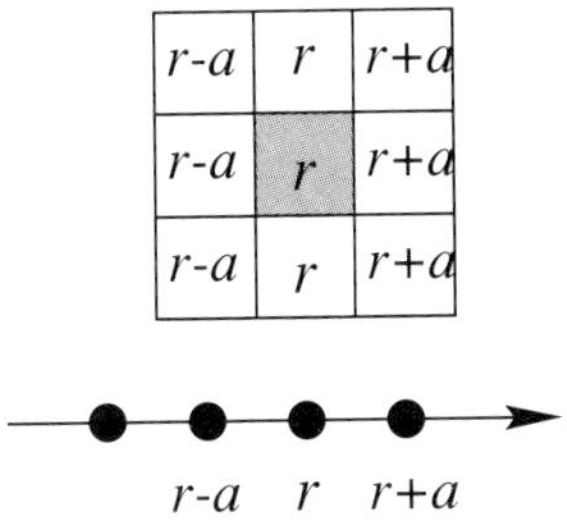

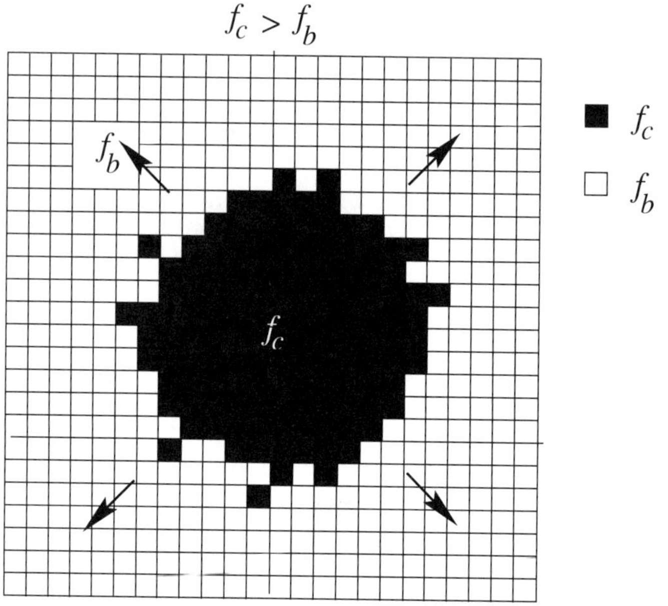

FIGURE 10.2 Circular wave of carrier genotype of fitness $f_c > f_b$ spreading over the background of fitness f_b. The boundary between the two genotypes moves with a speed v outwards.

We define $\rho(r, t)$ as the mean normalized population density of strings of the new genotype at a distance r from the center, at a time t measured from our initial seeding with the new genotype. We assume that the ages of cells near each other have roughly the same distribution and that this distribution is genotype-independent, ensuring that the selection of cells to be replaced does not depend on genotype either.

Then we can write a flux equation (the reaction-diffusion equation) that determines the change in the population density $\rho(r, t)$ of the carrier genotype, as a function of time. As we only have two genotypes in this description, the background genotype density will be $1 - \rho(r, t)$. Then, we have simply

$$\rho(r, t + 1) = \rho(r, t) + R_+(r, t) - R_-(r, t) \ , \qquad (10.6)$$

where

$$R_+(r,t) = f_c\left[\frac{3}{8}\rho(r-a,t) + \frac{1}{4}\rho(r,t) + \frac{3}{8}\rho(r+a,t)\right] \times (1-\rho(r,t)) \tag{10.7}$$

$$R_-(r,t) = f_b\left[\frac{3}{8}(1-\rho(r-a,t)) + \frac{1}{4}(1-\rho(r,t)) + \frac{3}{8}(1-\rho(r+a,t))\right] \times \rho(r,t). \tag{10.8}$$

Collecting these terms and writing $\rho(r,t+1) - \rho(t)$ as a time derivative, we can write

$$\frac{\partial\rho(r,t)}{\partial t} = f_c\left[\frac{3}{8}\rho(r-a,t) + \frac{1}{4}\rho(r,t) + \frac{3}{8}\rho(r+a,t)\right](1-\rho(r,t)) - f_b\left[\frac{3}{8}(1-\rho(r-a,t)) + \frac{1}{4}(1-\rho(r,t)) + \frac{3}{8}(1-\rho(r+a,t))\right]\rho(r,t)\,. \tag{10.9}$$

Since we are interested in the speed of the very front of the wave, we can assume ρ to be small. Also, from physical considerations we expect ρ to be reasonably smooth. Then, we can use a Taylor expansion for $\rho(r \pm a, t)$ and keep only the lowest order terms, to obtain

$$\frac{\partial\rho(r,t)}{\partial t} = \frac{3}{8}a^2 f_c \frac{\partial^2\rho(r,t)}{\partial r^2} + (f_c - f_b)\rho(r,t)\,. \tag{10.10}$$

This now has taken the form of a reaction diffusion equation just like we encountered earlier [in the chapter on SOC, Eq. (6.34)], with a reaction term given by just the difference between the replication rates. Note also that for the case where the replication rates are the same, the equation just reduces to the diffusion equation with diffusion coefficient (10.5).

Eq. (10.10) can be solved for the linear wavefront speed $v^{(b)}$. This is done essentially by parameterizing the wavefront of ρ with an exponential function (see Fig. 10.3 and Problem 10.2). Such a procedure yields the *Fisher* velocity:

$$v^{(b)} = a\sqrt{\frac{3}{2}}\sqrt{f_c(f_c - f_b)} \tag{10.11}$$

$$= 2\sqrt{D_c^{(b)}(f_c - f_b)}\,, \tag{10.12}$$

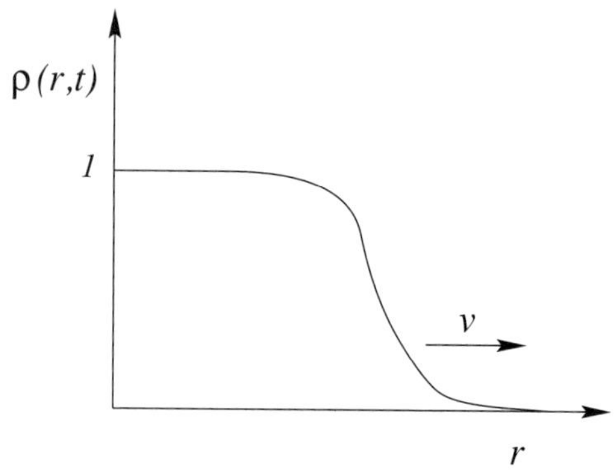

FIGURE 10.3 Profile of the solution of (10.10) and determination of the wavefront.

where $D_c^{(b)}$ is the diffusion coefficient of the carrier genotype when using a biased (by age) selection scheme, Eq. (10.5). (Consult Cross and Hohenberg, 1993, for details on how to obtain this wavefront speed.)

To study the case of wave-propagation with mutation, we shall make the assumption that all mutations are fatal. We can then calculate a steady state density of *nonviable* cells δ,

$$\delta = 1 - F^{1/8} , \tag{10.13}$$

where the fidelity F is the probability that a child will have the same genotype as its parent (*i.e.*, not be mutated). The fidelity is related to the mutation rate R by

$$F = (1 - R)^{\ell} , \tag{10.14}$$

where ℓ is the length of the particular string. Modifying our previous flux equation (10.10) to take into account these new factors and repeating our previous analysis yields

$$v^{(b)} = 2\,F\sqrt{D_c^{(b)}(f_c - F^{1/8}f_b)}\ . \tag{10.15}$$

Note that the speed of the wavefront is decreasing roughly linearly with the fidelity F, and vanishes for $F = 0$.

Let us now consider the effects of different selection schemes for choosing cells to be replaced. The relations we derived above hold true for the case in which we replace the oldest cell in the 8-cell neighborhood when replicating (age-based selection). Another method of choosing a cell for replacement is to choose a random neighboring cell regardless of age. This scheme, which we term *random* selection as opposed to the biased selection treated above, effectively halves the replication rate of all cells, as half of the cells never get the chance to produce an offspring before being removed (as Fig. 10.4 shows). It follows that the diffusion coefficient is also halved,

$$D^{(r)} = \frac{3}{16}a^2 f \tag{10.16}$$

$$= \frac{1}{2}D^{(b)} \tag{10.17}$$

and for the velocity of the wavefront (with no mutation), we find

$$v^{(r)} = 2\sqrt{D_c^{(r)}\,\frac{f_c - f_b}{2}}\ . \tag{10.18}$$

In Fig. 10.4, we show a histogram of the number of offspring that a cell obtains before being replaced by a neighbor's offspring for the biased selection case (left panel), and the random case (right panel). As expected from general arguments, half of the cells in the random selection scenario are replaced before having had a chance to produce their first offspring (resulting in a reduced diffusion coefficient), while biased selection ensures that most cells have exactly one child.

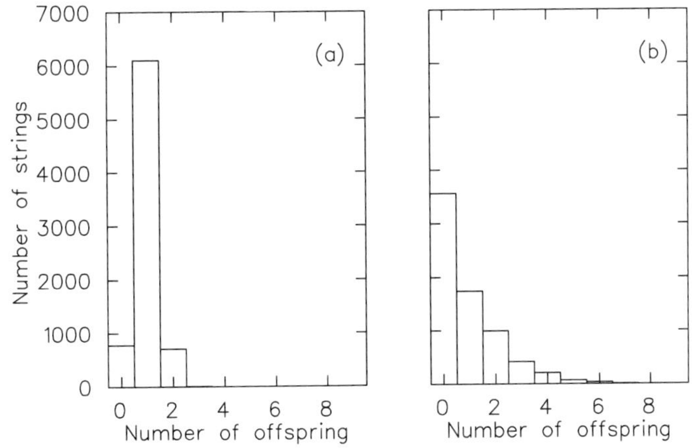

FIGURE 10.4 Distribution of number of strings generating different numbers of offspring, for the biased selection case (a), and the random selection scenario (b). The scale is arbitrary.

10.4 Comparison: Theory and Experiment

We carry out our experiments by first populating the grid with a single (background) genotype of fitness f_b. Then, a single string of the carrier genotype with fitness f_c is placed onto a point of the grid at time $t = 0$. We then observe the position and speed of the wavefronts formed, the mean squared displacement of the population of carrier genotypes, and various other parameters as a function of time.

With f_b kept constant,[2] we can vary f_b/f_c from 0.1 to 1.0 in increments of 0.1. Also, we vary the mutation rate R from 0 to 14×10^{-3} mutations per instruction, in increments of 1×10^{-3}.

Figure 10.5 shows a comparison of the theoretical vs. measured mean square displacement as a function of time for a genotype with no fitness advantage compared to its neighbors ($f_b/f_c = 1$). Because the process

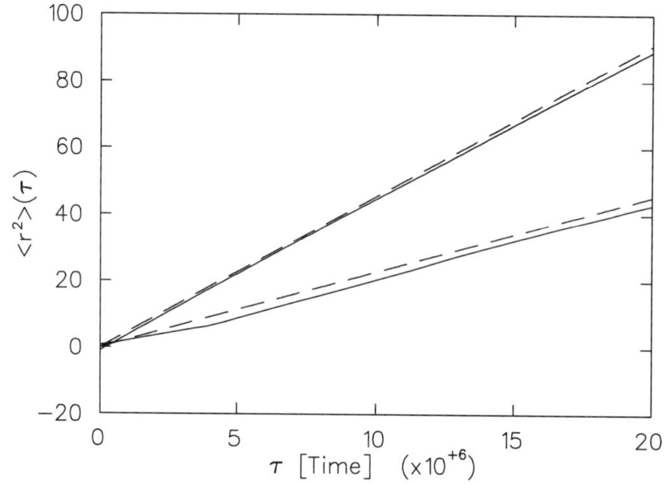

FIGURE 10.5 Mean squared displacement of genome as a function of time due to diffusion. Solid lines represent experimental results obtained from 1500 independent runs. Dashed lines are theoretical predictions. The upper curves are obtained with the biased selection scheme, while the lower curves result from the random selection scenario.

[2]The gestation time was approximately 330,000, where the base execution time for each instruction was (arbitrarily) set to 1000. Therefore, $f_b = \frac{1}{330000}$ here.

of diffusion is stochastic, many runs have to be superimposed in order to extract the linear law predicted by Eq. 10.4. The data shown here are extracted from approximately 1500 runs, obtained on a massively parallel supercomputer. The solid lines represent the (smoothed) averages of our measurements (for biased and random selection schemes), while the dashed lines are the theoretical predictions obtained from the diffusion coefficients (10.5) and (10.16), respectively. The slopes of the measured and predicted lines agree very well, confirming the validity of our random walk model and the diffusion coefficient predicted by it (without any free parameters). The slight discrepancy between thc experimental curves and the predicted ones at small times is due to a finite-size effect that can be traced back to the coarseness of the grid.

Fig. 10.6 shows the measured values of the wavefront speed for cases where $f_c > f_b$ and without mutation, with the corresponding predictions. Again, the higher curve is for biased and the lower for random selection. Note that the wavefront speed gain from an increase in fitness ratio is much better than linear. Note also that all predictions are again free of *any* adjustable parameters.

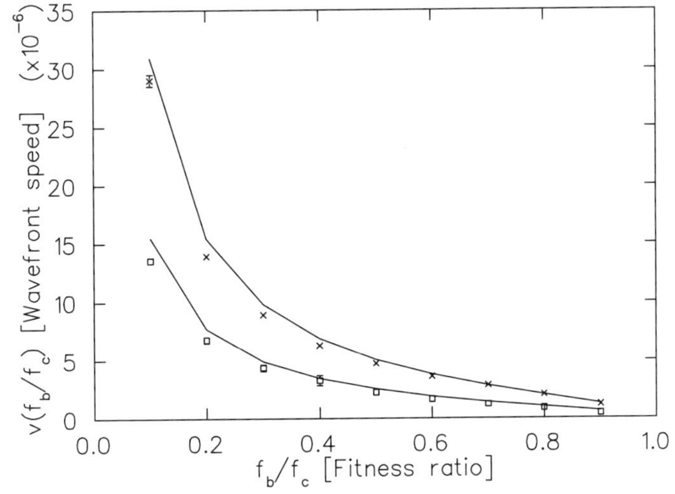

FIGURE 10.6 Wavefront speed of a genotype with fitness f_c propagating through a background of genotypes with fitness f_b, averaged over four runs for each data point. Upper curve: biased selection; lower curve: random selection. Solid lines are predictions of Eqs. (10.12) and (10.18).

The dependence of this curve on the mutation rate is shown in Fig. 10.7. Increasing the mutation rate tends to push the speed of the wave down. We should quickly note, however, that because we have only used copy mutations, there is no absolute cutoff point or error threshold F_c where all genotypes cease to be viable, with $F_c > 0$. Rather, genotypes can spread until F is very close to the limit $F_c = 0$. Error thresholds will be investigated in much more detail in Chapter 11.

Finally, we plot the dependence of the wavefront speed on the mutation rate for a fixed value of the fitness ratio ($f_b/f_c = 0.6$) in Fig. 10.8. Data were obtained from an average of four runs per point in the biased selection scheme. Again, the prediction based on the reaction-diffusion equation with mutation agrees well (within error bars) with our measurements.

The experiments shown here suggest that information propagation via replication into physically adjacent sites can be succinctly described by a reaction-diffusion equation. Such a model has been used in the description of in vitro evolution of RNA replicating in a solution of the bacterial enzyme Qβ-replicase [Bauer et al., 1989; McCaskill and Bauer, 1993], as well as in experiments involving the replication of viruses in

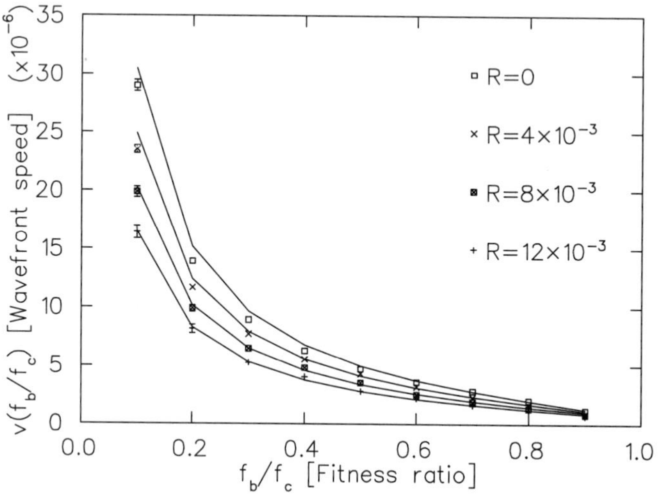

FIGURE 10.7 Measured wavefront speeds versus fitness ratio for selected mutation rates R (symbols) are plotted with the theoretical predictions from Eq. (10.15) (for the biased selection scheme only).

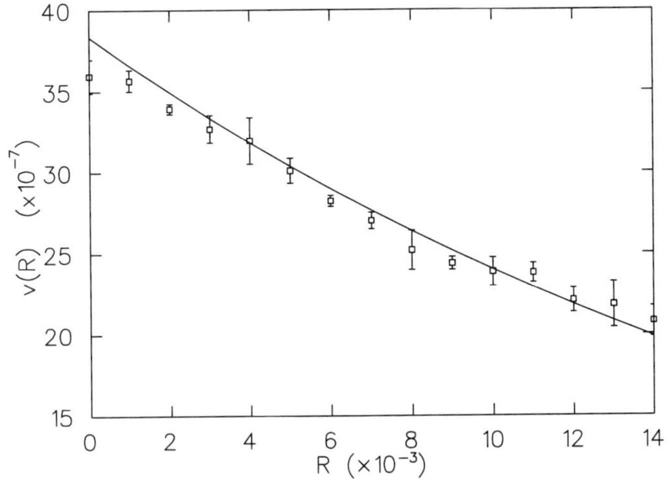

FIGURE 10.8 Wavefront speed of a genotype (biased selection) with relative fitness $f_b/f_c = 0.6$ as a function of mutation rate (symbols). Solid line is prediction of Eq. (10.15).

a host environment [Yin and McCaskill, 1992]. The same equation is used to describe the wave behavior of different strains of *E. coli* bacteria propagating in a petri dish [Agladze et al., 1993], even though the means of propagation in this case is motility rather than replication.

For artificial living systems such as the one we have investigated here, it is possible to formulate an approximate condition that ensures that it will (on average) *never* equilibrate, but rather consist of regions of local equilibrium that never come into informational contact. From the time scales mentioned above, we determine that the number of cells N in such a system must exceed a critical value:

$$N > \left(\frac{2\,v(f)}{R_\star\,a}\right)^{2/3}, \qquad (10.19)$$

where $R_\star$ is the rate of *nonlethal* mutations, $v(f)$ the Fisher velocity, and a the lattice spacing (assuming a mean time between nonlethal mutations $t_\star \approx (N\,R_\star)^{-1}$).

Beyond the obvious advantages of a nonequilibrium regime for genomic diversity and the origin of species, such circumstances offer the fascinating opportunity to investigate the possibility of nonequilibrium pattern formation in (artificial) living systems. As emphasized earlier, since it is widely believed that many of the processes that define life, including evolution, occur in a state that is far from equilibrium, to study such processes it is necessary to have systems that exhibit the properties of life we are interested in *and* that can be quantitatively studied in a rigorous manner in this regime.

10.5 Overview

In this chapter we used a *parallel* implementation of avida—the Artificial Life system sanda, which allows the investigation of large populations of self-replicating strings of code—for the purpose of observing nonequilibrium effects. The propagation of information was observed for a broad spectrum of relative fitness, ranging from the diffusion regime where the fitnesses are the same, through regimes where the difference in fitness leads to sharply defined wavefronts propagating at constant speed. The dynamics of information propagation leads to the determination of a crucial time scale of the system, which represents the average time for the system to return to an equilibrium state after a perturbation. This relaxation time depends primarily on the size of the system, and the speed of information propagation within it. Equilibration can only be achieved if the mean time between (nonlethal) mutations is larger than the mean relaxation time. Thus, a *sufficiently* large system will never be in equilibrium. Rather, it is inexorably driven far from equilibrium by persistent mutation pressure.

Problems

10.1 Derive Eq. (10.12), the speed of the wavefront, from the solution to Eq. (10.10) by approximating the exponential tail of $\rho(r - vt)$. The speed is obtained by examining the condition under which there is only *one* wavefront speed. (For a more rigorous derivation, consult Cross and Hohenberg, 1993, pp. 928–931).

10.2 Derive Eq. (10.19) and examine the assumptions underlying it. With reasonable estimates for f and $R_\star$, estimate the minimal size of the lattice that results in nonequilibrium effects.

CHAPTER ELEVEN

Adaptive Learning at the Error Threshold

Fluctuat nec mergitur.[1]

Motto of the city of Paris

In Chapter 9 we examined experiments where the population adapts to a user-defined landscape, and thus learns how to perform simple computational tasks that result in a speed-up of the code's metabolism. Here, we will not focus on *how* this happens (as before) but rather on two different, but related, questions. First we shall address the *circumstances* under which this happens most successfully, and secondly we will ponder how and why living systems seem to adjust the circumstances in such a manner that the adaptive powers of the population are maximal. As we shall see, this is a regime where the population is precariously perched at the edge of an *error catastrophe*, i.e., the probability for a string of code to acquire a mutation before being able to gestate a copy of itself is close to 100 percent. To address the first question, we shall conduct experiments to determine the advantage of such a regime, while secondly we detail the *approach* to this threshold, an undertaking that involves both theory and experiment.

---

[1]She floats, but sinketh not.

11.1 Information Processing at the Edge of Chaos

That something important is going on at the boundary between determinism and chaos in learning systems was established by Chris Langton in a study of computation and information processing in cellular automata [Langton, 1992]. Using his λ-parameter that characterizes CA rules (see Section 2.1), Langton could show that in between class II (deterministic) and class III (chaotic) rules, could be found the class IV rules, which display complex behavior and an ability to *store and transmit* information most efficiently. In order to make this determination, Langton studied the *mutual entropy* between a site at time t and a site at $t+1$, as a function of the λ-parameter, in two-dimensional automata with eight states and a von Neumann neighborhood. The mutual entropy here plays the same role as in Information Theory: it is the amount of information processed by the site. Langton determined that the information processed by a site is maximal in the region of the λ parameter characteristic of class IV automata: just at the edge between determinism and chaos. Let us briefly review his analysis.

Recall that the λ parameter represents the probability that a rule maps a neighborhood to the *nonquiescent* (active) state, which in the lingo of percolation theory we might dub the *occupied* state. Thus, if $\lambda = 0.5$, half of the rules will map any state into an active state, while the other half maps into the quiescent state. If we assume ergodicity, i.e., if we assume that the time average of the dynamics is well-represented by an average over the *configurations*, λ may be thought of as the *occupation probability* in two-dimensional percolation, as studied in Chapter 7.

The marginal (unconditional) entropy of a site j is obtained simply by calculating

$$H(x_j) = -\sum_{i=1}^{8} p_j(i) \log p_j(i) \,, \tag{11.1}$$

where $p_j(i)$ is the probability for site x_j to be in state i. These probabilities can be obtained by running the CA long enough and sampling the probability distribution. Finally, the *average* of this entropy over all sites is

$$\langle H\rangle = \frac{1}{N}\sum_{j}^{N} H(x_j) \,. \tag{11.2}$$

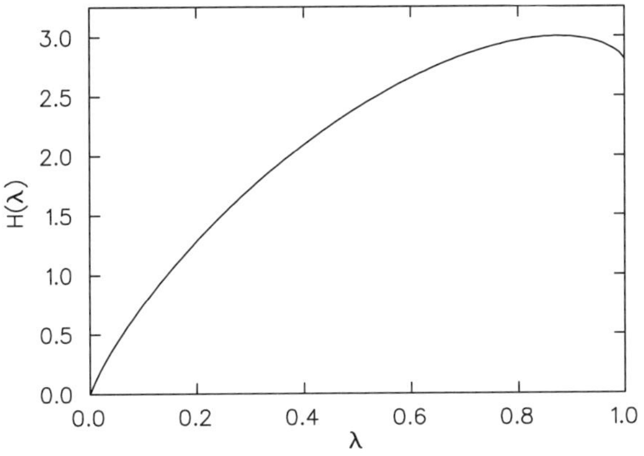

FIGURE 11.1 Bound for the average per-site entropy as a function of λ (in bits).

The entropy per site must be bounded by the entropy that is obtained if it is assumed that for a random rule each of the nonquiescent states is taken on with equal probability. Thus, the entropy of any rule must always be smaller or equal to this, and we find for the average per-site entropy as a function of the λ parameter

$$\langle H \rangle \leq -(1-\lambda)\log(1-\lambda) - \lambda\log(\lambda/7) \equiv H(\lambda)\ . \tag{11.3}$$

Indeed, the latter expression is just Eq. (11.1), with $p(0) = 1 - \lambda$ (the probability to map to the quiescent state) and all other $p(i) = \lambda/7$. The bound in Eq. (11.3) is shown in Fig. 11.1. While the entropy of random CA rules is not uniform below this curve, it shows quite obviously that at high λ the value of a site at any point in time is indeed completely random: the entropy is close to its maximum of 3 bits (every one of the eight states is equally probable).

Langton's entropy graph reveals more than this, though. It can be seen that there are essentially *two* kinds of dynamics that occur in the CA, one with quite low entropy $H \approx 0$, and the other with approximately $H \geq 1$ (see Fig. 11.2). Very few rule tables produce dynamics with $0 \leq H \leq 1$, and indeed an analysis of a trajectory of H, by changing λ via the table-walk-through method, reveals that the entropy usually *discontinually* jumps from zero to $H \approx 1$, in a manner akin to first-order phase transitions. This behavior stops if the starting point of the trajectory is $\lambda \approx 0.6$, very close to the percolation threshold $p_c \approx 0.59$ for two-dimensional regular lattices.

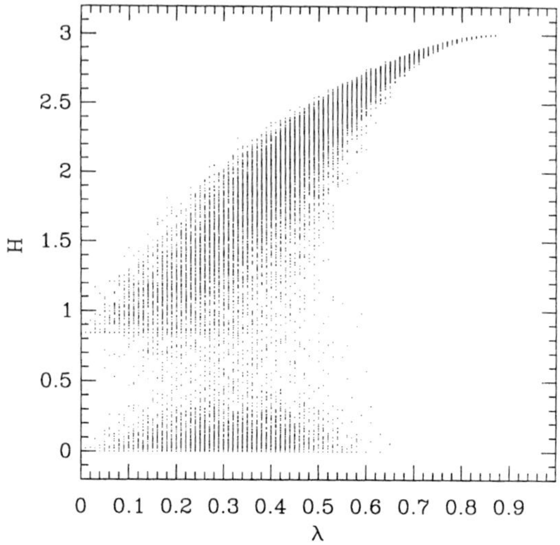

FIGURE 11.2 Average entropy of 10,000 randomly chosen rule tables of the eight state two-dimensional CA with von Neumann neighborhood (from Langton, 1992).

This seems to suggest that once the dynamics allow for an infinite cluster, no rule can be found anymore that keeps the entropy of the lattice low, as all sites are effectively connected.

Most interesting is the behavior of the (average) mutual entropy between a site at time t and a site at time $t+1$, or else a site x_j and its adjacent site x_{j+1}, for example. In both cases we can think of a site as a sort of information transmission channel, either in time or in space. The data collected in Fig. 11.3 (for the temporal channel) reveals that there is a maximum in the mutual entropy at around a value of $\lambda = 0.25$, a value on the boundary of the deterministic regime and the complex region that lies between $0.2 < \lambda < 0.4$, and not far away from the chaotic regime. This data can also be interpreted in a slightly different way, however. Let us calculate the mutual entropy which would result if the correlations that are detected are *entirely* due to the quiescence condition, which dictates that a quiescent site surrounded by quiescent sites only should return to the quiescent state. In this case, we can calculate the conditional entropy exactly, as follows. We denote an active state by i or j in general, whereas the quiescent state is denoted by 0. Then, the channel matrix (matrix of

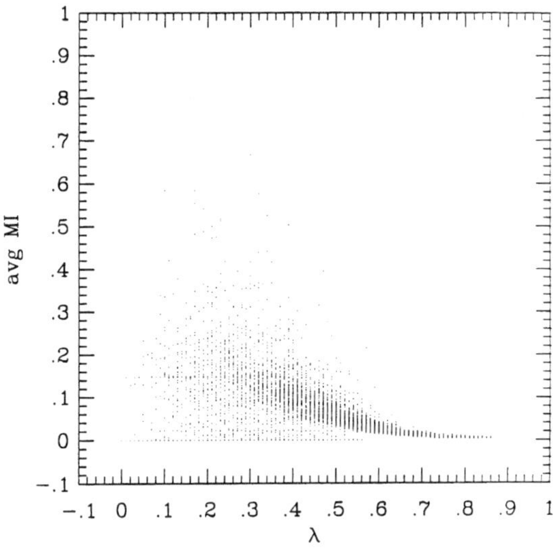

FIGURE 11.3 Average *mutual* entropy of 10,000 randomly chosen rule tables (from Langton, 1992).

conditional probabilities) can be written as

$$\begin{pmatrix} p_{0|0} & p_{i|0} \\ p_{0|j} & p_{i|j} \end{pmatrix} = \begin{pmatrix} 1-\lambda+\lambda(1-\lambda)^4 & [\lambda-\lambda(1-\lambda)^4]/7 \\ 1-\lambda & \lambda/7 \end{pmatrix}, \qquad (11.4)$$

where the first index refers to the site at $t+1$ and the second index to the site at time t. These probabilities can be obtained with a little reflection. For example, $p_{0|0}$, the probability that the site at $t+1$ is quiescent given that it was quiescent at time t, is the sum of two terms: $1-\lambda$ is the probability that this happened randomly, while $\lambda(1-\lambda)^4$ is the probability that this happened as a result of the quiescence condition.[2] All other probabilities are obtained in a similar manner. Then, we can calculate

$$H(x_{t+1}\,|\,x_t=0) = -(1-\sigma)\log(1-\sigma) - \sigma\log(\sigma/7)\,, \qquad (11.5)$$

where we defined

$$\sigma = \lambda - \lambda(1-\lambda)^4\,. \qquad (11.6)$$

[2]Obtained as $(1-\lambda)^4 - (1-\lambda)(1-\lambda)^4$. The second term removes from the probability that the state returns to quiescence *randomly* the probability that it did so as a result of the quiescence condition, to avoid double counting.

Also,

$$H(x_{t+1}\,|\,x_t = i) = -(1-\lambda)\log(1-\lambda) - \lambda\log(\lambda/7)\;, \qquad (11.7)$$

i.e., the entropy of a site at $t+1$, given that the site was previously an active site, is just the unconditional entropy (11.1). Then, the conditional entropy $H(t+1|t) \equiv H(x_{t+1}|x_t)$ is

$$H(t+1|t) = (1-\lambda)H(x_{t+1}\,|\,x_t = 0) + \lambda H(x_{t+1}\,|\,x_t = i) \qquad (11.8)$$

such that

$$\begin{aligned} H(t:t+1) &= H(t+1) - H(t+1|t) \\ &= (1-\lambda)\big(H(x_{t+1}\,|\,x_t = i) - H(x_{t+1}\,|\,x_t = 0)\big) \qquad (11.9) \\ &\equiv I(\lambda)\;. \end{aligned}$$

This just expresses the correlation between generations due to the quiescence condition *only*, which results in the difference between σ and λ. Eq. (11.9) is shown in Fig. 11.4, which agrees well with the envelope of the data shown in Fig. 11.3, even though the maximum of the distribution seems to be at a slightly smaller λ than the experimental results suggest. Thus, we might surmise from this that some important correlations exist in random CA at the edge of chaos that are *not* due to the quiescence condition. We conclude from this example that in a computational chemistry, the optimal regime requires a tradeoff between information storage and information transmission. Accurate information transmission

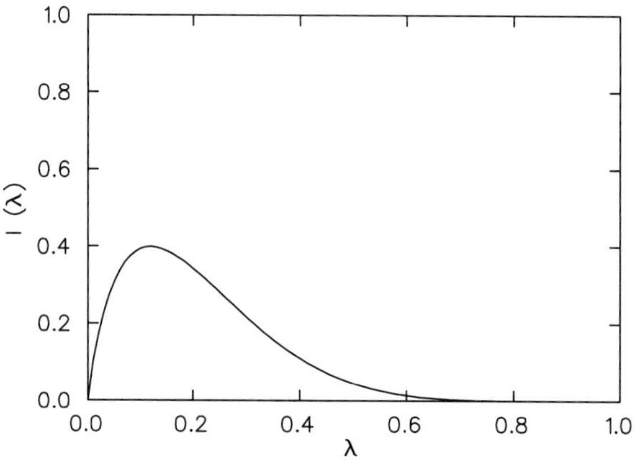

FIGURE 11.4 Mutual entropy calculated from Eq. (11.9).

requires low levels of noise: a low entropy environment. In order to store as much information as possible on the other hand, entropy should be as high as possible, as only a system with many states can carry much information. This tradeoff is apparently ideal in the computationally complex regime, a regime that allows computation universality. We shall see this signature even more clearly when we investigate the learning capability of populations of self-replicating strings.

11.2 Adaptation to Computation in avida

The question that Langton asked with respect to the computational abilities of cellular automata can also be asked in avida. What is the mutation probability at which a population of self-replicating strings best develops computational capabilities? The tradeoff between entropy and information will again become apparent: short programs have a relatively smaller mutation probability, and as a consequence a lower entropy. Thus, in order to store more information, the strings will tend to *increase* their length. Yet, there is a limit to this: the strings can only successfully self-replicate if the information stored in the genome can be *transmitted* accurately enough into the next generation. Thus, we would like the *mutual* entropy between one generation and the next to be maximal while also maximizing the entropy per string.

Rather than focusing on the entropy in the following experiment, we shall instead analyze the *adaptability* of the population, which can roughly be defined as the rate at which the population *absorbs* information from the environment. If this rate is maximal, we presume that the information transmission rate is also maximal, as the process by which information enters the genome is the same as that which provides the noise in the channel. In the lingo of Information Theory, the noise is also the message *source* (see Section 3.8).

Learning with Copy Mutations

We can investigate the *learning rate* of the population for two different kinds of mutation: we shall see that these lead to qualitatively different results. Let us first investigate *copy mutations*. Intuitively, we expect that the population learns rather slowly at low mutation rates, whereas very high rates should make the storage of any acquired information tremendously taxing. Thus, we expect to find a window in mutation rates at which learning is optimal. As adaptation is an intrinsically *stochastic* process, it is not straightforward to find a good measure for the learning capabilities of a population. Indeed, in experiments with tierra it was found that the average *time* for the population to learn a specific task was not a good measure, as the average and standard deviation were of the same order [Adami, 1995a]. This of course just reflects that the distribution of learning events has a long tail (power-law behavior). Still, a good measure for the learning capability can be found in the form of the learning *fraction*. This is a statistical measure that counts the fraction of runs in which a particular task has been achieved before a certain cutoff time. Imagine, for example, that we conduct ten runs, identical in all respects save the random number seed, at a specific mutation rate. Then, we can ask what fraction of the ten runs succeeded in acquiring, e.g., the NOT task before a fixed number of updates have elapsed. Thus, if the cutoff time is X updates, we define the learning fraction

$$f_X(R) = \frac{m}{n} \, , \tag{11.10}$$

where n is the total number of runs of this type, and m is the number of runs that were successful. In this manner, a learning fraction can be defined for each task, and the adaptive characteristics of the population mapped out.

To conduct this experiment, we have to decide on a world size, a suitable ancestor, and a number of mutation rates to test. The dependence of the adaptive capabilities on the population size is an interesting topic in itself, and shall be left as a project (Problem 11.1). Here, we shall test only one world, a grid of 40×40 programs. As we expect to take many runs per mutation rate, and let each of those go on for a sufficient number of updates such that the population has a good chance to acquire the relevant task, it is imperative that an experiment such as this one is prepared diligently in order not to waste CPU cycles. Thus, it is important to run *tests* to determine the range of mutation rates to investigate, as well as to *time* such test runs in order to estimate the CPU time necessary to accomplish all planned runs. Finally, we have to decide on a criterion to determine at what point in time a population has acquired a task. At our disposition we have such measures as

- the time at which the dominant genome can perform the task.
- the time at which a threshold number (say 10) figured it out.
- the time at which more than 50 percent of the population can do it.
- the time at which *almost all* programs are proficient in this task.

In most situations, all these measures are triggered in a relatively short period of time, as Fig. 9.3 shows. However, sometimes tasks (such as `NOT` in that figure) are mastered by only a background number of programs, and take over only at a much later time. For example, the `NOT` task in Figure 9.3 was learned by ≈ 3 percent of the population starting at update 4,000, enjoyed a brief burst of dominance at update 6,000, but withered in the background up until update 26,000 (not shown in that figure), when it suddenly jumped to 80 percent. In such a case, the measures proposed above differ substantially in their assessment of when a task was learned. In most cases, a task is only unlearned if it is superseded by a more powerful task. Thus, we choose as a criterion that a task is learned when the dominant genotype acquires the task for the *first* time.

For the `NOT` task, we set up 20 runs per mutation rate with rates ranging from low ($R = 0.0025$) to high ($R = 0.06$). As a cutoff, we set 10,000 updates, since `NOT` is usually learned before that at the optimal mutation rate. What this optimum is, and how it is determined, is the object of this analysis. Indeed, what we find is that there is an optimal mutation rate only for a fixed *length* of the code, as the mutation probability for each string is $1-(1-R)^\ell = 1-F$, where F is the fidelity of copying for a string of length ℓ. Thus, instead of plotting the learning fraction as a function of R, it is much more instructive to show it as a function of $R\ell$, as the genomes can change their length during evolution.

In Fig. 11.5 we show the learning fraction $f_{10K}(R\ell)$ for the `NOT` task in an experiment where we took 20 runs per mutation rate.[3] After collecting the data, each run is assigned an error probability $R\ell$ obtained from the mutation rate and the *average* length of the strings in that particular run. Note that the average length does not necessarily reflect the length of those programs that acquired the task accurately, but is a good measure

[3]This data was obtained by L. Borissov as part of a homework project.

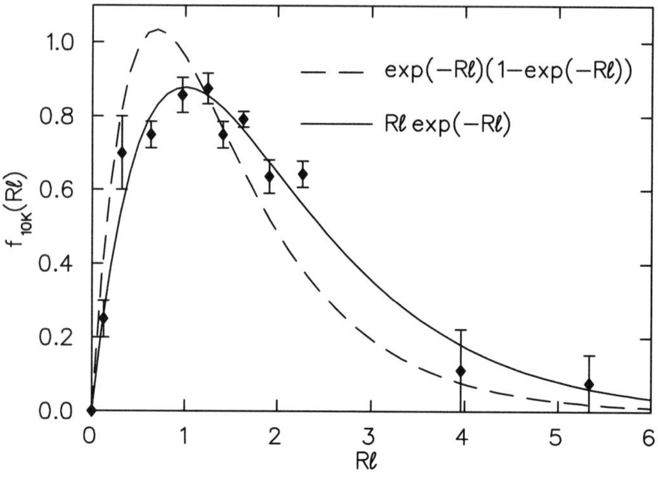

FIGURE 11.5 Learning fraction $f_{10K}(R)$ for the NOT task. Error bars are ±1 run per bin. The data are fitted with Eq. (11.11), with an overall factor as the only fitting parameter. The dashed line is a fit to the capacity of the "equipartitioned" genetic channel, Eq. (3.47).

of this in most cases. Finally, the runs were binned according to their $R\ell$ value, so that in each bin there were enough runs to estimate the learning fraction. The error bars in Fig. 11.5 are simply $1/n(R\ell)$, where $n(R\ell)$ is the number of runs (in each bin) centered around $R\ell$. Interestingly, most runs manage to adjust their $R\ell$ to a value between one and three, whereas runs with an $R\ell$ much below or much above these values are correspondingly rare. This type of behavior will be the center of our attention in the next section.

The data suggests that the learning fraction follows a geometric law of the form

$$f(R) \sim R\ell e^{-R\ell} , \tag{11.11}$$

which is maximal at $R\ell = 1$. It is tempting to believe that the learning fraction is proportional to the amount of information shared between the population and the environment. As this information, in order to be preserved, must also be shared between *generations* of strings, we might surmise that the learning fraction is just proportional to the mutual entropy between generations in the genetic channel introduced in Section 3.8. However, this entropy does not fit the data well (see dashed line in Fig. 11.5), which can be traced back to the simplifying assumptions made in its derivation. In realistic channels, the maximum of the distribution appears to be shifted from $R\ell = \ln 2$ to $R\ell = 1$. The form (11.11), while not obtained from a purely information-theoretic argument, can be interpreted as the Poisson probability that a daughter string has exactly one mutation. Thus, the experiments suggest that the most effective learning seems to occur when the strings are of such a length that there is, on average, exactly one mutation per string. As Fig. 11.5 shows, the population can still exist (albeit strenuously) up to rates significantly higher than the optimal one. This is different if the mutation affects not only the daughter (as in the copy process), but all programs indiscriminately. This is the case if the population is subject to Poisson-random bit-flip, or point mutation events, that we can liken to cosmic-ray mutations. What we shall see there is that most effective learning occurs right at the edge of chaos.

Learning with Cosmic-Ray Mutations

Cosmic-ray mutations can be turned on in avida by editing the appropriate line in the "Mutations" section of the `genesis` file (see Section 9.2 and the Appendix). Here, we shall study their effect on learning in the *absence* of copy mutations.

In contrast to copy mutations, cosmic ray mutations are characterized by a *flux* rather than a rate, as its units are

$$R_\star = \frac{\text{events}}{\text{unit time} \times \text{site}} \,. \tag{11.12}$$

Thus, the flux is determined only if we choose an appropriate unit of time in avida, which is the *update*. As explained in more detail in the Appendix, an update is the amount of time necessary for each program to execute a fixed slice of CPU time, here 30 instructions. The latter choice is, of course, arbitrary. With such a rate, the probability that two mutations are spaced by a time t is exponentially distributed, i.e.,

$$P(t) = R_\star \ell \, e^{-R_\star \ell \, t} \,, \tag{11.13}$$

and the average time between two mutations hitting a particular string of length ℓ is

$$\langle t \rangle = \frac{1}{R_\star \ell} \,. \tag{11.14}$$

For the experiments we analyze here [Adami and Brown, 1994], the only task to be acquired is the addition of two numbers provided in the input buffers. In order to get there, we reward (as earlier) intermediate steps that lead to the establishment of I/O operations culminating in programs that echo. For a population of 40×40 programs, the learning fraction as a function of mutation rate is depicted in Fig. 11.6 for three different cutoffs: 10K, 20K, and 50K updates. Note the sharp dropoff of the learning fraction as the critical mutation rate is reached. This was to be expected, as the population can no longer survive at a rate higher than the critical one, and learning is throttled trivially. Note also that the unit for the mutation rate in this graph is mutations per executed instruction, which turns out to be larger by about a factor of 10 than the rate defined by Eq. (11.12).

This situation is clearly qualitatively different from the case of copy mutations examined earlier, where the mother programs could still function even if all the daughters were flawed, and learning decreased exponentially fast away from thc optimal rate. For point mutations, the optimal rate appears to be right next to the lethal one: learning happens most effectively if the population is almost but not quite melting from the mutational pressure, a population balanced at the edge of chaos—swayed, but holding on.

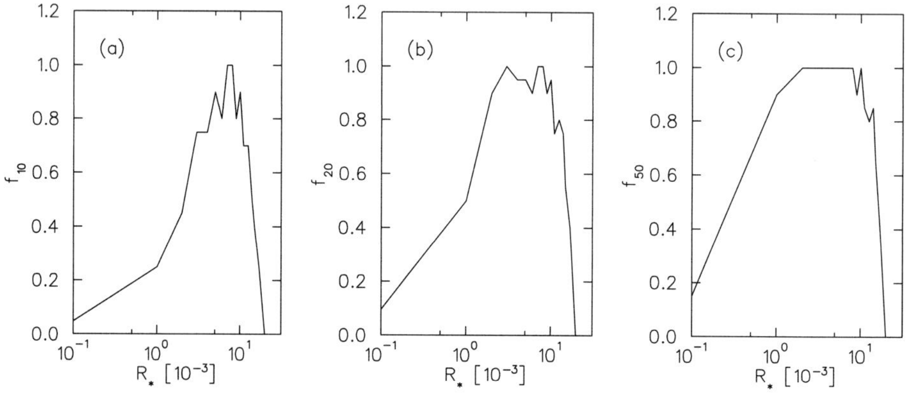

FIGURE 11.6 Learning fraction vs. mutation rate for point mutations. (a): $f_{10K}(R)$; (b): $f_{20K}(R)$; c: $f_{50K}(R)$.

11.3 Eigen's Error Threshold

While we have now observed the conditions that are most conducive to information acquisition by genomes, we have not talked about what actually happens in adapting populations as far as the average string length is concerned. Our goal will be to study the *dynamical* evolution of populations subject to a fixed mutation rate R. It was suggested by Eigen (1971) that real evolving populations actually manage to evolve *towards* this critical point, that genomes literally *adjust* the landscape they evolve in in such a way that the learning rate is near maximal. How this can, and does, happen in molecular evolution in general, and avida in particular, is the subject of the following sections. Before we focus on the dynamics of evolving populations, however, we need to review the concept of the error threshold developed by Eigen [Eigen and Schuster, 1979; Eigen et al., 1989a].

Imagine a population of self-replicating molecules under the constraint that their total number N is fixed (this is not necessary for the analysis, but will be assumed throughout here). Let each type of molecule (the genotype) be denoted by an index i, so that the abundances of genotypes n_i satisfy the conservation law

$$\frac{d}{dt}N = \frac{d}{dt}\sum_{i=1}^{N_g} n_i = 0, \tag{11.15}$$

where N_g is the total number of genotypes in the population. This number is generally not fixed, but can change during evolution. Equivalently, we could have summed over all *possible* genotypes $\mathcal{N} = \mathcal{D}^{\ell}$, where most of the n_i in that case would be zero. Let us write down a rate equation for each genotype, taking into account its production (by self-replication with rate ϵ_i and fidelity F), and its removal due to the condition (11.15), with rate $\langle\epsilon\rangle$. Thus,

$$\dot{n}_i = (F\epsilon_i - \langle\epsilon\rangle)\, n_i + \sum_{i\neq k} \phi_{ik} n_k\,, \tag{11.16}$$

where F (which usually only depends on the length of the molecule) is given in terms of the error rate per copy R as usual

$$F = (1-R)^{\ell}\,, \tag{11.17}$$

and ϕ_{ik} is a *transition matrix* element which is the probability for a mutation from genotype k to i, and that is necessary in order to achieve conservation of error copies. Indeed, every mis-copy of a genotype i (due to $F < 1$) must appear as some other genotype k:

$$\sum_i \epsilon_i(1-F)n_i \equiv N\langle\epsilon\rangle(1-F) = \sum_i\sum_{k\neq i}\phi_{ik}n_k\,. \tag{11.18}$$

This equation, together with Eq. (11.16), ensures that $\sum_i \dot{n}_i = 0$, as required by (11.15). We can use it also to write a "mean-field" equation for the abundances without referring to the mutation matrix ϕ explicitly. Assuming that ϕ_{ik} is independent of and equal for each genotype i (an assumption that is certainly wrong for each genotype, as mutations do not lead to arbitrary genotypes but only to those close in genotype space), we can write

$$\sum_{k\neq i}\phi_{ik}n_k \approx \frac{1}{N_g}N\langle\epsilon\rangle(1-F) \tag{11.19}$$

such that

$$\dot{n}_i(t) = \big(F\epsilon_i - \langle\epsilon\rangle\big)n_i(t) + \frac{N}{N_g}\langle\epsilon\rangle(1-F). \tag{11.20}$$

This equation cannot be solved straightforwardly, as $\langle\epsilon\rangle$ depends on time itself. However, it does allow us to estimate the genotype distribution function of *average* genotypes, i.e., genotypes of fitness about $\langle\epsilon\rangle$, for which $\dot{n}_i \approx 0$. For those,

$$n_i \approx \frac{N}{N_g}\frac{1-F}{1-\sigma_i F}\,, \tag{11.21}$$

where we introduced the *superiority* parameter

$$\sigma_i = \frac{\epsilon_i}{\langle\epsilon\rangle}\,. \tag{11.22}$$

Note that this distribution does not hold for fitter genotypes with $\sigma_i > 1$, as for those the condition $\dot{n}_i = 0$ does not hold. Rather, we can use Eq. (11.20) to formulate a condition for successful survival of the information stored by the population. If we assume that most of the information is stored in those genotypes that replicate most successfully, i.e., the dominant (most abundant) genotypes, this information can only be stored accurately if the genomes that bear it can replicate this information with a rate exceeding zero, not counting the average influx from mis-copies. In other words, the precise message stored in the most successful genomes will be perpetuated only if these genomes have a nonvanishing *growth rate*. From Eq. (11.20), we see that this is the case if

$$F\epsilon_{\text{best}} - \langle\epsilon\rangle > 0 \,, \tag{11.23}$$

or

$$\sigma_{\text{best}}F > 1 \,, \tag{11.24}$$

where we introduced the superiority parameter σ_{best} for the most abundant genotype, termed the quasispecies by Eigen. In terms of the mutation rate R and the length of the code, this condition can approximately be written as[4] (using Eq. 11.17)

$$R\ell_{\text{best}} < \log_e \sigma_{\text{best}} \,, \tag{11.25}$$

which puts a limit on the length of the code as a function of the mutation rate R: this is Eigen's error-threshold condition. Essentially, it implies that the concentration n_{best}/N of the quasispecies, i.e., the dominant genotype, decreases for R and σ fixed, up until it reaches zero if $R\ell = \log_e \sigma_{\text{best}}$, i.e., at the error threshold.

11.4 Molecular Evolution as an Ising Model

Eigen suggested thus that natural populations automatically move *towards* the error threshold $R\ell = \log\sigma$, which, if true, would also imply that the populations maximize their adaptive capabilities, as usually $\log\sigma \sim 1$ (as we saw in Section 11.2). This behavior has been observed experimentally in populations of bacteriophage $Q\beta$ RNA replicating in $Q\beta$ replicase [Domingo et al., 1976; Domingo et al. 1978], as well as populations of RNA viruses [Domingo et al., 1980]. The data show, for example, that phage $Q\beta$ operates *exactly* at its error threshold, as evidenced by

---

[4]In the last equation, the logarithm is the natural one and we made use of the approximation $-\log_e(1 - R) \approx R + \mathcal{O}(R^2)$, which is quite accurate for usual mutation rates.

the fact that the concentration of the master sequence is virtually undetectable. The way that adapting populations can influence the error probabilities affecting their genome is, of course, by changing the *length* of the sequence, and it appears that adapting populations tend to increase their length up to the limit imposed by condition (11.25). From our earlier investigations, we are aware that this is a necessary behavior in order to store as much information about the environment as possible. The question that we would like to ask in the following concerns the structure of the fitness landscape that enables such a behavior.

In order to investigate such a question, we need to construct a *dynamical* model that treats the evolution of molecular strings *directly*, and where we can ask about the most probable state of an ensemble of strings in equilibrium. The model described in the previous section can serve such a purpose, if we refine it by including a fitness landscape and solve it explicitly in the limit $t \to \infty$, rather than settling for a mean-field treatment as we did above. A particularly enlightening treatment of this model in terms of the statistical mechanics of the *Ising* model is due to Leuthaüsser [Leuthäusser, 1986; Leuthäusser, 1987]. There, the binary genome of a string is mapped to a *spin chain*: a string of symbols x_i, where each x_i can take on the values -1 and $+1$ (to be thought of as little magnets that can take on only two orientations) of length ℓ. For such strings, the mutation matrix ϕ_{ij} introduced earlier can be written in a simple manner, using the genetic distance between sequences i and j. With d denoting this distance, the probability of obtaining sequence i from sequence j can be obtained in terms of the probability to take d mutational steps. Thus, ϕ_{ij} will be proportional to the probability that exactly d out of the ℓ spins are flipped, which is

$$P(j,i) = (1-R)^{\ell-d} R^d \,, \tag{11.26}$$

where R is again the probability to flip one spin (i.e., one symbol). ϕ_{ij} is then obtained by multiplying this probability by the replication rate of genotype i (as this will determine how many erroneous copies will be produced per unit time) such that

$$\phi_{ij} = \epsilon_j (1-R)^\ell \left[\frac{R}{1-R}\right]^{d(i,j)} . \tag{11.27}$$

For these two-valued chains, d is given by the Hamming distance between the two chains, which can be written as

$$d(i,j) = \frac{1}{2}\left(\ell - \sum_{k=1}^{\ell} x_k^{(i)} x_k^{(j)}\right) . \tag{11.28}$$

This expression is easily verified, as two positions $x_k^{(j)}$ and $x_k^{(i)}$ that are different will result in a product equal to -1, whereas if they are the same, the product is $+1$. Thus, for example, if two sequences differ only in one position,

$$\sum_{k=1}^{\ell} x_k^{(i)} x_k^{(j)} = (\ell - 1) \times 1 + 1 \times (-1) = \ell - 2 \, , \tag{11.29}$$

such that $d = 1$. We can now return to Equation (11.16) and write it as a *matrix* equation. But first, we return to the *discretized* form

$$n_i(t+1) = (F\epsilon_i + 1 - \langle\epsilon\rangle)\, n_i + \sum_{i \neq j} F\epsilon_i \left[\frac{R}{1-R}\right]^{d(i,j)} n_j \, . \tag{11.30}$$

A redefinition of $n_i(t)$ (with N the total number of molecules),

$$n_i(t) \to \rho_i(t) = \frac{n_i(t)}{N} \exp\left\{\int_0^t dt' \langle\epsilon\rangle(t')\right\} , \tag{11.31}$$

simplifies this equation considerably, leading to

$$\rho_i(t+1) = \sum_j W_{ij} \rho_j(t) \tag{11.32}$$

and

$$W_{ij} = F\epsilon_j \left[\frac{R}{1-R}\right]^{d(i,j)} . \tag{11.33}$$

Note that the sum over j in Eq. (11.32) includes the term $i = j$ [unlike the sum in (11.16)], which just gives the diagonal element $F\epsilon_i$, as $d(i = j) = 0$.

Let us now follow Leuthäusser and write W_{ij} in a form that is analogous to matrices that arise in the physics of real spin systems: magnets. First, we recognize that

$$\left[\frac{R}{1-R}\right]^{d(i,j)} = \left[\frac{R}{1-R}\right]^{\ell/2} \exp\left(-\log\left[\frac{R}{1-R}\right] \frac{\vec{x}^{(i)} \cdot \vec{x}^{(j)}}{2}\right), \tag{11.34}$$

with obvious notation for the vector product of $\vec{x}^{(i)}$ and $\vec{x}^{(j)}$. Finally, we can introduce the parameter

$$\beta = \log \frac{1-R}{R} \tag{11.35}$$

[which is positive for the mutation rates of interest ($R < 0.5$)], so that

$$W_{ij} = e^{-\beta H_{ij}} \,, \tag{11.36}$$

with

$$H_{ij} = -\frac{1}{2}\vec{x}^{(i)} \cdot \vec{x}^{(j)} + \frac{\log \epsilon(i)}{\beta} - \frac{\ell}{2\beta} \log R(1-R) \,. \tag{11.37}$$

Then, the time evolution equation can be written as

$$\vec{\rho}\,(t+1) = \mathbf{W}\vec{\rho}\,(t) \,, \tag{11.38}$$

where

$$(\mathbf{W})_{ij} = W_{ij}(t+1,t) \tag{11.39}$$

is the *transfer matrix* that takes generation t into generation $t+1$, and therefore

$$\vec{\rho}\,(t=t_n) = \mathbf{W}^n \vec{\rho}(0) \,. \tag{11.40}$$

The mapping of concentrations $\vec{\rho}$ from one generation to the next can be described pictorially as in Fig. 11.7, where in the horizontal dimension we spread out the spin chain, whereas the time evolution of the concentrations takes place vertically. In order to see the analogy with the two-dimensional Ising model, and to analyze the time development of the concentrations $\vec{\rho}(t)$ explicitly, we need to specify a replication landscape $\epsilon(\vec{x})$. A simple example landscape can be constructed by defining a function that returns the fitness of a string $\vec{x}$ as a function of the *distance* of that string to a master sequence $\vec{\xi}$, the *fittest* string in the landscape

$$\epsilon(\vec{x}^{(i)}) = e^{\frac{K}{2\ell^2}(\vec{x}^{(i)} \cdot \vec{\xi})^2} \,. \tag{11.41}$$

Note that because of the redefinition (11.31), the fitness ϵ for a string with no replicative ability is 1 rather than 0, while the maximal fitness ($\vec{x} \equiv \vec{\xi}$) is

$$\epsilon_{\max} = e^{K/2} \,. \tag{11.42}$$

The dynamics of the spin chain, or alternatively the time evolution of the molecular string, can now succinctly be described by the equation

$$\vec{\rho}\,(t_n) = e^{-\beta H}\vec{\rho}\,(t=0)\ , \tag{11.43}$$

with the Hamiltonian

$$-\beta H = -\beta \sum_{i=0}^{n-1}\Big(J\,\vec{x}^{(i)}\cdot\vec{x}^{(i+1)} - \frac{K}{2\beta\ell^2}\sum_{k\neq k'}^{\ell}\xi_k\xi_{k'}x_k^{(i)}x_{k'}^{(i)}\Big), \tag{11.44}$$

and $J = \frac{1}{2}$. With a redefinition $x_k \to \xi_k x_k$, the Hamiltonian can be rewritten in precise analogy to the Hamiltonian of the two-dimensional Ising model,

$$H = \sum_{i=0}^{n-1}\Big(J_x\,\vec{x}^{(i)}\cdot\vec{x}^{(i+1)} - J_y\sum_{k\neq k'}^{\ell}x_k^{(i)}x_{k'}^{(i)}\Big), \tag{11.45}$$

with J_x the interaction in the horizontal direction equal to $J = \frac{1}{2}$, and J_y the interaction in the vertical direction given by $K/2\beta\ell^2$. The interactions J_x and J_y are indicated in Fig. 11.7.

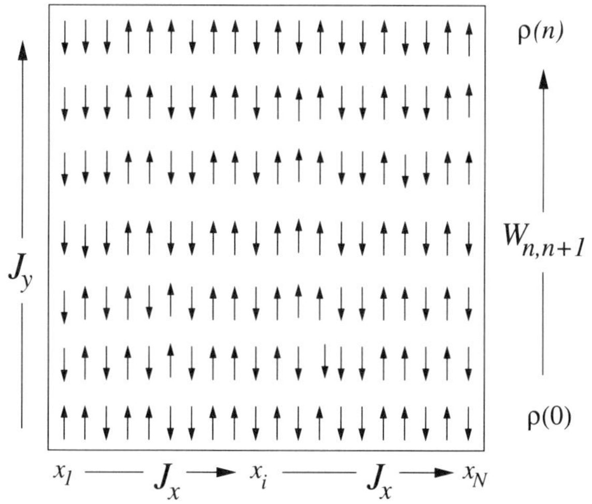

FIGURE 11.7 Time-evolution of the spin chain, with the sequence depicted horizontally, while the time-evolution of the concentrations proceeds vertically. The lattice may be viewed as a two-dimensional Ising crystal, with interactions in the horizontal directions (interaction between adjacent spins) mediated by J_x, while interactions in the vertical direction are specified by J_y.

Note that we left out the constant term $\ell/2 \log R(1-R)$ from the Hamiltonian, as it is independent of $\vec{x}$, and therefore unimportant for the dynamics or the analysis that follows. For such simple landscapes as defined by (11.41) (sometimes called Mattis landscapes [Mattis, 1976]), where the number of degenerate maxima does not grow exponentially with the length of the molecule (in the landscape above, there is only *one* maximum, independent of ℓ), adaptation will result in molecules that tend to *minimize* their length. This theoretical behavior is borne out by the early experiments with replicating RNA of Spiegelman [Mills et al., 1967]. There, RNA molecules taken out of their natural metabolism and allowed to optimize via repeated sequences of selection and replication shrink in length and shed all unnecessary genes. Intuitively, this can be easily understood by considering the competition between the speed of the search and the size of the space to be searched, as a function of sequence length ℓ.

For constant string length ℓ, the number of possible strings is 2^ℓ, whereas there is only one maximum $\vec{\xi}$. On the other hand, as the string length increases, the speed of search through the space increases, as the probability for mutation is $1-(1-R)^\ell$ per string. For the landscape described above, this gain in search speed (when increasing the length ℓ) is swamped by the corresponding increase in search space, if we assume that there is one solution ξ_ℓ (with fitness independent of ℓ) in every sub-landscape of fixed ℓ. Then it would be of no advantage for molecules to grow in length, as the maxima waiting to be discovered in the landscapes with high ℓ are so sparse that they might as well be absent. Thus, the landscape looks essentially flat, a situation described in Section 8.6. Indeed, we noticed there (Fig. 8.7) that the length of the strings drops consistently during evolution, much like in the Spiegelman experiments. In order for there to be an evolutionary advantage to a higher search speed, the maxima in high-dimensional genetic spaces need to be correspondingly frequent; in other words, there ought to be a *degeneracy* of maxima that increases *exponentially* with the string length. Such a situation is intuitively appealing. What this implies is that a sequence with an advantageous chemistry has many equivalent sequences that may or may not be very close in genetic space.

With respect to our computational chemistry in avida, this amounts to saying that for any string that triggers the bonus for a computational task, there are many equivalent strings (but with different genomes) that trigger the same bonus. In a sense the landscape must be of a form that there are many *overlapping* maxima. Before constructing such a landscape for the model described above, we should pause to discuss other results obtained with the statistical treatment of molecular evolution.

In the above discussion, we have stopped short of actually solving the Ising model in two dimensions with a Mattis landscape. These steps have been carried out by Leuthäusser and by Tarazona (1992) and we need not repeat them here. What transpires from the solution of this model is that the error threshold discovered by Eigen is in fact, in the context of the Ising model, related to a second-order phase transition separating an ordered regime (below the critical error probability) from a disordered one. As the ordered phase looks much like a *frozen* system, this transition is often called a *freezing* transition. Indeed, the critical parameter in this model is the "temperature"

$$\theta = \frac{R\ell}{(1-R)K}\,, \tag{11.46}$$

and the critical point is reached as $\theta \to 1$, which we recognize as the error threshold (in the limit of small R) as $K \sim \log\sigma$.

The disordered phase is characterized by a vanishing *magnetization*

$$\mathbf{m} = \frac{1}{\ell}\sum_{k=1}^{\ell}\xi_k\langle x_k\rangle \approx \sqrt{1-\theta^2}\,, \tag{11.47}$$

where $\langle\ldots\rangle$ denotes the statistical average over the population at a given generation. Thus, if the magnetization vanishes, it implies that the strings take on all possible instructions at each site, i.e., all correlations have vanished and the population is essentially random. On the other hand, in a population replicating with high fidelity, $\mathbf{m}$ is approaching 1, as almost all sequences are identical to the master sequence $\vec{\xi}$.

In the next section, we describe the Ising model with a fitness landscape different from the Mattis landscape, and that features degenerate, overlapping maxima.

11.5 The Race to the Error Threshold

Examples of landscapes for which the number of degenerate maxima is exponential in the length of the sequence are spin glasses [Anderson 1983; Mézard et al., 1987]. Such landscapes have a number of attractive features, and are also known to be fractal. As a consequence, they appear to be ideal candidates for evolutionary landscapes. However, explicit models for molecular evolution such as the Eigen-Ising model described above cannot be solved analytically, even approximately, for such landscapes, and thus this prospect has not as yet been realized. Still, other landscapes that feature overlapping exponentially degenerate maxima that *are* amenable to analytical calculation do exist [Adami and Schuster, 1997], and we shall investigate their consequences in the following.

The idea is, quite generally, to investigate the Eigen-Ising model for *variable* string length ℓ, and to let the dynamics pick out the most favorable sequence length. There are two scenarios that we can imagine here. On the one hand, we may start with a population that is monotypic and with a single length, and allow mutations to create strings with shorter and longer sequences than this ancestor. The mutants in the tail of such a distribution then could adapt to become fitter than the original sequence, in which case the original sequence length will be driven to extinction and the new length will dominate. Of course, mutants around this new quasispecies will again provide the possibility for strings of differing length to compete with the dominant length. In such a scenario, a pressure to increase or decrease length will result in a sequence of takeovers that will monotonically increase or decrease the length of the dominant sequence. An equivalent scenario is obtained without assuming that mutations can change the length of sequence, if we start the evolution with a uniform distribution over sequence lengths of equally fit sequences. The dynamics will then pick out the optimal one, without this optimal length having to arise out of a mutational event. While in both scenarios the optimal sequence length is reached, the dynamics of the population itself may be quite different.

The competition between sequence lengths in a landscape turns out to be a delicate tradeoff between search speed and space as mentioned earlier, and can be likened to a race. Indeed, because of selection imposed on the landscape, the first string to reach a higher plateau of fitness will, in general, take all, i.e., it will drive all other strings into extinction even if they have a much higher potential of fitness in the long run, and thus end the competition. In the language of critical transitions in the Ising model introduced above, such a winner-takes-all event is akin to a freezing transition. If the overall landscape is a superposition of landscapes with different ℓ, and there is only one maximum per ℓ, then the short strings will invariably win as they find the maximum the fastest. What if the *density* of maxima does *not* drop exponentially with rising string length ℓ, but stays roughly constant? In that case, the increased search speed due to the higher mutation probability $p(\ell) \approx 1 - \exp(-R\ell)$ will result in longer strings finding a maximum first, and the dynamics should therefore favor an increase in sequence length. Note that this implies that such fitness landscapes are in effect fractal, as it is assumed that new maxima can be found for *any* ℓ, and the landscape is effectively infinite. Still, we have seen that there is a limit to the length of self-replicating sequences, which is the Eigen threshold. An increase in length over this limit would result in the information stored in the sequence to be dissipated, and the adaptive powers of the population to be lost. Thus, the scenario described above can be described as a *race towards the error threshold.*

Mathematically, we can obtain such a fractal, exponentially degenerate landscape by summing the Mattis landscape over q different $\vec{\xi}$ vectors with $q = 2^{\ell\mu}$ and μ a parameter $0 < \mu < 1$, i.e., the number is exponential in the string length:

$$\epsilon(\vec{x}) = \sum_{\xi_1 \cdots \xi_q} \left[e^{\frac{K}{2\ell^2}(\vec{\xi}\cdot\vec{x})^2} - 1 \right] + 1 \,. \tag{11.48}$$

The important aspect of this landscape is its $2^{\ell\mu}$-fold degeneracy, which is quite unlike the degeneracy in the corresponding *Hopfield* landscape [Hopfield, 1982],

$$\epsilon(\vec{x}) = \exp\left[\frac{K}{2\ell^2} \sum_{\xi_1 \cdots \xi_q} (\vec{\xi}\cdot\vec{x})^2 \right] , \tag{11.49}$$

where the sum over the maxima is *inside* the exponential. Such a landscape does not lead to exponential degeneracy.

An explicit solution of the Eigen-Ising model with a particular fitness landscape involves obtaining the equilibrium distribution $\vec{\rho}\,(t_n)$ for t_n large, by calculating the right-hand side of Eq. (11.40) given an initial distribution $\vec{\rho}\,(0)$, the idea being that as $t \to \infty$, only the distribution with the *largest* eigenvalue of the transfer matrix $\mathbf{W}^n$ [see Eq. (11.40)] will survive this limit. The corresponding eigenvector is the most probable equilibrium distribution we are after. A calculation along the lines of Leuthäusser shows that with the landscape (11.48), the critical parameter

θ is

$$\theta = \frac{R\ell}{K(1-\mu)}, \tag{11.50}$$

and the largest eigenvalue λ_ℓ becomes (in the vicinity of the $\theta = 1$) [Adami and Schuster, 1997]

$$e^{\lambda_\ell} \approx 2^{\mu\ell} K(1-\theta) \exp\left[\frac{1}{2} K(1-\mu)^2 (1-\theta)^2\right] . \tag{11.51}$$

It is this function, therefore, that determines which length ℓ will dominate when we calculate the average length

$$\langle \ell \rangle = \frac{\sum_{\ell=1}^{\infty} \langle \ell \vec{\rho}\,(t)\rangle_{\vec{x}}}{\sum_{\ell=1}^{\infty} \langle \vec{\rho}\,(t)\rangle_{\vec{x}}} . \tag{11.52}$$

This average is akin to the Gibbs average Eq. (4.51), and is maximal where the largest eigenvalue λ_ℓ of Eq. (11.51) is maximal. Let us plot λ_ℓ as a function of θ for different values of the error threshold, and with the parameters $\mu = 0.1$ and $K = 1$, versus the critical parameter $\theta = \ell/\ell_c$ [implicitly defining ℓ_c via Eq. (11.50)]. As mentioned before, the average length of the sequences is determined by the maximum of λ_ℓ, so the fact that λ_ℓ is maximal exactly *at* the error threshold $\theta = 1$, i.e., $\ell = \ell_c$ in Fig. 11.8, shows that evolution in a landscape with exponential

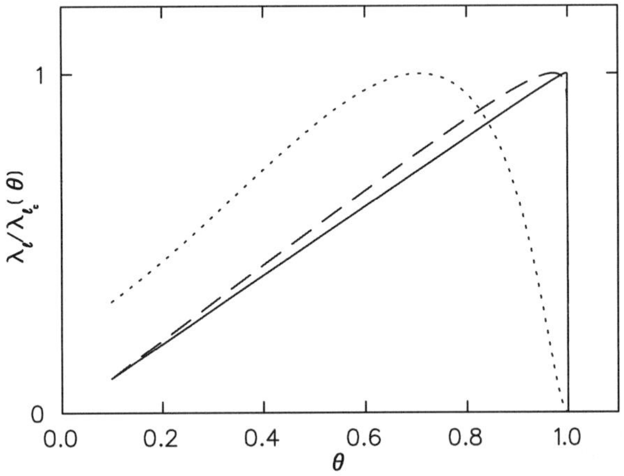

FIGURE 11.8 Normalized eigenvalue $\lambda_\ell/\lambda_{\ell_c}$ as a function of the normalized molecular length $\theta = \ell/\ell_c$ for three different values of the error threshold: $\ell_c = 100$ (dotted line), $\ell_c = 1000$ (dashed), and $\ell_c = 10^4$ (solid line).

degeneracy drives the average length of sequences *towards* the critical length ℓ_c, which is determined by the error rate.

11.6 Approach to Error Threshold in avida

While the evidence garnered from this theoretical analysis is compelling, it would be interesting to observe these dynamics directly, in real or artificial evolving populations. In avida, we can monitor both the fitness gain σ as well as the length of the dominant genotype ℓ, such that an approach to the error threshold, should it take place in avida, can be monitored.

First, let us take a look at some important characteristics of *single* runs in avida, before embarking on a statistical analysis. As we are interested in the dynamics of length changes of the genome, we should establish whether a particular genome length usually dominates the population, or whether the distribution of lengths is so broad that the dynamics can pick out the optimal one from a population at any time. We shall find that the former is the case: due to the selection effect, the genome length distribution is usually very narrow, and an evolution to longer genomes can only occur if these lengths are created before they can be populated. For this reason, we turn on *insert* and *delete* mutations at a rate of 5 percent, which should provide a certain amount of natural width to the genotype length distribution. This is not strictly necessary (as genome length changes occur spontaneously also without insert and delete mutations), but it speeds up evolution considerably.

A typical run at a mutation rate $R = 0.02$ is shown in Fig. 11.9. The fitness increases very fast at the beginning of the run, during which the unadapted ancestor of length $\ell = 30$ adapts to the landscape. Around update 10,000, the length of the dominant program increases dramatically to absorb the information that gives rise to a higher metabolic rate. At about update 20,000, most of the information has been absorbed, and the landscape essentially appears flat to the population. Consequently, the fitness stays constant and the length slowly decreases as the population optimizes its code. In Fig. 11.10, we can see a cross-section of the length distribution of threshold (replicating) genotypes at three different times. In the first panel, the cross-section at an early stage is depicted, with most genotypes

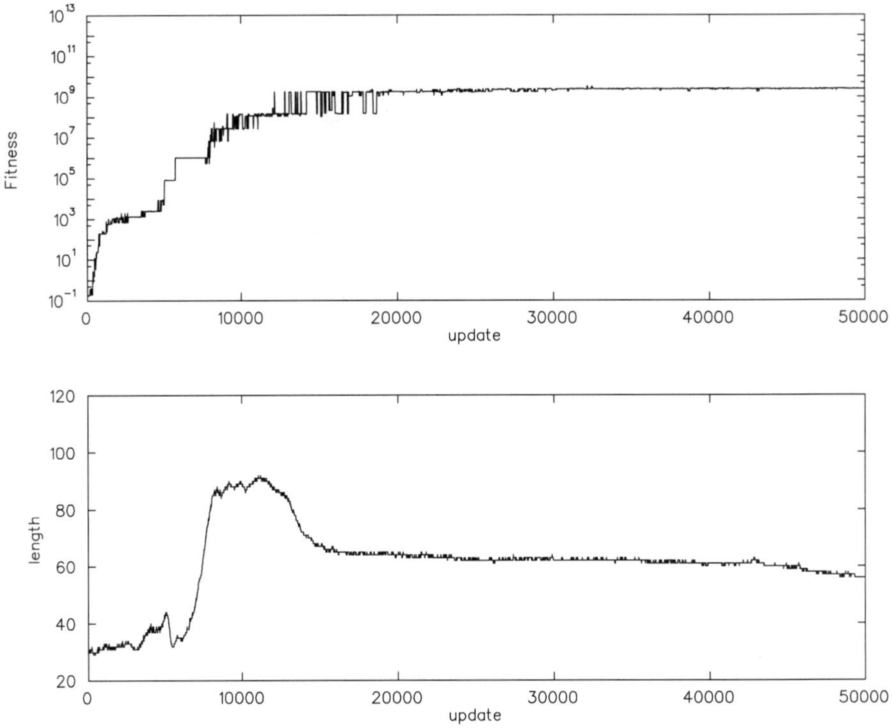

FIGURE 11.9 Fitness of dominant genotype (quasispecies) as a function of time in updates (upper panel) and length of average genotype (lower panel).

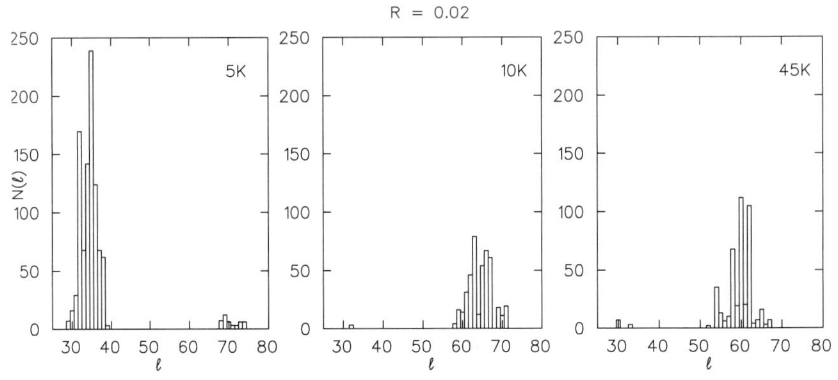

FIGURE 11.10 Length distribution of replicating genotypes at three stages in the evolution depicted in Fig. 11.9. Early (5,000 updates, left panel), shortly after the major size transition (10,000 updates, middle panel), and late (45,000 updates, right panel).

only marginally larger than the $\ell = 30$ class that was used to impregnate the population. However, a small contingent of considerably longer genotypes of size $\ell \approx 70$ is already visible. At 10,000 updates, these genotypes have taken over and completely wiped out the ancestral class of genotypes, which could not compete with the newly emerged programs with advanced computational chemistries. Note that the average length of genotypes is larger than the class that dominates around update 10,000, as can be seen from the lower panel of Fig. 11.9. Thus, many longer genotypes that do not replicate faithfully are present right after the transition, but are weeded out in a slow process of purification as time goes on. At update 45,000, the length distribution of the dominating genotype is roughly unchanged (right panel in Fig. 11.10), whereas the average length has dropped considerably.

The theory that predicts that the length of evolving genotypes approaches the bound given by the ratio between the logarithm of the superiority σ and the mutation rate R,

$$\ell \to \ell_{\rm max} = \frac{\log \sigma}{R}, \tag{11.53}$$

where

$$\sigma = \frac{\epsilon_{\rm best}}{\langle \epsilon \rangle} \tag{11.54}$$

can be tested by comparing the length of the dominant genotype ℓ to $\log \sigma$ after equilibration, i.e., after all or most of the information available in the environment has been absorbed, for different mutation rates. Let us first check that Eigen's threshold inequality is never violated by the population, i.e., that the product $R\ell$ never exceeds the value $\log \sigma$, which of course changes during the run as more and more fit genotypes emerge. In Fig. 11.11, we plot $R\ell$ against $\log \sigma$ in a *scatter plot*, where each point is a pair $(R\ell, \log \sigma)$ at one point in time in the evolution of a run at a fixed mutation rate $R = 0.01$ that continued for 50,000 updates with data taken every 10 updates. While an *approach* to the threshold is not apparent in this view (as the time evolution is not evident), it is clear that

$$\log \sigma \geq R\ell \tag{11.55}$$

at all times, just as Eigen theorized (and as we showed in the previous sections), as points do not venture below the solid line that demarcates the region allowed by Eq. (11.55).

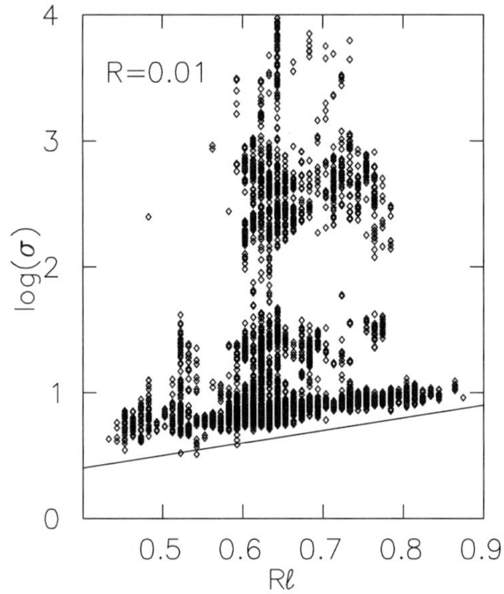

FIGURE 11.11 Scatter plot of $R\ell$ versus $\log\sigma$ for the evolution of a run at $R = 0.01$. The solid line represents the error threshold $R\ell = \log\sigma$.

The theory outlined in the previous section claims more, however. It predicts that after equilibration, i.e., at a time when all the information of the landscape has been absorbed, the population has managed to approach the threshold $R\ell = \log\sigma$ closely, thus maximizing its search speed and adaptive capabilities. To show this, we take a multitude of runs at fixed mutation rates and let them continue for 50,000 updates in the hope that equilibration is achieved. At this point, we average ℓ and $\log\sigma$ over the last 1,000 updates to obtain the *equilibrium* values for each run and their variance. A large variance in a quantity indicates that the run possibly was not quite at equilibrium, and instead was undergoing a nonequilibrium phase transition. For each mutation rate, we plot the final pairs ℓ and $\log\sigma$ against each other. If the population adapted to the error threshold, the points ought to lie on a straight line with a slope given by R, according to the relation

$$\log\sigma = R\ell\,. \tag{11.56}$$

In order to account for an insert/delete probability P (here $P = 5 \times 10^{-2}$) Eq. (11.56) must be replaced by

$$\log \sigma \approx R \cdot \ell + \frac{2P}{(1-R)^{\ell}} \qquad (11.57)$$

leading to a deviation from the linear law (11.56). Note that there is no free parameter in this prediction, and the accuracy of the approach to the error threshold can be measured by how well the *fitted* R agrees with the known mutation rate. In Fig. 11.12, we can see this plot for four different mutation rates: $R = 0.0075$, $R = 0.01$, $R = 0.015$, and $R = 0.02$. In each case, the points do, for the most part, lie on the predicted curve, within the error bars which are taken to be the standard deviation over the last 1000 updates. The fitted R is within 5 percent of the actual value used in the runs. Deviations from the predicted behavior can occur for a

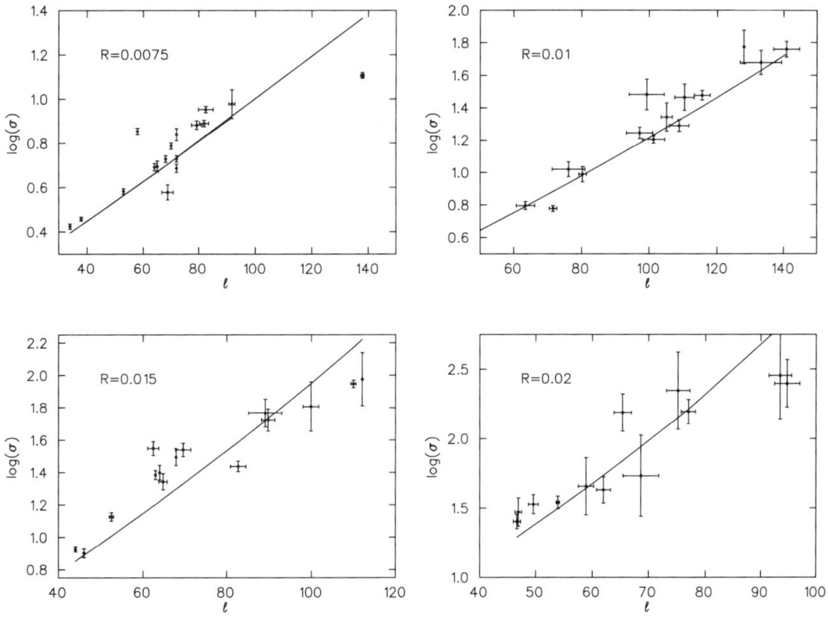

FIGURE 11.12 Log σ plotted against ℓ averaged over the last 1,000 updates for runs at fixed mutation rates. If the population has approached the error threshold, the points must lie on a straight line (for $P = 0$) with a slope given by the mutation rate. The solid lines are fit to the data via Eq. (11.57) with R of 0.0078, 0.0095, 0.0150, and 0.0217, respectively.

number of systematic reasons, foremost among them the possibility that the population is not at equilibrium when the data are taken. Usually, this translates in larger error bars in $\log \sigma$, while the average itself is off the line. Another systematic error is associated with the measurement of $\log \sigma$ proper. As the runs progress, the best genotype is usually very rare, and ϵ_{best} becomes harder and harder to measure (the vanishing of the quasispecies at the error threshold is predicted by Eigen's theory). Also, the average replication rate is difficult to measure in a population subjected to high mutation rates, as many genotypes do not replicate correctly, or else do so inconsistently. For example, a genotype may come to dominate the population that does *not* breed true (make exact copies of itself) due to a faulty algorithm rather than an error rate. The effective mutation probability for such a genotype is much higher than the rate R might suggest, an effect that is beyond the scope of Eigen's theory of the error threshold.

Finally, let us observe how the mutation rate affects the lengths of programs *on average*. According to the theory laid out here, the length of the programs should grow steadily (as long as there still is information to be discovered) until reaching the maximum length, which is limited by the mutation rate. Indeed, Fig. 11.13 seems to confirm this. There,

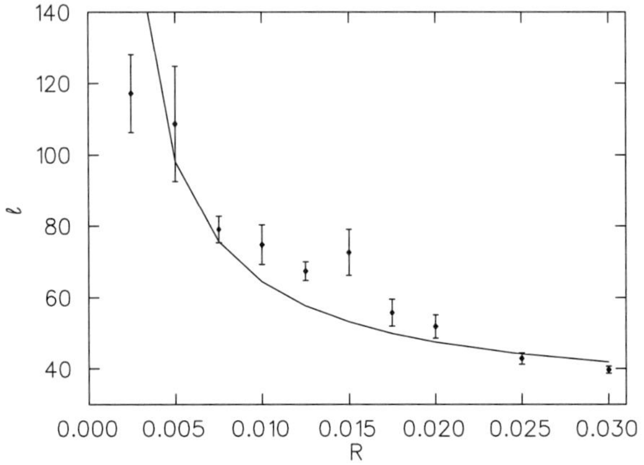

FIGURE 11.13 Average length of programs after adaptation ℓ plotted versus the mutation rate R prevailing during the adaptive process, for between 10 and 20 runs at each mutation rate. The solid line is a hyperbolic fit to the data to lead the eye.

the average length of the dominant program at the end of the run is plotted against the mutation rate, showing clearly that runs at low error rates develop much longer codes on average, while those evolving in the high-noise environment cannot progress past a maximum length. Even though the variation of lengths at each rate is large (reflecting the different fitnesses achieved) the figure convincingly demonstrates that $R\ell \approx$ const.

Notwithstanding the possible systematic trends, the data presented in Fig. 11.12 and Fig. 11.13 are consistent in suggesting that evolution, at equilibrium, does indeed adjust the length of the strings in such a way as to be right *at* the error threshold, where learning and adaptation take place the fastest and information transmission is almost optimal (recall the discussion in Section 11.2). A further increase in the mutation rate would turn the information transmission into a *useless* channel, all information present in one generation would be lost in the next, and the population would cease to occupy a coherent cloud in genotype space and drift apart.

11.7 Overview

Adapting populations must make a tradeoff between the accuracy of information transmission and the storage of information, which is translated into a balance between the deterministic and the chaotic regime. Experiments with avida show that, with copy mutations, populations adapt the most effectively if $R\ell \approx 1$, unlike the naive genetic channel for which the optimum is at $R\ell \approx \ln 2$. Eigen's theory of self-replicating macromolecules suggests that there is a threshold for $R\ell$: we must find $R\ell \leq \log \sigma$ at all times. This theory can be cast in terms of the equilibrium dynamics of two-dimensional *Ising* models, and such Eigen-Ising models can be solved for a number of simple model landscapes. A particular choice of an exponentially degenerate landscape with overlapping maxima suggests that populations of self-replicating macromolecules will always *approach* this threshold by adjusting their length to the maximum allowed by the condition. This behavior can be verified in avida, implicitly suggesting that the artificial chemistry in avida is indeed exponentially degenerate (up to the "exhaustion" of the landscape) and therefore fractal in the range of adaptation investigated.

Problems

11.1 Determine the behavior of the learning fraction for *fixed* mutation rate as a function of population size. As this kind of project can be very CPU-intensive, be sure to plan number of runs, durations, population sizes, etc., carefully before embarking on the project. Estimate the behavior *before* taking any data by making educated assumptions. Compare the result with these expectations.

11.2 (a) Verify in avida the *MacArthur-Wilson Law*, which relates the number of species N on an island to the area A of the island by the "universal" relation

$$N = c\,A^z\,. \tag{11.58}$$

The constant c may depend on the type of island, but the exponent z is independent of it, and usually in the range $0.2 < z < 0.4$ [MacArthur and Wilson, 1967]. If N is the number of genotypes, what is the exponent measured in avida? Can this result be understood from first principles?

(b) What is the exponent if you measure the number of *species* instead? What do you expect for the exponent as higher and higher levels of the taxonomic hierarchy are considered?

The avida User's Manual

The accompanying software was originally written and compiled by Charles Ofria in 1994–97, at the California Institute of Technology. Portions of the source code have also been written by Dennis Adler and Travis Collier. Previous incarnations of avida were written and compiled by C. Titus Brown. Please send any questions, comments or suggested patches to us at `avida@krl.caltech.edu`

A.1 Introduction

This manual is intended to be an operating guide and an investigative aide to the avida system. It is structured so that new readers need only concentrate on the introduction and the beginner's guide in order to gain a

By Charles Ofria, C. Titus Brown, and Christoph Adami.

working knowledge of avida. This introduction contains a quick overview of the system, and the following section contains a fuller description of both the theoretical design and this particular implementation. The next group of sections explore some of the design decisions made (which emphasize those we consider non-intuitive), and explain the specifics on configuring runs. Finally, the user's interactions with avida (installation, configuration, run-time use, and output files) are detailed with sample applications for research.

What Is avida?

The computer program avida is an auto-adaptive genetic system designed primarily for use as a platform in Artificial Life research. The avida system is based on concepts similar to those employed by the tierra program, that is to say it is a population of self-reproducing strings with a Turing-complete genetic basis subjected to Poisson-random mutations. The population adapts to the combination of an intrinsic fitness landscape (self-reproduction) and an externally imposed (extrinsic) fitness function provided by the researcher. By studying this system, one can examine evolutionary adaptation, general traits of living systems (such as self-organization), and other issues pertaining to theoretical or evolutionary biology and dynamic systems.

Since the predominant auto-adaptive system available is Ray's tierra, and avida is based on similar concepts, we present a short list of differences between avida and tierra. In avida we provide:

- A two-dimensional environment with local interactions *only* (as opposed to the global interactions in tierra.) This prevents information from propagating at an exponential rate and allows us to study localized phenomena. The slower information propagation causes the population to maintain a state outside of equilibrium (where learning is optimal) for a much longer period of time. Additionally (in future versions of avida) local interactions will permit nearly linear increases in execution efficiency on parallel computers.
- A detailed flexibility in configuring runs. Avida allows the user to select from a variety of time-slicing methods (which allow for synchronous or asynchronous execution of creatures), mutation schemes, methods for placement of newborn creatures, and even the flexibility to model

other auto-adaptive Artificial Life systems such as tierra (to establish baselines for comparisons.) In future versions, we plan to expand this emulation to standard adaptive systems such as genetic algorithms, and allow for other CPU structures.

- The ability to specify fitness landscapes for goals beyond the optimization of gestation time, along with the theoretical possibility for open-ended evolution (given a sufficiently complex fitness landscape.)
- Allowing precise measurements and the collection of statistical information (as is required for a scientific research platform.) Avida allows the user to configure a wide range of output files, and observe almost any aspect of the soup.

Brief Overview

The avida system creates an artificial (virtual) environment inside of a computer. The system implements a 2D grid of virtual processors which execute a limited assembly language; programs are stored as sequential strings of instructions in the system memory. Every program (typically termed *cell, organism, string* or *creature*) is associated with a processor, or grid point. Therefore, the maximum population of organisms is given by the dimensions of the grid, $N \times M$, and not by the size of the total genome space of the population, as in tierra. For purposes of Artificial Life research, the assembly language used must support self-reproduction; the assembly language instructions available are described in Section A.5.

The virtual environment is initially seeded with a human-designed program that self-replicates. This program and its descendants are then subjected to random mutations of various possible types which change instructions within their memory; resulting in unfavorable, neutral, and favorable program mutations. Mutations are qualified in a strictly Darwinian sense; any mutation which results in an increased ability to reproduce in the given environment is considered favorable. While it is clear that the vast majority of mutations will be unfavorable—typically causing the creature to fail to reproduce entirely—or else neutral, those few that *are* favorable will cause organisms to reproduce more effectively and thus thrive in the environment.

Over time, organisms which are better suited to the environment are generated that are derived from the initial (*ancestor*) creature. All that remains is the specification of an environment such that tasks not otherwise intrinsically useful to self-reproduction are assimilated. A method of altering the *time slice*, or amount of time apportioned to each processor, is described in Section A.3.

While avida is clearly a genetic algorithm (GA) variation (to which nearly all evolutionary systems with a genetic coding can be reduced), the presence of a computationally (Turing) complete genetic basis differentiates it from traditional genetic algorithms. In addition, selection in avida more closely resembles natural selection than most GA mechanisms; this is a result of the implicit (and dynamic) co-evolutionary fitness landscape automatically created by the reproductive requirement. This co-evolutionary pressure classifies avida as an *auto-adaptive* system, as opposed to standard genetic algorithms (or *adaptive*) systems, in which the creatures have no interaction with each other. Finally, avida is an evolutionary system that is easy to study quantitatively yet maintains the hallmark complexity of living systems.

Contacting the avida Group

The members of the avida group can be reached by email at `avida@krl.caltech.edu`. Consult the group's Web page for changes and notices at `http://www.krl.caltech.edu/avida/`. Research articles by members of the avida group are available via anonymous FTP at `ftp://ftp.krl.caltech.edu/pub/avida/`. Please feel free to contact us with any questions, comments, or ideas you may have about avida. For technical questions, contact `avida-help@krl.caltech.edu`. (Technical Support is *not* guaranteed with avida, but we will typically try to help with any problem you may have.)

A.2 A Beginner's Guide to avida

In this section, we give a general description of the structure of avida and examine a few of the implementation details.

The avida System

At the highest level, the avida system consists of a grid and a set of interactions between points on the grid. The grid is a two-dimensional lattice with natural Cartesian coordinates specifying the grid points and periodic boundary conditions at the edges, resulting in a toroidal geometry. Each grid point harbors a separate central processing unit (CPU).

This leads to a virtual representation of a standard multiple-instruction multiple-data (MIMD) parallel machine; the CPUs can be run either synchronously or asynchronously, closely approximating actual parallel machines. The closeness of this approximation depends largely on the *granularity*, or simplicity, of the instruction set. If a single instruction can have an extremely large effect on the surrounding environment, it will be correspondingly harder to approximate parallel execution using an asynchronous update technique. Section A.3 on time slicing details the various mechanisms by which these time slices are allocated and CPUs are scheduled for execution.

While this representation is in itself general enough to be used in many types of simulations, e.g., the study of MIMD adaptive computation, avida is designed explicitly for Artificial Life research. Hence, we term each grid point a *cell* or *organism*, and each associated list of instructions the *genome* of that organism. Naturally, active grid points are *alive*, and inactive (or empty) grid points are *dead*. The genome of each CPU is *circular*, as in most bacteria and viruses. As a consequence, the instruction pointer never leaves a cell unless it is forced by explicit command (see *parasitism* and the `jump-p` command in Section A.5).

We define the assembly language to be simple and to support the capacity for *reproduction*, where the program from one processor can copy itself and replace the program of a neighboring processor with its copy. We then seed the environment with a simple human-coded self-replicating program which soon spreads throughout the available lattice.

These active programs have a small probability of each command being randomly permuted in the copy of their genomes (see Section A.6 for a description of this process); this results in a Darwinian selection process. By using different time-slice allocation methods (see Section A.3) we can specify a fitness landscape such that the programs will evolve to exhibit useful behavior beyond that of simple self-reproduction.

A.3 Time Slicing and the Fitness Landscape

This section discusses the theory behind the time slicing schemes used in avida. Time slicing is the method by which the creatures in the grid are allocated CPU time to execute their code. Some creatures have faster CPUs than others, so the time slicing code must make sure they all run at the appropriate (relative) rates. Each creature has a *merit* which determines the speed of its CPU; the time slicer will adjust this merit as creatures perform tasks deemed desirable in this environment.

Time Slicing

The mechanism by which portions (or *slices*) of CPU time are distributed to the individual proccssors on the grid significantly influences the global behavior of the population; here we examine it in detail.

The idea behind a system such as avida is to explore the complexity that the chemistry of self-replication has introduced, without actually simulating chemical reactions. Rather, the chemistry of the polypeptides coded for in the DNA is replaced by executing the string of instructions. As in chemistry, we expect different genomes to have different properties, most of which are reflected by either a greater or lesser amount of CPU time triggered by the cell. This is the equivalent of endothermic or exothermic reactions. Replication and execution of instructions costs CPU time, thus consuming energy.

On the other hand, certain instructions executed in the right order can trigger large amounts of bonus CPU time, the equivalent of a catalyzed reaction which benefits replication. As in real chemistry, it is problematic to specify which sequence of instructions is beneficial; rather, we construct the environment by rewarding *effects*, such as we did in our experiments with integer computation (discussed below). In the remainder of this section, we point out how this mechanism can be used to breed desired traits.

In addition to this external bonus structure, which effectively distinguishes one environment from another, we specify basic systems of CPU-time distribution that describe the low-level aspects of the virtual chemistry we are constructing. This is the time-slicing system proper. To define the time slicer, we have to decide how much time a cell should be worth *by default*, i.e., without studying the effect this string has when executed in a population.

A simple choice would be to give each string a constant time slice regardless of its size. This is the primary mechanism used in the tierra system. With such a choice, all cells attempt to minimize the length of their code by shedding superfluous instructions. The shorter they are, the less they have to copy, and the more copies they can make of themselves in the fixed amount of time they are given. The gestation time is roughly linear in the length of the cell. The advantage gained by shrinking the code is so dramatic, however, that cells might even choose to shed sections of code that trigger moderate bonuses. Such a method certainly provides for very efficient optimization while discouraging the evolution of complex code by magnifying the barrier to neighboring local minima in the fitness landscape. As far as the structure of the fitness landscape for the strings is concerned, such a slicer increases the local slopes and thus accelerates convergence to a local energy minimum while reducing ergodicity.

Another possibility is to distribute CPU time in a manner proportional to the length of the code. This is the *size-neutral* scheme also used in tierra. The resulting fitness landscape is intuitively much smoother; strings that behave in the same way but differ in length of code are degenerate as far as their replication rate is concerned and far-lying regions in genotype-space can be accessed easily. Clearly this mechanism is much more conducive to the evolution of complexity. However, it has a certain disadvantage from a practical point of view, as the instruction set provides the possibility to *jump over* sections of code. The cells soon discover that they can earn free CPU time by developing code that is neither executed, nor properly copied. This does not exist in real chemistry, as even DNA that is not expressed still participates in chemical interactions.

We therefore developed two mechanisms to handle this. The first counts only those instructions that were copied into the creature, or alternatively (as defined during the configuration of the run) only those instructions executed by the creature, in evaluating the effective length. Under these conditions, *lean* cells are favored over those that carry sections of un-executed, un-copied code. The second mechanism forces a creature to copy at least 70 percent of itself into its daughter, and execute at least 70 percent of itself, or else all divides will fail. The precise parameters are, again, configurable in the `genesis` file.

The slicers discussed here can be distinguished by a simple formula that describes the mechanism. Specifically, the time doled out (allocated) to each cell *a priori* is proportional to the creature's *merit*, where merit is determined by the number of instructions copied into (or executed by, depending on the configuration) the creature, times any bonuses given to a creature through its interaction with the environment. This multiplication (as compared to the method of *addition* of bonuses used in previous incarnations of avida and in tierra) serves to ensure that there is no size bias in evolution. In the case where bonuses are additive, they will soon overshadow the size component of the merit thus bringing back the constant time-slice paradigm. Note that enforcing size neutrality is strictly speaking un-biological, as it is known that self-replicating strings will shed all unnecessary instructions if given the opportunity. In avida, size neutrality is necessary in order to jump start the evolution of complexity.

Carving a Landscape

Since the time slicer defines the landscape (and thus the "physics" and artificial chemistry) associated with self-replication, we can superimpose any landscape we deem interesting. This is done by specifying bonus CPU time associated with the *phenotype* of the string. By rewarding *actions* rather than a particular sequence of commands within a genotype, we introduce the possibility for open-ended evolution. As the set of possible strings that have the same phenotype is effectively infinite if no bounds are put on the length of strings, it is impossible to construct a string with maximum fitness given a complex enough environment. The complexity of the landscape (here identified roughly as proportional to the number of distinct local minima) increases exponentially with the number of distinct bonuses specified, as they can in principle be triggered simultaneously and in any order (often integrated so precisely that the same section of code will work on multiple tasks at once). As an example, consider the landscape we constructed in such a way that the adapted population would reflect a phenotype capable of adding integer numbers.

As a first step, we reward doing any form of input and output. We then further reward the correct input/output (I/O) structure (i.e., a minimum of two `get` and one `put` command, in that order). For each such step, we multiply the creature's merit by a fixed amount. This is admittedly a deviation of the principles just stated, as we reward a genotypical signature rather than a phenotype. However, since I/O is a low-level characteristic (achieved by executing the *single* instructions `get` and `put`), this deviation appears to be warranted. Also, any additions to the instruction set which also allow for manipulation of the input and output buffers should also trigger their bonuses.

Next, we reward the capability to *echo* numbers (which were read from the input) into the output. This reward is phenotypic, as it can be obtained in a number of distinct ways, all of which are rewarded as long as they properly complete the task. Finally, if a `put` command writes into the output buffer a number which is the sum of two previously read numbers, the string is rewarded with another bonus. Each of these rewards can be triggered multiple times each gestation period, typically to a maximum of three. By default, the bonuses multiply the creature's merit by a fixed factor the first time they are triggered, while only multiplying them by a smaller factor (1.25) thereafter. This is both to encourage diversity in ability, and because it is typically easier to perform a task multiple times than it was to learn it initially.

How such a bonus structure *carves* a landscape in the space of all fitness improvements becomes obvious (or at least intuitive) if we analyze the population shortly after it adapted to the *echo* bonus. At that point, mutated strings write all sorts of numbers into the output, numbers that are obtained via random manipulations. Among those we find sums, differences and multiples of the input numbers. The gene for addition is simply filtered out by rewarding this particular task out of all those currently being performed. Any other task can be filtered in the same manner. Quite literally, rewarding addition creates a valley that only those cells with the appropriate gene can occupy. Since cells in the lower regions of the landscape obtain more offspring than those higher up, they soon dominate the population and drive strings missing the gene into extinction. Once this is achieved, the adapted population spreads in diversity via the effects of mutation, to explore new regions of the landscape where perhaps more crevices can be found. Such a sequence of adaptations results in a fitness curve resembling a staircase, which can be a true fractal if the environment is complex enough.

Fitness

The fitness of a cell in such simple systems is given by the total effective number of offspring it can generate in its environment. Theoretically, this is given by the merit $\mathcal{M}$ earned by the cell, divided by the time it takes to generate an offspring, the gestation time t_g

$$\alpha = \frac{\mathcal{M}}{t_g}. \tag{A.1}$$

This fitness measurement is a unitless quantity which can be directly compared to the fitness of any other creature to determine their relative reproduction speeds. If one creature has twice the fitness of another then it will be able to have twice the number of offspring per unit time.

Choosing a Time Slicing Algorithm

These algorithms handle the details of grid cxccution; they ensure that cells are executed with simulated parallelism to minimize any advantage dependent on execution order which may occur.

This system provides a method of time-keeping independent of grid size; in tierra, the standard time measure is one million instructions, which depends entirely on the population size. In a large population, several measurement periods might pass before the population is entirely executed even once. In avida as the grid becomes larger, the entire grid is executed during each measurement period, and as the time slicer allocates more time (more energy) the cells run correspondingly *faster* in relation to the measurement periods. In addition, many of the algorithms provide a natural interface to distribute the avida system across multiple processors (see below).

In order to properly configure the time-slicing scheme, edit the "Time Slicing" section of the `genesis` file (see Section A.9). The first value, `AVE_TIME_SLICE` will set the average number of instructions a creature should execute every update. By default, this value is set to 30.

Next we have `SLICING_METHOD`; this is an important control which determines just how the time blocks are doled out. The options are:

- `CONSTANT`: This means that all of the CPUs get the same amount of CPU time, and their merit is ignored. This obviously encourages shrinking, and removes all incentive to learn environmental tasks (unless merit is taken into account in the placement of new offspring; see BIRTH_METHOD below.) CPU time is doled out evenly such that each creature executes a single instruction before any execute their second one.

- BLOCK: In this time slicing scheme, all CPUs are allocated a block of CPU time such that the size of each time-slice block is proportional to the creature's merit. The CPUs are executed in sequence for their entire block, quite similar to tierra.
- PROBABILISTIC: Instructions are executed in a semi-random fashion in this method, such that the probability of a single creature having an instruction executed is proportional to that creature's merit. Thus, on the average, each creature does obtain an amount of CPU time proportional to its merit. This method has the most realistic feel to it, but the random component does slow learning slightly, and the constant use of the random number generator makes avida run slightly slower.
- INTEGRATED: This is the default slicing method in avida. It has each CPU execute a single instruction at a time in a deterministic fashion such that the relative speeds of the individual CPUs are proportional to the merit of the creature using them. Effectively this comes as close to a perfectly synchronous parallel execution as possible.

The next variable that can be configured for time-slicing is the SIZE_MERIT_METHOD. This determines what the base value of the creature's merit is proportional to: its full size, its executed size, its copied size, the smallest of the latter two, or just a constant value (independent of size). In avida, the default method is to select the minimum of executed and copied size to determine base merit. The relative value of the different methods is briefly discussed above. These different choices allow for a varying amount of *junk* (i.e., unexecuted or even uncopied) code to develop in the creature's genome.

Finally, we have TASK_MERIT_METHOD, which simply determines if tasks should be used in determining merit. This is a binary switch, turned on by default.

A.4 Reproduction

The process of replication dominates the dynamics of the system. Here, we present an overview of the method by which programs reproduce, and we then discuss the exact implementation.

Self-Replication and Offspring

Reproduction in avida is typically carried out in four distinct processes:

- Allocation of new memory.
- Copying of the parent program into the new memory, instruction by instruction.
- Division of the program into parent and child programs.
- Placement of the child program into the lattice.

The first three processes are implemented in the instruction set (and are thus the responsibility of the individual program), while the fourth process is automatically handled by the environment when a successful division takes place.

In a correctly self-replicating program (see the example program in Section A.5), the size of the allocated memory is typically exactly the program's size (doubling the total memory from its original size), with division occurring at its midpoint after the copying process is finished. In principle, there is no reason that a program could not use a different method (such as tripling its size, and making two copies of itself, or creating a self-extracting smaller program); however, the instruction set (and the handwritten ancestor) are biased towards the first method.

The reproduction system is separated into two conceptual parts: first, the generation of offspring, which is handled by individual cells; and second, the placement of offspring, which is handled by the interaction lattice or the world structure.

Placement of the offspring is done in a *localized* manner; the offspring of a program can only be placed within the immediate neighborhood of that program's location (the near eight grid positions on a 2-D lattice). First, an empty location is sought; if no free location exists, the oldest cell (in the default scheme) is chosen and replaced with the offspring. The search for the oldest cell includes the parent program; thus, a creature can actually cause itself to be killed by the placement of its child.

The process of placement is entirely a function of the environment; as soon as a successful division occurs, the new program is automatically placed. It is intrinsically part of the "physics" of the system, while the replication process itself (the allocation, copying, and division) is part of the instruction set.

Choosing a Placement Method

In order to choose a method of placing a daughter cell, edit the "Reproduction" section of the `genesis` file (see Section A.9). The `BIRTH_METHOD` variable determines how new offspring are placed, and this can have a *large* effect on the soup. If there are any empty cells available, they will always have top priority as possible locations. If none of the eight immediate cells connected to the mother are vacant, one of the surrounding creatures (or the mother herself) must be removed to make room for this new child. The options are:

- **Choose Randomly:** a creature is chosen at random from the mother and its eight neighbors. This method is poor for evolution because approximately half of the creatures will be replaced before they have had any chance to have offspring, and hence will never have a chance to prove themselves.
- **Choose Eldest:** this is the default method in avida. The creatures in the neighborhood around the mother (including the mother itself) will be evaluated, and the oldest of them will be removed. In the case of a tie, the cell to be removed for the new creature will be chosen randomly from the eldest ones.
- **Choose max Age/Merit** (highest [age divided by merit]): this placement method favors creatures with a higher merit, and is an additional way to encourage creatures to learn specific tasks. With this birth method combined with the constant time-slicing scheme, the creatures will be given CPU time equivalent to other (time-slicing based) learning schemes. The difference here is that rather than having faster CPUs, the creatures will simply live for a longer period of time. See Section A.3 for more information on this.

- **Choose Empty:** this is a very limited birth method which currently is useful only when *death* is turned on (see below). In this mode cells are prevented from killing each other, and only new born creatures are allowed to move into empty cells.

Choosing a DEATH_METHOD for the soup is somewhat less complicated. In this version of avida, there are three methods available. The first is to simply turn death off such that creatures can only die from being replaced by a newborn, and *not* through old age. The second way is to fix the maximum number of instructions that any creature can execute, effectively giving them a finite lifetime akin to *decay*. Creatures that hit this age limit are replaced by empty space. This age limit is specified by the AGE_LIMIT variable. Finally, the last method is to have the maximum number of instructions a creature can execute be a multiple of that creature's length. If you select this method, the maximum number of instructions that a creature can live is its length *multiplied* by AGE_LIMIT.

A.5 The Virtual Computer

The virtual computer implemented in avida consists of a central processing unit (CPU) and an *instruction set*. These components define the low-level behavior of each program; the CPU and the instruction set together form the *hardware* of a Turing machine.

When a genome is loaded into the memory (as the *software*) of a CPU, the initial state of the Turing machine is set. The hardware, combined with the interaction with other CPUs, then governs the set of transitions between CPU states.

The CPU Structure

The CPU contains the following set of variables, as shown in Fig. 1.1:

- A *memory* which contains the assembly source code to be run. Each location in memory contains a single instruction, and a set of flags to denote if the instruction has ever been executed, copied, mutated, etc. Additionally the memory has an *instruction pointer* (IP) which indicates the next position to be executed.

- Three *registers* that are used by the program. These are often operated upon by the various instructions, and can contain arbitrary 32-bit values.
- Two *stacks* that are used for storage. These are of variable (though finite) size. The default size of the stack is 10.
- An *input buffer* and an *output buffer* which the creatures use to receive information, and return the processed results. Each buffer also has a pointer to indicate the active position within it.
- A *facing* which determines which of the CPU's neighbors it is currently pointing towards.

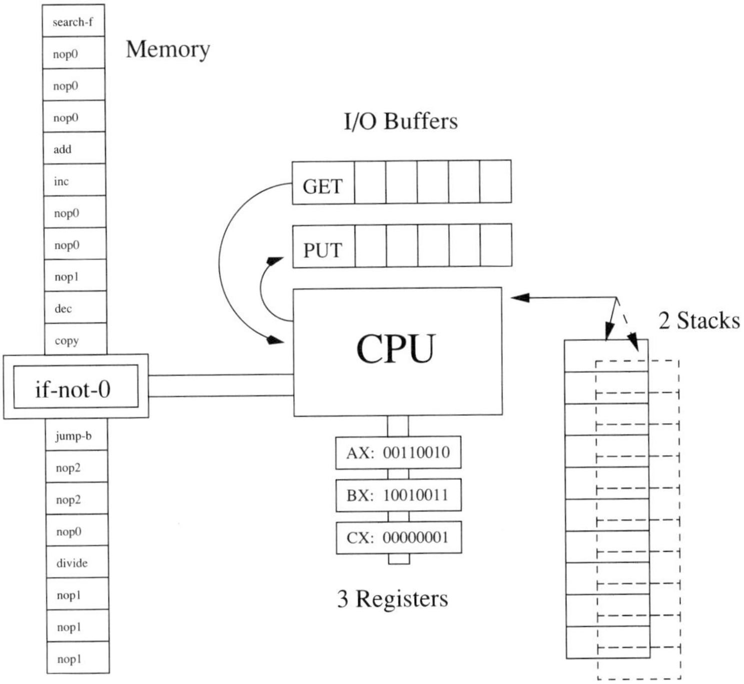

FIGURE A.1 Structure of the virtual CPU in avida.

The Instruction Set Implementation

The instruction set in avida is loaded on startup from a configuration file (`inst_set.24.base` by default). This allows selection of different instruction sets without recompilation, as well as allowing different numbers of instructions to be specified. However, it is not possible to alter the behavior of individual instructions or add new instructions without recompiling avida; this has to be done directly in the source code.

All of the available instructions are listed in the `inst_set.*` files with a `1` or a `0` next to an instruction to indicate if it should or should not be included. Changing the instruction set to be used simply involves adjusting these flags.

The instructions were created with three things in mind:

- To be as complete as possible (both in a Turing complete sense, and, more practically, to ensure that simple operations only require a few instructions).
- For each instruction to be as robust and versatile as possible; all instructions should take an appropriate action in any situation where they are executed.
- To have as little redundancy as possible between instructions. (Those instructions which are redundant will typically not be turned on simultaneously for a run.)

One major concept which differentiates this assembly language from its real-world counterparts is in the additional uses of `nops` (no-operation commands). These have no effect on the CPU when executed, but may modify the behavior of any instruction which precedes them. This occurs in two ways; most of the time it will change the register affected by a command. For example, an `inc` command followed by the instruction `nop-A` would cause the contents of the `AX` register to be incremented, while an `inc` command followed by a `nop-B` would increment `BX`.

Below, whenever a register name is surrounded by `?`'s in an instruction description, it is the default register to be used. If a `nop` follows the command, the register it represents will replace this default.

The second way `nops` can be used is as labels (reference points) for a search or a jump as in tierra. If `nop-A` follows a jump-forward command, it scans forward for the first complementary label (nop-B) and moves the instruction pointer there. Labels may be composed of more than a single nop instruction.

The label system used in avida allows for an arbitrary number of different nops. By default, we have three nop instructions, nop-A's complement is nop-B, nop-B's is nop-C, and nop-C's is nop-A.

A description of all of the instructions implemented in avida follows below. Those with a "*" next to them are part of the default instruction set.

No-operations

There are three nops in the default instruction set, and a fourth which is a "pure" no-operation, as follows:

* `nop-A`

* `nop-B`

* `nop-C`

`nop-X` : A pure no-operation instruction. It will do nothing in all cases.

Flow control operations

`if-not-0` : If the ?BX? register is not 0, execute the next instruction, otherwise skip it.

* `if-n-equ` : If the ?BX? register does not equal its complement, execute the next instruction, otherwise skip it. (Thus a nop-A following this command causes AX and BX to be compared; nop-B—the default—compares BX and CX, and finally, a nop-C compares CX and AX.)

`if-bit-1` : Execute the next instruction if the last bit of ?BX? is one.

* `jump-b` and `jump-f` : If a label follows, search for its complement in the backwards/forwards direction; if a match is found, jump to it.

If there is no label, jump by BX instructions in the proper direction. If there is a label, but its complement is not found, the jump will fail.

`jump-p` : Jump specifically into the memory of another CPU, decided by the current facing. Jump to the position in the faced creature at an instruction after the first occurrence of the complement label. If no complement label can be found, this instruction fails. If no label is initially provided to the instruction, the IP (instruction pointer) will move to line BX in the faced CPU's memory. A creature's IP may only move to an immediate neighbor and no further (local interactions only).

* `call` : Put the location of the next instruction on the stack, and jump forward to the complement of the label which follows. If there is no label, jump BX instructions.

* `return` : Pop the top value from the stack and jump to that index in the creature's memory.

Single argument math operations

All of these affect the ?BX? register.

* `shift-r` and `shift-l` : Rotate the bits of the ?BX? register in the appropriate direction.

`bit-1` : Set the last bit of ?BX? to 1.

* `inc` and `dec` : Increment or decrement ?BX?.

`zero` : Set ?BX? to zero.

* `push` and `pop` : Push ?BX? onto the stack or pop the stack into ?BX?.

`set-num` : Set BX to the ternary equivalent of the label which follows. To do this, take nop-A as a 0, nop-B as a 1 and nop-C as a 2. Thus nop-C nop-A nop-B would translate to 2 0 1 in ternary, or 19 in decimal. If there is no label, set it to 0.

Double argument math operations

* `add` : Set ?BX? equal to the sum of the BX and the CX registers : ?BX? = BX + CX.

* `sub` : ?BX? = BX - CX.

* `nand` : ?BX? = BX NAND CX (in bitwise fashion).

  `nor` : ?BX? = BX NOR CX (bitwise).

  `order` : Place BX and CX in the proper order, i.e., such that CX > BX.

"Biological" operations

* `allocate` : Allocate ?BX? instructions of memory at the end of the memory for this CPU and return the start location of this memory into AX. Only one allocate may occur between successful divides; any additional ones will automatically fail. Additionally, not more than twice or less than half of the current memory size can successfully be allocated.

* `divide` : Split the memory in this CPU at ?AX?, placing the instructions beyond the dividing point into a new cell. There are a number of conditions under which a divide will fail. Those are:

  (a) If either the mother or the daughter would have less than 10 instructions.
  (b) If the creature has not completed a successful allocation of memory.
  (c) If less than 70 percent of the mother was executed.
  (d) If less than 70 percent of the daughter's memory was copied into.
  (e) If the daughter would be less than half or more than double the mother's size.

* `copy` : Copy a command from the memory location pointed to by the BX register to the memory location pointed to by AX + BX, i.e., copy the instruction at location BX into a location *offset* by AX. If a location is out of range of the memory, then it will be cycled back into range.

  `read` : Copy a command from memory at BX into the CX register.

  `write` : Copy a command from the CX register into the memory location at AX + BX.

  `if-n-cpy` : Only execute the next line if the contents of memory locations BX and AX + BX are *identical*; otherwise skip it. This command has an error rate equal to the copy mutation rate. (It can be used to do some level of error checking.)

I/O and sensory operations

* `get` : Read the next value from the input buffer into ?CX?.

* `put` : Place ?BX? into the output buffer and set the register used to 0.

* `search-f` and `search-b` : Search in the appropriate direction for the complement label and return its distance. The returned value is placed in the BX register, and the size of the label that followed is put in CX. If a complement label is not found, a distance of 0 is returned.

Additional instructions

`switch_stack` : Toggles the active stack.

`rotate-l` and `rotate-r` : Rotate the current facing of the CPU in the appropriate direction.

`inject` : This instruction acts somewhat similarly to divide, but rather than killing another creature and replacing it with its offspring, the code is instead injected into the middle of a running CPU's memory. The CPU which the code is to be injected into is chosen by the facing of the active CPU and the position is found by using complement labels. If a complement label can not be found in the memory of the CPU faced, the instruction fails.

`set-cmut` : This command allows a CPU to set its own copy mutation rate. When a command is executed, it simply takes the value in `?BX?` and uses that for its new copy mutation rate ($\times 10^{-4}$). The minimal value which can be set is, of course, 1.

`mod-cmut` : This instruction modifies the copy mutation rate of a CPU. When executed, the copy mutation rate of the CPU has `?BX?` $\times 10^{-4}$ added to it. The minimal value it can be modified by is 0.0001.

An Example Program

What follows is one of the simpler ancestor organisms distributed with avida; it is contained in the file `creature.small` in the `work/genebank/` directory. It is a simple self-replicator, and very short compared to other ancestors included in this distribution. Due to its efficiency in self-replication, this creature is not well suited as an ancestor for adaptation experiments. Note that when an instruction pointer runs off of the end of the program, it will automatically loop back to its beginning, in order to ensure that the chemistry of the program never halts.

This simple program contains two label pairs (α and β), one for the purpose of calculating the length, and the other for the implementation of a copy loop. The `search-f` followed by label α searches forward in the genome for its complement, and returns its distance (from the end of the first label to the end of the second) into the `BX` register, and the size of the label into `CX`. The program then adds `CX` to the `BX` register, to account for the length of the label itself, and finally increments `BX` to account for the single instruction before the first label. The creature now has its own size in `BX`. When the instruction on line five (`allocate`) is called, the program is doubled in length and the absolute address of the new chunk of memory is put in `AX`. Now, `AX` contains the offset of the newly allocated section, and `BX` contains the length of the cell (which, if a creature is copying itself properly, will always be the same). Lines 6 through 11 move the size of the creature (via the stack) into the `CX` register, and clear out `BX`.

00	search-f	find distance to the end label
01	nop-A	label α
02	nop-A	
03	add	account for the end label's size
04	inc	account for the initial `search-f`
05	allocate	allocate space for daughter.
06	push	move size from BX onto the stack.
07	nop-B	
08	pop	move size off of the stack into CX
09	nop-C	
10	pop	since the stack is empty, pop 0 into BX
11	nop-B	label $\bar{\beta}$ (Copy Loop start)
12	nop-C	
13	copy	copy the current line...
14	inc	move onto the next line.
15	if-n-equ	if we aren't done copying...
16	jump-b	...jump back to the loop's beginning.
17	nop-A	label β
18	nop-B	
19	divide	done copying; separate the daughter!
20	nop-B	label $\bar{\alpha}$
21	nop-B	

The copy loop follows. It starts by copying from line `BX` and uses `BX + AX` as the destination. Initially, `BX` is 0 and `AX` is the size (22). This means the first time through the copy loop, line 0 is copied to line 22 (the first line in the newly allocated memory). Then line 14 is executed and `BX` is incremented. Finally, the `if-n-equ` tests to see if whether `BX` and `CX` are different, and jumps back to the beginning of the loop if this is the case. The loop will continue (copying a new line each time) until `BX` equals `CX` and hence all the lines have been copied. Finally, the `divide` instruction is reached.

The `divide` instruction divides the cell at the offset specified in the `AX` register, creating a parent and daughter cell. The daughter is placed in a neighboring cell (the exact mechanism of which is described in Section A.4). At the end of the program, the instruction pointer is looped back to the beginning.

Here is a trace of the execution of the program, with the values of the registers and stack at each moment in time.

line	instruction	AX	BX	CX	stack	comments
00	search-f	0	19	2		
03	add	0	21	2		
04	inc	0	22	2		We have size!
05	allocate	22	22	2		Allocate space
06	push	22	22	2	22	push `BX`
08	pop	22	22	22		pop `CX`
10	pop	22	0	22		pop `BX`
12	nop-C	22	0	22		No-Operation
13	copy	22	0	22		Copy line 0...
14	inc	22	1	22		
15	if-n-equ	22	1	22		
16	jump-b	22	1	22		... to start of loop
13	copy	22	1	22		Copy line 1...
14	inc	22	2	22		
15	if-n-equ	22	2	22		
16	jump-b	22	2	22		...to start of loop
13	copy	22	2	22		Copy line 2...
	...					...
13	copy	22	21	22		Copy line 21..
14	inc	22	22	22		
15	if-n-equ	22	22	22		
17	nop-A	22	22	22		
18	nop-B	22	22	22		
19	divide	22	22	22		Divided!
20	nop-B	22	22	22		
21	nop-B	22	22	22		

For this program, the *gestation time* (the number of instructions required to reproduce) varies between 98 and 100 instructions. The first time through requires 98 instructions: the portion before the copy loop consists of 8 executed instructions; the copy loop contains another 4 instructions that are each executed 22 times each, except for the last time the copy loop is executed, where the last `jump-b` is skipped over; and from there it is another 3 instructions until the `divide` is issued. So, $8 + (4 \times 22 - 1) + 3 = 98$. However, the second time through, an additional 2 instructions are executed because of the α label after the `divide` instruction, and a gestation time of 100, rather than 98, is averaged into the genotype record.

A.6 Mutations

Avida has a range of both *explicit* and *implicit* mutations. Five forms of explicit mutations have been implemented in avida.

There are three intuitive ways in which mutations can be triggered in avida. The first is the most basic cosmic-ray or *point* mutation, which is an external random process independent of the action of the cell—it will hit randomly chosen points in the soup at Poisson-randomly distributed times. Next we have *copy* mutations, which can occur whenever a creature tries to copy a line. There is a small probability (fixed by the user) that this copy will be flawed. Finally we have divide mutations which occur at the time of birth of the child, and which modifies the parent's genome in a predetermined way, such as inserting or deleting a random instruction from a random location.

The primary type of these explicit mutations are copy mutations. They are used most commonly in avida runs as they also are the most prevalent in natural biological systems. Overall, the types of mutations implemented in avida are

- Copy mutations
- Point mutations
- Divide mutations
- Divide insertions
- Divide deletions

Naturally, the rates for all these mutations must be below a certain threshold in order to avoid killing the population, while a rate that is too low slows evolution to a crawl (see Chapter 11 of the book for a study of learning vs. mutation rate.)

It is also important to note that each of the different forms of mutations have different rates that are expressed in different units. Copy mutations are *per-site*, point mutations are *per-site per-update*, and divide mutations are *per-creature*. What this means is that for the former two, longer creatures stand a better chance of being mutated so they put a pressure on the soup for creature size to shrink. On the other hand, with divide mutations longer creatures can often better survive a single mutation, and hence put a slight pressure for sequence growth on the population.

Implicit mutations in avida typically involve mistakes (due to incomplete or faulty *algorithms*) committed in the act of self-copying, usually instigated by code corrupted by mutations. There are a wide assortment of these, many of which have not been categorized due to the ability of creatures to always find new and surprising methods of operation.

One of the most common forms of implicit mutations is the duplication of code within a creature; sometimes the flow of execution will be distorted such that a section of code will be replicated multiple times within a single daughter. A second (similar) form occurs when a creature only partially copies itself over a dead creature which previously occupied the CPU. The two creatures (old and new) have effectively merged into a single one—a process which has been dubbed *necrophilia*. Other implicit mechanisms are certainly possible. As a result, the *effective* copy-fidelity of a program can be significantly lower than the one calculated with mutation rates only.

Setting Mutation Rates

Mutation rates can be configured in the `genesis` file in the "Mutations" section. There are five settings; all of these are scaled independently, depending on the type of mutation, and are explained in more detail below.

- `POINT_MUT_RATE`: This value is the probability of mutation *per-site per-update*, in units of 10^{-6}. Typically, a population can survive values up to approximately 500, assuming that no other mutations are turned on.
- `COPY_MUT_RATE`: The probability of a copy instruction writing the incorrect instruction. These are the most common mutations used in avida since they best model simple biological systems. The values are specified in units of 10^{-4}. Depending on the length of the genomes,

soups can sometimes survive values up to 350 or so. Typically, 30 is considered a low mutation rate, while 200 is high.

- DIVIDE_MUT_RATE: The probability of mutation every time a divide occurs. This causes a single instruction, chosen randomly from the code, to mutate. All divide-based mutations are specified in units of 10^{-2}. Thus a soup can handle divide mutation rates as high as 85 (i.e., one instruction is mutated 85 percent of the time its host divides).
- DIVIDE_INS_RATE and DIVIDE_DEL_RATE: The probability to inject or remove a random instruction to (or from) a random location is triggered in a similar way to DIVIDE_MUT_RATE (and also specified in unts of 10^{-2}). A typical population appears to tolerate up to about a value of 85.

By default, COPY_MUT_RATE is set to a value of 50 (which translates to a probability of 0.005 per copy, and insert and delete mutations are turned on to 5 percent, i.e., DIVIDE_INS_RATE and DIVIDE_DEL_RATE are set to 5. The insert and delete mutations facilitate size changes in the population, which is helpful in the development of complexity.

Additionally, there are two other places where experimentation with mutations is possible. The first is in the instruction set; instructions are available which allow programs to alter their own copy mutation rate. These are `set-cmut` and `mod-cmut`. (See Section A.5 for more information on initializing and using these instructions.)

The other place where mutation rate can be controlled is the `event-list` file. There is a single event called `set_copy_mut` which resets the copy mutation rate of all the CPUs *during* a run. See Section A.9 or the `event_file` itself for more information on this.

A.7 Installing avida

Version 1.0 of avida is working under Microsoft Windows 95, NT, and most Unix platforms (specifically tested under HPUX, linux, OSF1, SunOS, Solaris, IRIX, and IBM AIX). Separate versions of the program are included for both Win95/NT and Unix (with text-based user-interfaces). All of the needed files are on the CD-ROM which comes with this book, or at the TELOS web page at `http://www.telospub.com/PHYSICS/Alife.html`, where the most up-to-date version of both avida and this manual can be found. New versions of the software, as well as updated versions of the manual, will also be placed on the avida ftp site at ftp://ftp.krl.caltech.edu/pub/avida/.

Windows 95 and NT

The Windows version of avida is the easiest to work with. Retrieve the file Avida1_0.EXE and execute it. This is a self-extracting file that will create a directory called Avida with two subdirectories: The avida subdirectory contains all of the source code for the program, and the work subdirectory is where all of the configuration files are located, and where avida should be run. Upon unpacking, there will be a precompiled avida executable in the work subdirectory, which can be run immediately.

UNIX Systems

The file avida-1.0.0.tgz on the CD-ROM and the ftp site contains the source code for avida. Compiling avida from this source code should be straightforward on most systems.

The command "gunzip <filename>" will uncompress the file, then "tar -xvf <filename>" will extract the actual avida files. If this does not work, consult your systems administrator for the specifics of your system. Note that on the CD-ROM, there is in addition a shell script unix-avida which unpacks avida upon executing it.

Extracting the program creates an avida-1.0.0 directory with the source and work sub-directories. In order to produce an executable file, avida has to be compiled by going into the work directory and executing "configure" and then "make". This should create the executable avida in the sub-directory work, where all the configuration files, such as genesis, etc., are also located. Here, it is sufficient to type avida to run the program.

If this build should fail, the actual Makefile may be edited directly to select compiler, user-interface, and include directories.

A.8 The Text Interface

For the most part, the text interface to avida is simple and straightforward to use; most of the options available are listed on the screen. For example, at all times there is a menu-bar at the top of the screen which lists the current update, and the keys to press to go to any of the most used avida screens. This section further defines the function of each of these options.

The Map Screen

The Map screen displays the spatial representation of the population of organisms in avida. The grid itself is toroidal and typically will not fit entirely on the screen. The arrow keys will allow the users to adjust the portion of the map viewed. On computers using a "curses" interface, the arrow keys will often not function properly, in which case the number keys 8, 2, 4, and 6 can be used to move up, down, left, and right respectively. Here is an example map screen (showing genotype mode on a noncolor terminal):

```
+----------------+----------------------------------------------+-----------+
| Update: 97     |  [M]ap  [S]tats  [O]ptions  [Z]oom  [Q]uit   |   Avida   |
+----------------+----------------------------------------------+-----------+
                          A A A A . I A A A A K . . . K K A A + K K K
                        A A A A A I I A A . A K K . . + K A K K K . K A A A
                      A . A A A A A A A . A A K A A + K K A A A K K A A A
                        A A A A A A A I . . A K A . A K K K K A K K K A A
                        A A . A . A A I A A A A A . A A K K K K A . A .
                      A A A A A A A A A A A . . A A A A A A . . K A . A .
                      A A . A A A A A A A A A . A K K A . A A . A . K A A
                        A A A A A A . A A A A A A A . A A . . K A A A A + A
                      A A A A A . A A A A . A . A F A A A A A A A F A A A +
                    A   A A . A . A A A A A . A F A A A A . . A F A A A +
                        A A A A + A A A A A A A A F A A A A A F F F F A A A
                      A A A . . A A A . A A A A F A . + A A A A + F F F . .
                        A A A . . A A A A A A A A . A . + F A F A F F F F F
                      A . A A A A . A A A A A A A A A F A F F F . + . A . .
                      . A A A A A A . A A . A A A A F . . F F F F . F . A
                        A A + A . A A . H H H A A A . F A . F F F F F A A
                          A A A A A A + A A A H . . A . A F F . F F A F +
                        A A A A . A A A A A A A A A A A A . A A A F F F A
                        A A A . . A A A A A . A A A A . A A A A F F F   .
                          A   A A A A A A A A . . A A . . B G G A G .
* Clipping last 40 line(s) *                              [<]  Genotype View  [>]
```

Many different features can be displayed by the different map modes (which can be cycled through using the '<' and '>' keys). The available map viewing styles are:

- **Genotype mode:** This view of of the soup will display the genotype at each location. The most abundant genotypes each have a unique color (or letter if colors are not available) assigned to them; all sites which are of the same color (or letter) harbor the same genotype. Creatures which have passed the genotype threshold, but are not abundant enough to have a color of their own, are represented by white (or a +) on the map. Sub-threshold genotypes are denoted by gray (or a .).
- **Species mode:** This viewing method is only available when speciation has been turned on. It works in a manner very similar to the genotype mode, except single colors (or letters) are assigned to entire species. Additional species (those which are less abundant) are denoted by the color white (or marked by a +). Species mode tends to be much cleaner than genotype mode for observing the current dynamics taking place.
- **Age mode:** The age of each creature is displayed using a similar scheme where a 0-9 represents the exact age (in updates) if possible; otherwise a range of Roman numerals are used to approximate the actual age (X = 10–20, L = 21–80, C = 81–200, + = 201 or greater...)
- **Breed True mode:** This mode simply displays a binary gray or white (- or * if no colors) indicating whether a program at that grid point is an exact copy of its mother. White (or a *) stands for a true-breeding program.
- **Parasite mode:** This mode works identically to Breed True, with white (or *) indicating creatures which are parasites.
- **PointMut mode:** Again, this mode works like Breed True, where white (or *) marks those creatures that have been hit by point mutations.

The Stats Screen

This screen displays all of the current statistics about the ongoing run. A typical snapshot of the screen looks like this:

```
+---------------+------------------------------------------+-----------+
| Update: 97    | [M]ap  [S]tats  [O]ptions  [Z]oom  [Q]uit |   Avida   |
+---------------+------------------------------------------+-----------+

Tot Births.:        2   -- Dominant Genotype --             Dominant  Average
Breed True.:      475   Name........: 031-aaaaa  Fitness..:   0.2213   0.1893
Parasites..:        0   ID..........:         1  Merit....:       31       31
Energy.....:     0.16   Species ID..:         0  Gestation:      140    140.3
Max Fitness:   0.3115   Age.........:        97  Size.....:       31     30.8
Max Merit..: 3.8e+01                             Copy Size:       31     30.8
                                                 Exec Size:       25     24.9
            Current    Total  Ave Age  Entropy   Abundance:      415     5.85
Creatures:      655  2.1e+03      4.5     6.48   Births...:        0    0.018
Genotypes:      112  3.2e+02     71.3     0.68   BirthRate:    0.214    0.183
Threshold:       15  1.6e+01     83.8
Species..:        9  9.0e+00     89.0     0.41

+----------------------------------------------------------------------+
| Input...:     19       Not.....:       0       Nor.....:      0       |
| Output..:      1       And.....:       0       Xor.....:      0       |
| I/O.....:      0       ~A Or B.:       0       Equals..:      0       |
| Echo....:      0       ~A And B:       0                              |
| Nand....:      0       Or......:       0                              |
+----------------------------------------------------------------------+
```

Starting at the upper left column of the screen, the first block of statistics describes the current state of the soup. These statistics are defined as follows:

- `Tot Births` indicates the number of creatures which have been born during the past update.
- `Breed True` is the total number of creatures currently in the soup which are exact copies of their parents.
- `Parasites` is the total number of creatures which have displayed parasitic behavior, i.e., the number which have executed instructions outside of their own memory.

- `Energy` is the log of the ratio between the fitness of the dominant genotype and the average fitness in the soup. (See Glossary.)
- `Max Fitness` is the highest fitness that can be found in the soup. During equilibrium this will usually be approximately the same as the fitness of the dominant genotype.
- `Max Merit` is the highest merit that can be found in the soup. Since this gives no indication as to the replication abilities of this creature, it is typically not a very revealing quantity.

To the right of the soup status column, we have some information about the dominant genotype. This section simply lists the name of this genotype, its ID, its species ID, and its age (how many updates it has existed for). The first three of these statistics are purely for identification purposes.

On the right side of the screen, more statistics are given for a number of common measurements on the dominant genotype, as well as the average across all genotypes. All of these statistics are described in the Glossary.

In the middle left side of the screen, we have information about the various taxonomic levels in avida; we give the current abundance of each in the `Current` column, the total number of each that have existed over the entire run in the `Total` column, the average number of updates each have existed in the `Ave Age` column, and finally the entropy of each in the `Entropy` column.

Finally, along the bottom of this screen, we list the total number of creatures which have completed each of the assortment of tasks available in avida. These numbers reflect only those creatures which have actually finished the task, so even if every creature in the soup is capable of completing a task, not all of them may be listed because the newborns would not have finished it for the first time.

The Histogram Screen

This screen is a histogram of the most abundant genotypes in the population. A typical screen looks like this:

```
+---------------+--------------------------------------------+-----------+
| Update: 97    | [M]ap [S]tats [O]ptions [Z]oom [Q]uit       |  Avida    |
+---------------+--------------------------------------------+-----------+

Fitness Name        Histogram: [<] Genotype Abundance [>]
 0.2213 031-aaaaa: AAAAAAAAAAAAAAAAAAAAAAAAAAAAAAAAAAAAAAAAAAAAAAAAAA  415
 0.2204 031-aaaad: A                                                     4
 0.1845 031-aaaah: A                                                     5
 0.2727 029-aaaac: H                                                     3
 0.2251 031-aaaaf: A                                                     5
 0.2227 031-aaaab: AAAAAA                                               47
 0.2236 031-aaaac: C                                                     4
 0.2214 031-aaaag: A                                                     4
 0.2192 029-aaaaa: D                                                     5

 0.2190 030-aaaaa: BBBBBB                                               44
```

The first number here represents the fitness of the creature; this is the relative replication rate as compared to the other creatures in the population.

Next comes the name of the genotype (for example 109-aaaad). This is an identifier for the genotype, and the name of the file it will be saved under if it is extracted. The number portion (before the dash) of the name is the length of the code for that genotype, and the letter sequence after the dash gives a unique identifier for it. These are never repeated throughout a single run.

The repeated letter after the name is the actual histogram; the number of letters which appear here is the relative current abundance of the genotype. This allows quick recognition of which genotypes are dominant in the soup. The letter itself represents the species of this genotype, so any two lines with the same letter are of the same species.

Finally, each line ends with a number which is the exact abundance of creatures currently within this genotype.

The Options Screen

This screen lists all of the options which are both currently available, and were used to initialize this run. A typical screen looks like this:

```
+---------------+-----------------------------------------+-----------+
| Update: 97    | [M]ap  [S]tats  [O]ptions  [Z]oom  [Q]uit |   Avida   |
+---------------+-----------------------------------------+-----------+

Current CPU..: (0, 0)                 Time Slicing.: Constant
Genotype.....: 031-aaaaa              Task Merit...: Exponential
ID #.........: 0                      Size Merit...: Full Size
                                      Birth Method.: Replace max age
Max Updates..: 50000                  Ave TimeSlice: 30
World Size...: 60x60
Random Seed..: 1                      Point  Mut: 0
Threshold....: 3                      Copy   Mut: 50
                                      Divide Mut: 0    Ins: 0    Del: 0
Debug Level..: OFF
Inst Set.....: inst_set.24.base
Task Set.....: task_set
Events File..: event_list


+---------------------------------------------------------------------+
| [H]istogram Screen          [C]hoose New CPU        Un-[P]ause      |
| [B]lank Screen              [E]xtract Creature      [N]ext Update   |
| [R]edraw Screen                                                     |
+---------------------------------------------------------------------+
```

The upper left corner of this screen gives information about the active genotype in the soup, and the remainder of the upper portions of the screen list values from the `genesis` file, and what they were initialized to.

The lower part of the screen (within a box) shows the special options available to the user. They are:

- **[H]istogram Screen:** This will go to the histogram screen described above.
- **[B]lank Screen:** This option will clear the screen making avida run marginally faster (since it will not be wasting much CPU time on the display)

- **[R]edraw Screen:** If the screen gets garbled, this will erase it and refresh all text which is supposed to appear.
- **[C]hoose New CPU:** This option will put avida in map mode with the cursor on the screen. Position the cursor over the CPU you would like to select, and press `<enter>`. Additionally, in the Windows version of avida, the mouse can be used to select CPUs while in this mode. A single click highlights the CPU targeted, and a double click selects it as the new active CPU, and exits from this mode. The selected CPUs inner workings and genome can then be viewed in the Zoom screen (see below).
- **[E]xtract Creature:** This will save the genotype of the active creature (the one currently selected) to a file by the same name as the genotype. An extracted creature will include all of its statistics as comments (gestation time, fitness, tasks completed, etc.) and can be loaded into another soup without modifying the file.
- **[P]ause:** This freezes activity in the soup, but still allows navigation through the interface and examination of the soup. Additionally, many modes have additional options when the soup is paused. To Un[P]ause, press P again.
- **[N]ext Update:** When paused, this will advance the soup a single update.

The Zoom Screen

This screen contains all of the information about the state of the active CPU. Here is a typical screenshot:

```
+---------------+------------------------------------------+-----------+
| Update: 97    | [M]ap  [S]tats  [O]ptions  [Z]oom  [Q]uit |   Avida   |
+---------------+------------------------------------------+-----------+

Current CPU.: (0, 0)         +-------------+-------------+-----------------+
Genotype....: 031-aaaaa      | Memory:  62 | Stack A     |  A A A A A A A  |
Species.....: spec-0         +-------------+-------------+  A A . A . A F  |
                             | 16:   nop-A |           0 |  A A A . A F A  |
Gestation...:     140        | 17:   nop-B |           0 |  A A A[A]A A F  |
CurrentMerit:      31        | 18:    copy |           0 |  . A A A A F A  |
LastMerit...:      31        | 19:     inc |           0 |  A A A A A A .  |
Fitness.....:  0.2214        | 20: if-n-equ |          0 |  A A A A A A A  |
Offspring...:       0        | 21:  jump-b |           0 |[<] Genotypes [>]|
Errors......:       0        | 22:   nop-C |           0 +-----------------+
Age.........:       3        | 23:   nop-A |           0 | AX:          31 |
Executed....:      69        | 24:   nop-C |           0 | BX:          15 |
Last Divide.:      69        | 25:   nop-A |           0 | CX:          31 |
Flags.......:  A             +-------------+-------------+-----------------+
Facing......: (59, 58)       | Inputs      | Get.:  0  Not.:  0  Nor.:  0  |
                             +-------------+ Put.:  0  And.:  0  Xor.:  0  |
Un-[P]ause                   |   343139087 | GGP.:  0  ~Or.:  0  Equ.:  0  |
[N]ext Update                |   314146099 | Echo:  0  ~And:  0            |
[Space] Next Instruction     |    81633365 | Nand:  0  Or..:  0            |
[-] and [+] Scroll Memory    +-------------+-------------------------------+
```

The column on the left of the screen gives all of the current statistics for this CPU, and the right of the screen contains information about the actual hardware in the CPU; the memory, the stack, the registers, and the I/O buffers.

The execution statistics recorded here are:

- `Gestation`: This is the gestation time for the creature (the number of instructions it needs to execute in order to copy itself.) If the creature has not copied itself, a zero appears here.
- `CurrentMerit`: This is the merit which the creature is currently building during this gestation cycle. Every time a new task is completed, this value will increase appropriately. When a divide occurs, the cur-

rent merit will be reset to its base value (typically the creature size), and the value it was at before the divide will be used to determine how much CPU time it should get.

- `LastMerit`: This is the merit which determines how much CPU time this creature gets. It is locked in when a creature divides (transferred from the current merit). When a creature is first born, it is initialized to the merit of its parent.
- `Fitness`: As described elsewhere, fitness is the relative replication rate of this creature.
- `Offspring`: The number of offspring this creature has produced.
- `Errors`: Every time the creature tries to execute an instruction and fails (e.g., tries to allocate negative space) this value is incremented. This has no feedback into the soup; it is only for the reference of the user.
- `Age`: The number of updates that this creature has lived.
- `Executed`: The number of instructions this creature has executed since it was born.
- `Last Divide`: The number of instructions this creature has executed since its last divide.
- `Flags`: There are a number of flags within a CPU which can be set; a corresponding letter will appear in this field when this is the case. The flags are - A: Allocated (the creature has allocated memory for itself which it has not yet divided off); I: Injected (the creature was injected into the soup by the user); M: Mutated (an instruction has been struck by a point mutation); P: Parasite (the creature has executed code from within another's memory); T: True Copy (the creature is an exact copy of its parent).
- `Facing` or `Executing`: This attribute gives the direction in which the creature is pointed. *Facing* indicates that instructions which involve other creatures will use this creature, while *Executing* means that the code in the specified creature is actually being run (parasitically) by this creature's CPU.

This screen is especially useful while avida is paused. The space bar will cause the active creature to advance a single instruction, and the return key will cause it to advance a full update. In this way, the execution of the creature can be fully examined.

A.9 Configuring avida Runs

Avida runs can be setup in detail through the use of several different files; a `genesis` file, a file defining the instruction set, one listing the events to occur during the run, a creature file (which is used to initialize the soup), and finally a file which indicates bonus levels for tasks. By default, these files are (respectively)

```
genesis
inst_set.24.base
event_list
creature.base
task_set
```

A description of each of them follows.

The Genesis File

The `genesis` file initializes many variables within an avida run. The basic format for the file is: `<variable name> <value>`

This file is split up in a number of sections, each of which contain a selection of variables (described briefly within the actual file). Here we attempt to fully detail the use of each variable.

The first section in the `genesis` file, "Architecture Variables," determines the overall structure of the run. These settings are:

- `MODE`: This variable sets the geometry of the soup. Only two options are available in version 1.0 of avida: mode 1 (a tierra emulator), and mode 2 (the default avida).
- `MAX_UPDATES`: Determines the number of updates the run should last. For a baseline, the default setup (50001 updates with a 60 × 60 world and an average time slice of 30) will take about 5 hours on an unloaded 200MHz Pentium Pro. A value of 0 for this variable will prevent the run from ever terminating by itself.
- `WORLD_SIZE`: This is used *only* in tierra mode. It will determine the maximum number of creatures which should be in the population. By default, this is set to 3600.

- WORLD-X and WORLD-Y: These determine the dimensions of the lattice in avida mode. By default, these are both 60, hence the population has 3600 creatures in it. The minimum for each of these (due to some algorithms used to speed up portions of avida) is 3.
- RANDOM_SEED: This is a number off of which all of the randomness in avida is based. If the seed is altered from one run to the next, the runs will progress in different manners. On the other hand, a single run can be repeated exactly by keeping the seed the same. For a seed based on the current time, use 0 here.

The "Configuration Files" section of the genesis file determines which other files should be used to configure the avida run. Note that to change the genesis file used, use the command line option: -g <genesis_filename>

- DEFAULT_DIR: This is the directory where avida will look for the remaining configuration files. By default, ../work/ will be used (thus any directory on the same level as work will find the configuration files.) This should be changed to an absolute path, so that the configurations can be found if you run avida from anywhere else on your system. Note that the genesis file must be in the same directory as avida unless the -g option is used to specify its location on startup.
- INST_SET: The instruction set file to be used for this run. This file configures the assembly language used by the virtual CPUs. The default is inst_set.24.base. See later in this section for more details on how the files are set up and which ones are available with the avida distribution.
- TASK_SET: The task set file to be used. This file configures the rewards given to CPUs for performing specific tasks. The default is "task_set." See later in this section for more details on modifying this file.
- EVENT_FILE: The event file contains a list of specific actions which should occur during the run. The default file used is called "event_list." See later in this section for more details.
- START_CREATURE: This option designates the file of the ancestor program which should be used to seed the soup. The default ancestor is located in creature.base. See later in this section for information on how creatures are stored in files.

The "Viewer" section has only a single variable; `VIEW_MODE`. This determines the screen which the user interface initializes to when an avida run is started. By default, it is blank to maximize the speed of the run, but the options are 0 = BLANK, 1 = MAP, 2 = STATS, 3 = HIST, 4 = OPTIONS, and 5 = ZOOM, each of which directly correspond to a view in avida.

The next three sections in the **genesis** file ("Reproduction," "Mutations," and "Time Slicing") are all quite important, as the variables there can have a *very* large effect on runs, and as such each are described in more detail in their own manual sections (A.4, A.6, and A.3 respectively).

The "Genotype Info" section controls what information we record about genotypes and species. The variables are:

- `THRESHOLD`: This value determines the number of creatures which must be present in a genotype in order for it to be statistically interesting. Typically this is set to 3; if a genotype does not have at least 3 members it probably cannot reproduce itself, and a non-replicating genotype will be unlikely to have a full 3 members just by chance.
- `GENOTYPE_PRINT`: This is a 0/1 on/off switch. When it is turned on, all genotypes that become threshold will automatically be extracted to the **genebank** directory. Each of these files will be small, but if a run continues for a long time there can literally be millions of them, so be careful!
- `SPECIES_PRINT`: This works similar to the previous variable, but only prints out the representative genotype every time a new *species* is created. The files names are in the format: spec-<number>.
- `GENOTYPE_PRINT_DOM`: Again, this works similar to the previous two variables, but it causes only the genotype which is currently dominating the population to be extracted into the genebank directory. Additionally, if a value larger than one is placed here, the genotype would have to be dominant for at least that many consecutive updates before it is printed. Hence a 10 would only have a genotype printed after it was the most abundant in the soup for a full ten updates.
- `SPECIES_RECORDING`: This setting determines if we should keep species information during a run. Setting this variable to 0 turns this feature off; switching to 1 keeps full information (i.e., every time a new threshold genotype is created, it is checked against *all* current species to see into which, if any, it best fits). Finally, setting this switch to 2 keeps

limited information; it only compares the new threshold genotype to the species of its parent genotype. In practice, settings 1 and 2 produce similar results as it is rare for multiple genotypes to evolve into the same species independently. As setting 2 is significantly faster, it is the default.

- `SPECIES_THRESHOLD`: When comparing two different genotypes to determine if they are the same species, we cross them over at all possible points and then count the number of times the hybrid creature fails to reproduce. The value of `SPECIES_THRESHOLD` determines the maximum number of failures allowed for them to still be considered the same species.

Finally, the "Data and Log Files" section determines if and how often information should be added to various output files. For all files, a 0 next to the appropriate variable indicates that the file should not be printed at all. For the *data* and *status* files, a positive number indicates that the files should have information added to them *this* set interval of updates. Most of the data files have a default value of ten, and hence are updated every tenth update. For log files, a tag other than 0 implies that this file should be produced. See Section A.10 for more information on exactly what is recorded in each of these files.

The Instruction Set File

The instruction set file is one of the simplest to modify. It consists of a list of possible commands for the virtual assembly language, each with a 1 or 0 next to them determining if they should be included. Any instruction not listed here will automatically be assumed to have a 0 next to it (so, theoretically only the included instructions need to be in the list.)

Avida comes with a selection of default instruction set files that are each for a specific purpose. Those files are:

- `inst_set.24.base` : This is the default instruction set, as described in Section A.5.
- `inst_set.24.const` : The `const` instruction set prevents creatures from changing their size. Instead of the `allocate` and `divide` instructions from the default set, they use `c_alloc` and `c_divide` which prevent any size changes in the creatures.

- `inst_set.25.error_check` : The error-checking instruction set adds the single instruction `if-n-cpy` to those available in the default set. Higher mutation rates will actually cause creatures to learn to double check all of their copies, and re-do those which have failed.
- `inst_set.27.parasite` : The parasite instruction set adds the commands `jump-p`, `rotate-r`, and `rotate-l` to the default one. This allows creatures to select an adjacent creature as host (using the `rotate` instructions) and jump into its memory to execute its code. Do not use the *Death* function (`DEATH_METHOD` = 0) with this instruction set.

For more information on how each instruction actually works, see Section A.5.

The Event File

This file configures the events to occur during an avida run, by listing the update it should happen at, the name of the event, and any relevant arguments for it. The format is:

```
<update> <event-name> [<args>...]
```

As an example, if a creature named `creature.happy` is to be injected at update number 42 into CPU 100, the following line should be added to the file:

```
42 inject creature.happy 100
```

Only four types of events are enabled in avida version 1.0. Those are:

- `cycle <cycle_length> <event>`: Execute the `<event>` listed every `<cycle_length>` updates.
- `inject <filename> <cpu_num>`: Put the creature located in `<filename>` into CPU number `<cpu_num>`.
- `pause`: Freeze the avida run in the viewer. This will not work with viewers that do not have the ability to pause.
- `set_copy_mut <new_mut_rate>`: Reset the copy mutation rates in all the CPUs to `<new_mut_rate>`.

The Genebank Files

Avida comes with a selection of possible ancestor creatures to initialize (or inject into) runs, located in the directory `genebank`. These files include the code of the creatures, and are often commented. Comments all start with a # symbol; it and all of the text following are ignored when the creatures are loaded into avida (as are blank lines). The creature files which are included in avida are:

- `creature.base` : The default creature used in avida; 31 lines long.
- `creature.error_check` : This creature uses the error checking instruction set to decrease its effective copy mutation rate.
- `creature.host` and `creature.parasite` : These creatures are a pair such that the host is capable of self replication on its own while the parasite only calculates its own size and then jumps into the host to finish the remainder of its replication.
- `creature.small` : This is a small creature capable of self-replication as described in Section A.5.

The Task File

The `task_set` file lists all of the possible bonuses that avida creatures can receive, and the relative level (on an exponential scale) of that bonus. To be exact, a bonus b multiplies a creature's merit by $1 + 2^{b-3}$. Thus a bonus of 3 would double the merit, and 4 would multiply it by 3.

The file is set up so that each task in avida is listed, followed by the bonus it should trigger for this run. After this (as comments), the meaning of this task is given, as well as the minimum number of `nands` a program would need to use to be able to complete the operation. The `task_set` file is located by default in the `work` directory.

Command Line Options

There are five command line options implemented in version 1.0 of avida. They are:

- `-g[enesis] <filename>`: This option allows the use of a genesis file other than the default `genesis`.

- `-h[elp]`: This option simply lists all of the command line options available in avida.
- `-s[eed] <value>`: Set random seed to the value given.
- `-v[ersion]`: This will give the version number for this avida. This manual is written for avida version 1.0.0. The version ID is broken up such that the first number is only changed if there is a *major* structural revision, the second number is changed for all significant changes to code, and the final number is changed every time there are bug fixes implemented (thus, this manual should be good for all avida versions 1.0.x)

A.10 Guide to Output Files

The avida program generates a number of output files for the analysis of the run. Which of the many possible output files are generated is specified in the `genesis` file (see Section A.9). There are three different kinds of output files produced by avida; *data* files, *log* files, and *status* files, all of which will be discussed. The following overview is a guide to what the data in these files represent.

Data Files

Data files are printed every fixed number updates (where the number is configurable in the `genesis` file for each output file independently). Each column in the data file represents a specific variable as defined below. In the lists below, the number in front of a variable indicates the column of the file where the particular measure can be found. Note that the *fitness* (column 4 in the following two files) rises exponentially, so it is advised to plot the logarithm of this quantity instead.

average.dat

All of the variables in this file are averages taken over all strings in the population.

1. update number
2. avg. merit

3. avg. gestation time
4. avg. fitness
5. avg. replication rate
6. avg. size
7. avg. copied size
8. avg. executed size
9. avg. births for a genotype per update
10. avg. genotype abundance

dominant.dat

The variables in this file represent metabolic data of the genotype currently dominating the soup. In many cases this genotype is simply the most abundant in the soup and does not necessarily have a very clear dominance.

1. update number
2. dom. merit
3. dom. gestation time
4. dom. fitness
5. dom. replication rate
6. dom. size
7. dom. copied size
8. dom. executed size
9. dom. births per update
10. dom. abundance
11. highest fitness
12. highest replication rate
13. dominant name

stats.dat

This file collects various statistics of the population; typically they involve all strings, but sometimes only the dominant.

1. update number
2. energy (average inferiority) = $\log \alpha_{\text{dom}}/\langle\alpha\rangle$
3. effective mutation probability (ave. creature)
   $1-(1-R)^{\langle\ell\rangle}$
4. effective mutation probability (dom. creature)
   $1-(1-R)^{\ell_{\text{dom}}}$
5. $-\langle\ell\rangle\log(1-R)\approx R\langle\ell\rangle$
6. $-\ell_{\text{dom}}\log(1-R)\approx R\ell_{\text{dom}}$
7. genotype change (in number from last update to this)
8. entropy of genotypes
9. entropy of species

tasks.dat

This file counts the number of creatures in the soup which have completed the particular task. Columns represent specific tasks as listed below.

1. update number
2. `get` count
3. `put` count
4. GGP (`get get put`) count

5+. task counts (tasks are in same order as on Stats screen).

count.dat

The file keeps a count of how many of each event or measure there are *at this update*. Each column represents a different count.

1. update number
2. num. instructions executed
3. num. creatures
4. num. genotypes
5. num. threshold genotypes
6. num. species
7. num. threshold species

8. num. deaths
9. num. breed true
10. num. parasites

totals.dat

This file is similar to count.dat, but rather than per update, it measures the total number of each event or measure over the course of the entire run.

1. update number
2. total instructions executed
3. total creatures
4. total genotypes
5. total threshold genotypes
6. total species

Log Files

Log files are printed whenever the event they are set to record occurs. Each column in the log file has a specific meaning.

creature.log

This file is used to record every time a new creature is born or a creature dies, and can be used to perfectly playback an avida run. The columns are

1. update number
2. cell ID# (location)
3. genotype ID#

genotype.log

Entries are appended to this file every time a genotype dies. The following information is included about each genotype:

1. update number
2. genotype ID
3. parent genotype ID (-1 for injected creatures)
4. parent genotype distance (-1 for injected creatures)
5. genotype abundance (from birth to extinction)
6. parasite abundance within genotype
7. genotype age
8. sequence length

threshold.log

New data are printed to this file each time a genotype reaches threshold abundance. The specific information is:

1. update number
2. genotype ID
3. species ID
4. name

species.log

New information is printed to this file every time a *species* dies.

1. update number
2. species ID
3. parent species ID
4. total threshold genotypes in species
5. total creatures in species
6. species age

breed.log

Entries are appended to this file every time a *creature* dies.

1. update number
2. genotype ID
3. num. divides
4. age of creature

phylogeny.log

This file records information every time a creature divides.

1. update number
2. child genotype ID
3. parent genotype ID
4. child-parent genetic distance

Status Files

Information is added to these files at time intervals set in the `genesis` file.

genotype.status

The status files are printed in a different format than the data or log files. The file `genotype.status` prints "cross-sections" of the population at each requested update. For each genotype, its ID number, current number of living members, species ID, and sequence length is printed, separated from another genotype by a colon. If there were only two genotypes at update 50 for example, the entry would be

```
50 : 1 12 1 30 : 17 4 1 30
```

which would indicate that at update 50 there were 12 creatures of genotype #1 (with sequence length 30) and 4 creatures of genotype #14, also of sequence length 30, with both genotypes belonging to species # 1.

diversity.status

In this status file, genetic distances between members of the population are listed. For each requested update (the frequency of output is selected in the `genesis` file), the abundance of pairs of threshold creatures a genetic distance d apart is listed in ascending order of genetic distance. Thus, an entry

```
65 6695 348 348 1318 72 0 0 9 0 366
```

reports that at update 65, there were 6695 pairs a distance 0 apart, 248 pairs which were one mutation distant from each other, another 348 pairs two steps removed, etc., and finally 366 pairs at a genetic distance $d = 9$.

A.11 Summary of Variables

α_i	fitness of genotype i
α_{dom}	fitness of dominant genotype
$\langle\alpha\rangle$	fitness of average genotype
A_i	birth rate of genotype i
H	entropy
I	inferiority
N	number of creatures in population
n_i	number of creatures in genotype i
t_g	gestation time
t_a	allocated time
$\mathcal{M}$	merit

A.12 Glossary

Abundance: The total number of sub-taxa within a taxon. For example, we will commonly look at the total *abundance* of creatures within a genotype, or the abundance of genotypes within a species.

Adaptive System: A system in which the population of programs will evolve to optimize an *extrinsic* fitness function imposed on their environment. Typically these programs will have no direct interactions with each other; they will only be evaluated for their fitness, and those with the maximal fitness will be chosen to survive and propagate.

Ancestor: The creature used to initialize the population in an avida run.

Auto-Adaptive System: A system of *self-replicating* agents in an environment with an *implicit* fitness function. An agent's ability to interact with both the environment and other agents will determine how well it will be able to reproduce. Only indirect control over that environment can be used to direct the evolution of these agents.

Birth Rate: The number of offspring per update a creature (or genotype) is expected to have. This can be calculated approximately by

$$A = \frac{\text{fitness}}{\text{ave\_merit}} * \text{ave\_time\_slice} = \frac{\alpha}{\langle \mathcal{M} \rangle} \langle t_a \rangle \tag{A.2}$$

This value depends on the average merit $\langle \mathcal{M} \rangle$, and hence on other creatures currently in the environment. The inverse of this value is the expected number of updates it would take for the creature to have a child, given its current competition.

Cell: A single organism located at a lattice point in avida. A cell consists primarily of a genome, and a CPU executing that genome.

Copied Size: The number of lines in a creature which were actually copied into it from its parent. All lines which were *not* copied from the parent are typically random.

Copy Mutation: A stochastic event occurring when copying a single line of code from one point in memory space to another. Typically this event affecting the `copy` instruction results in the instruction being written to be different from the one that was read (while still being a legal instruction). Other instructions that perform similar functions (such as `write`) are subject to similar errors.

Cosmic-Ray Mutation: See Point Mutation.

CPU (Central Processing Unit): This is the machine that processes the genome of a creature. It consists of a memory space, three registers, two stacks, a facing, I/O buffers, and an instruction pointer. The CPU will move through the instructions in memory, executing each and then advancing. Most instructions (unless defined otherwise) will deterministically alter the state of the CPU.

Creature: See Cell.

Effective Mutation Probability: The probability of a specific creature (or its child) to be mutated in its attempt to copy itself.

Energy: A measure of the average *inferiority* in the soup (see Inferiority).

Entropy: A measure used to determine the *disorder* in the population, according to Shannon Information Theory. In this measure, the probability of occurrence of a single genotype i, p_i, is approximated by n_i/N:

$$H = -\sum_i \frac{n_i}{N} \log \frac{n_i}{N} \tag{A.3}$$

where n_i is the current abundance of this genotype and N is the total number of strings in the population.

Executed Size: The number of instructions in the genome of a creature which are actually executed at least once during the course of its lifetime. A single `nop` used to modify the register an instruction interacts with *does* count as an executed instruction, but full labels only have their first `nop` counted (if these were counted in full, it could cause creatures to have very long labels in order to increase their executed size).

Fidelity: The probability for a string to correctly transmit its code to its offspring. The fidelity F is just 1-P, where P is the error probability. If only copy errors arise with probability R, the fidelity is

$$F = (1 - R)^{\ell} \,, \tag{A.4}$$

where ℓ is the length of the code. If insert and delete mutation occur with probability P_i and P_d respectively, the effective fidelity is

$$F = (1 - R)^{\ell} - P_i - P_d \,. \tag{A.5}$$

Fitness: A unit-less measurement of the replication ability of a particular creature in a specified environment. By itself, fitness has little intrinsic meaning, but when compared to that of another creature it gives a ratio of their respective replication rates. Specifically, to calculate fitness, we take a creature's merit and divide it by its gestation time. ($\alpha = \mathcal{M}/t_g$). Since merit increases exponentially with the number of tasks acquired, fitness is best described by the log of its actual value (see also *Inferiority*).

Genebank: A directory where the genome of hand-written as well as *extracted* creatures is deposited.

Genome: The assembly language program used to define a creature. The genome seeds the memory component of the CPU when a creature is executed.

Genotype: A taxonomic level recorded in avida which represents all creatures with a totally identical genome. Genotype is one of the standard tools used to study avida, as all creatures of a particular genotype should behave similarly given a fixed environment.

Gestation Time: The number of instructions a creature must execute to produce a single offspring. This is typically proportional to the length of the creature.

Inferiority: A measure which determines how much *worse* a particular genotype is than the genotype which is currently dominating the system. If $\alpha_i \geq 0$ represents the fitness of genotype i, its inferiority is (this is the case in tierra)

$$I_i = \alpha_{\rm best} - \alpha_i \; . \tag{A.6}$$

In avida, fitness is taken as the logarithm of the actual (purely computational) replication rate because merits are doled out using an exponential scheme (see *Fitness*). In that case, the measure of inferiority is

$$I_i = \log \alpha_{\rm best} - \log \alpha_i \; . \tag{A.7}$$

Instruction: A single command in the assembly language of the CPU. When executed, an instruction modifies some of the parts of the CPU in a deterministic fashion.

Instruction Set: The collection of instructions in the assembly language the creatures are written in. Whenever an instruction is mutated, the new instruction is chosen at random from the instruction set (with all instructions given an equal probability of being selected).

Label: A sequence of nops (no-operation instructions) in the genome that are used to modify the instruction that precedes them. Typically they are used to reference another point in the code where the *complement* label is located.

Merit: A value indicating the amount of CPU time a particular creature deserves (or has earned) taking into account its length and the tasks that it has successfully completed.

Necrophilia: A term used to describe a form of crossover which goes on in avida. This occurs when a creature only manages to copy part of itself into space already containing the genome of a dead creature. In effect,the two genomes are merged into a single unit.

Phenotype: A classification system that measures *what* a creature can do without ever checking *how* it is done. In other words, the phenotype reflects gestation time, tasks completed, and the like, but never takes into account the actual source code (the genotype).

Point Mutation: (Also called cosmic ray mutations). This form of mutation is a random change from one instruction to another in the memory space of a creature. This can occur at any time and is not limited to whether the creature is executing a particular task, or even executing at all.

Population: The collection of all of the active creatures on the lattice in an avida run. This is sometimes also referred to as the *soup*.

Quasispecies: Also called the *consensus* sequence: the genotype obtained by picking at each location the allele (instruction) which is the most frequent in the population. This measure can strictly only be defined for sequences of the same length. If the most abundant genotype has more than 50 percent of the population, this genotype automatically becomes the quasispecies. After equilibration, the consensus sequence usually has zero or close to zero representatives in the population (approach to the error threshold).

Replication Rate: The absolute speed at which a creature can self-replicate, i.e., the number of offspring per unit time. This is simply the inverse of the creature's gestation time.

Self-Replication: The process a creature goes through in making an exact copy of its genotype in a daughter cell.

Soup: See Population.

Species: A taxonomic level above genotype. All creatures in a species are similar on a functional and structural level, but not necessarily in all instruction positions on their genome. Species can be used to study clouds around an archetype (*quasispecies*) in genome space.

Task: A feature imposed on the environment that can be triggered by certain actions of a creature, which will in turn cause the merit (and hence fitness) of that creature to increase.

Template: See Label; this term was more commonly used in tierra.

Threshold Genotype: A genotype which has reached a minimum abundance specified in the `genesis` file. (By default, this minimum is three). This is used to determine if the genotype is properly self-replicating (since it would be very unlikely to observe this many copies of a creature that could not properly copy itself). For this reason, many statistics are only taken on threshold genotypes.

Time-slice: The number of instructions executed in a particular CPU during a single update. By default (and in most avida configurations), this is proportional to the *merit* of the organism in that CPU.

Time-slicer: The portion of code in avida which doles out time slices to CPUs, and is responsible for executing the proper number of instructions in those CPUs.

Unrolling the Loop: An evolutionary step the population will sometimes take to lower their gestation times. This process involves copying *two or more* instructions each time through the copy loop to minimize the effect of loop overhead.

Update: An artificial unit of time during which all creatures execute their time-slice. All statistics about a creature are taken at the end of each update.

References

Adami, C. (1994). On modeling life. *Artificial Life* **1**, 429.

Adami, C. (1995a). Learning and complexity in genetic auto-adaptive systems. *Physica* **D 80**, 154.

Adami, C. (1995b). Self-organized criticality in living systems. *Phys. Lett.* **A 203**, 23.

Adami, C. and C.T. Brown (1994). Evolutionary learning in the 2D Artificial Life system "Avida." Proc. of *Artificial Life IV*; MIT July 6–8, 1994. R. Brooks and P. Maes, Eds. (MIT Press, Cambridge, MA), p. 377.

Adami, C., C.T. Brown, and M.R. Haggerty (1995). Abundance distributions in Artificial Life and stochastic models: "Age and Area" revisited. In *Advances in Artificial Life*. Proc. of 3d Europ. Conf. on Artificial Life, Granada, Spain, June 4–6, 1995. Lecture Notes in Artificial Intelligence Vol. 929. F. Morán, A. Moreno, J.J. Merelo, P. Chacón, Eds. (Springer-Verlag, New York), p. 503.

Adami, C. and N.J. Cerf (1996). Complexity, computation, and measurement. Proc. of 4th Workshop on Physics and Computation, Boston Univ. Nov. 22–24, 1996. T. Toffoli, M. Biafore, and J. Leao, Eds. (New England Complex Systems Institute), p. 7.

Adami, C. and N.J. Cerf (1997). A practical lower bound on the complexity of symbolic sequences. *Physica* **D**, to be published.

Adami, C. and H.G. Schuster (1997). Driving molecular evolution to the error threshold. *Phys. Rev.* **E**, to be published.

Agladze, K. *et al.* (1993). Wave mechanisms of pattern formation in microbial populations. *Proc. Roy. Soc. Lond. B* **253**, 131.

Anderson, P.W. (1983). Suggested model for prebiotic evolution: The use of chaos. *Proc. Natl. Acad. Sci. USA* **80**, 3386.

Badii, R. and A. Politi (1997). *Complexity* (Cambridge University Press).

Bak, P. (1996). *How Nature Works: The Science of Self-Organized Criticality* (Springer-Verlag, New York).

Bak, P., K. Chen and C. Tang (1990). A forest-fire model and some thoughts on turbulence. *Phys. Lett.* **A 147**, 297.

Bak, P. and K. Sneppen (1993). Punctuated equilibrium and criticality in a simple model of evolution. *Phys. Rev. Lett.* **71**, 4083.

Bak, P., C. Tang, and K. Wiesenfeld (1987). Self-organized criticality: An explanation of $1/f$ noise. *Phys. Rev. Lett.* **59**, 381.

Bak, P., C. Tang, and K. Wiesenfeld (1988). Self-organized criticality. *Phys. Rev.* **A 38**, 364.

Bartel, D.P. and J. Szostak (1993). Isolation of new ribozymes from a large pool of random sequences. *Science* **261**, 1411.

Basharin, G.P. (1959). On a statistical estimate for the entropy of a sequence of independent random variables. *Theory Probability Appl.* **4**, 333.

Bauer, G.J., J.S. McCaskill, and H. Otten (1989). Travelling waves of in vitro evolving RNA. *Proc. Natl. Acad. Sci. USA* **86**, 7937.

Becker, T., H. de Vries, and B. Eckhardt (1995). Dynamics of a stochastically driven running sandpile. *J. Nonlin Sci.* **5**, 167.

Bennett, C.H. (1973). Logical reversibility of computation. *IBM J. Res. Dev.* **17**, 525–532.

Bennett, C.H. (1982). The thermodynamics of computation—A review. *Int. J. Theor. Phys.* **21**, 905–940.

Bennett, C.H. (1995). Universal computation and physical dynamics. *Physica* **D 86**, 268.

Berlekamp, E., J.H. Conway, and R. Guy (1982). *Winning Ways for Your Mathematical Plays* (Academic Press, New York).

Bonabeau, E.W. and G. Theraulaz (1994). Why do we need Artificial Life? *Artificial Life* **1**, 303.

Brown, T.A., Ed. (1991). *Molecular Biology LABFAX* (BIOS Scientific, Oxford, England).

Burlando, B. (1990). The fractal dimension of taxonomic systems. *J. Theor. Biol.* **146**, 99.

Burlando, B. (1993). The fractal geometry of evolution. *J. Theor. Biol.* **163**, 161.

Campi, X. (1987). Introduction à la théorie des modèles de formation d'amas. In *Au-delà du champ moyen*, Notes de cours de l'école Joliot-Curie de Physique Nucléaire, Maubuisson, Gironde, 14–18 Septembre, 1987, 166.

Carlson, J.M, J.T. Chayes, E.R. Grannan, and G.H. Swindle (1990). Self-organized criticality and singular diffusion. *Phys. Rev. Lett.* **65**, 2547.

Carlson, J.M, E.R. Grannan, C. Singh, and G.H. Swindle (1993). Fluctuations in self-organizing systems. *Phys. Rev. E* **48**, 668.

Cerf, N.J. and C. Adami (1997). Negative entropy and information in quantum mechanics. *Phys. Rev. Lett.* **79**, to be published.

Chaitin, G.J. (1966). On the length of programs for computing finite binary sequences. *J. ACM.* **13**, 547.

Chu, J. and C. Adami (1997). Propagation of information in populations of self-replicating code. Proc. of *Artificial Life V*, Nara, Japan, May 16–18, 1996. C.G. Langton and K. Shimohara, Eds. (MIT Press, Cambridge, MA), p. 462.

Codd, E.F. (1968). *Cellular Automata* (Academic Press, New York).

Cross, M.C. and P.C. Hohenberg (1993). Pattern formation outside of equilibrium. *Rev. Mod. Phys.* **65**, 851.

Derrida, B. (1981). Random-energy model: An exactly solvable model of disordered systems. *Phys. Rev. B* **24**, 2613.

Dewdney, A. (1984). In the game called Core War hostile programs engage in a battle of bits. *Sci. Amer.* **250/5**, 14.

Dobzhansky, T. and S. Wright (1943). Genetics of natural populations. X. Dispersion rate in *Drosophila pseudoobscura*. *Genetics* **28**, 304.

Domingo, E., R.A. Flavell, and C. Weissmann (1976). In vitro site-directed mutagenesis: Generation and properties of an infectious extracistronic mutant of bacteria *Qβ*. *Gene* **1**, 3.

Domingo, E., E. Sabo, T. Taniguchi, and C. Weissmann (1978). Nucleotide sequence homogeneity of an RNA phage population. *Cell* **13**, 735.

Domingo, E., M. Dávila, and J. Ortin (1980). Nucleotide sequence heterogeneity of the RNA from a natural population of foot-and-mouth-disease virus. *Gene* **11**, 333.

Drossel, B. and F. Schwabl (1992). Self-organized criticality in a forest-fire model. *Physica A* **191**, 47.

Edwards, S.F. and P.W. Anderson (1977). Theory of spin-glasses. *J. Phys.* **F 5**, 965.

Eigen, M. (1971). Self-organization of matter and the evolution of biological macromolccules. *Naturwissenschaften* **28**, 465.

Eigen, M. (1978). How does information originate? Principles of biological self-organization. *Adv. in Chem. Phys.* **33**, 211.

Eigen, M. (1985). In *Disordered Systems and Biological Organization*. E. Bienenstock, F. Fogelman, and G. Weisbuch, Eds. (Springer-Verlag, Berlin), p. 25.

Eigen, M. (1986). The physics of molecular evolution. *Chemica Scripta* **26B**, 13.

Eigen, M., B.F. Lindemann, M. Tietze, R. Winkler-Oswatitsch, A. Dress, and A. von Haseler (1989). How old is the genetic code? Statistical geometry of tRNA provides an answer. *Science* **244**, 673.

Eigen, M., J. McCaskill, and P. Schuster (1989). The molecular quasi-species. *Adv. in Chem. Phys.* **75**, 149.

Eigen, M. and P. Schuster (1979). *The Hypercycle—A Principle of Natural Self-Organization* (Springer-Verlag, Berlin).

Ekland, E.H., J.W. Szostak, and D.P. Bartel (1995). Structurally complex and highly-active RNA ligases derived from random RNA sequences. *Science* **269**, 364.

Fisher, R.A. (1937). The wave of advance of advantageous genes. *Ann. Eugen.* **7**, 355.

Fontana, W. and L.W. Buss (1994). "The arrival of the fittest": Toward a theory of biological organization. *Bull. Math. Biol.* **56**, 1.

Fontana, W., P.F. Stadler, E.G. Bornberg-Bauer, T. Griesmacher, Ivo L. Hofacker, M. Tacker, P. Tarazona, E.W. Weinberger, and P. Schuster (1993). RNA folding and combinatory landscapes. *Phys. Rev.* **E 47**, 2083.

Foster, P.L. and J. Cairns (1992). Mechanisms of directed mutation. *Genetics* **131**, 783.

Gaylord, R.J. and P.R. Wellin (1995). *Computer Simulations with Mathematica* (Springer-Verlag, New York).

Gaylord, R.J. and K. Nishidate (1996). *Modeling Nature: Cellular Automata Simulations with Mathematica* (Springer-Verlag, New York).

Gil, L. and D. Sornette (1996). Landau-Ginzburg theory of self-organized criticality. *Phys. Rev. Lett.* **76**, 3991.

Goss S., J.-L. Deneubourg, R. Beckers, and J.-L. Henrotte, Eds. (1993). Recipes for collective movement. In *Proc. of the 2d Conf. on Artificial Life*, Université Libre de Bruxelles, Brussels, p. 400.

Gould, S.J. and N. Eldredge (1977). Punctuated equilibria: The tempo and mode of evolution reconsidered. *Paleobiology* **3**, 115.

Gould, S.J. and N. Eldredge (1993). Punctuated equilibrium comes of age. *Nature* **366**, 223.

Grassberger, P. and H. Kantz (1991). On a forest fire model with supposed self-organized criticality. *J. Stat. Phys.* **63**, 685.

Hager, A.J., J.D. Pollard, and J.W. Szostak (1996). Ribozymes—Aiming at RNA replication and protein synthesis. *Chem. Biol.* **3**, 717.

Haldane, J.B.S. (1929). *The Origin of Life* (Rationalist Annual).

Hofacker, I.L., W. Fontana, P.F. Stadler, and P. Schuster. Vienna RNA Package. Available by FTP from `ftp.itc.univie.ac.at`

Hopfield, J.J. (1982). Neural networks and physical systems with emergent collective computational abilities. *Proc. Nat. Acad. Sci. USA* **79**, 2554.

Hull, D.B. (1995). Naming and classifying computer viruses. Unpublished.

Huynen, M., P.F. Stadler, and W. Fontana (1996). Smoothness within ruggedness: The role of neutrality in adaptation. *Proc. Nat. Acad. Sci. USA* **93**, 397.

Kadanoff, L.P., S.R. Nagel, L. Wu, and S. Zhou (1989). Scaling and universality in avalanches. *Phys. Rev.* **A 39**, 6524.

Kauffman, S.A. (1993). *The Origins of Order* (Oxford University Press, Oxford).

Kauffman, S.A. and S. Johnsen (1991). Coevolution to the edge of chaos—Coupled fitness landscapes, poised states, and coevolutionary avalanches. *J. Theor. Biol.* **149**, 467.

Kauffman, S. and S. Levin (1987). Towards a general theory of adaptive walks on rugged landscapes. *J. Theor. Biol.* **128**, 11.

Klafter, J., M.F. Shlesinger, and G. Zumofen (1996). Beyond Brownian motion. *Physics Today* **49/2**, 33.

Koch, A.L. (1996). What size should a bacterium be—A question of scale. *Ann. Rev. Microbiol.* **50**, 317.

Kolmogorov, A.N. (1965). Three approaches to the definition of the concept "quantity of information." *Probl. Inform. Transmisssion* **1**, 1.

Kolmogorov, A.N. (1983). Combinatorial foundations of information theory and the calculus of probabilities. *Russian Math. Surveys* **38**, 29.

Landau, L.D. and E.M. Lifshitz (1980). *Statistical Physics, 3d Edition Part 1* (Pergamon Press).

Landauer, R. (1961). Irreversibility and heat generation in the computing process. *IBM J. Res. Dev.* **5**, 183.

Landauer, R. (1991). Information is physical. *Physics Today* **44(5)**, 23–29.

Langton, C.G. (1984). Self-reproduction in cellular automata. *Physica* **D 10**, 135.

Langton, C.G. (1986). Studying Artificial Life with cellular automata. *Physica* **D 22**, 120.

Langton, C.G., Ed. (1989). *Artificial Life*. Proc. of an interdisciplinary workshop on the synthesis and simulation of living systems, Los Alamos, NM 1988 (Addison-Wesley, Redwood City, CA).

Langton, C.G. (1992). Life at the edge of chaos. In *Artificial Life II*, Ref. [Langton et al., 1992a], p. 41.

Langton, C.G., C. Taylor, J.D. Farmer, and S. Rasmussen, Eds. (1992a). *Artificial Life II*, Proc. of an interdisciplinary workshop on the synthesis and simulation of living systems, Los Alamos, NM 1990 (Addison-Wesley, Redwood City, CA).

Langton, C.G., Ed. (1995). *Artificial Life: An Overview* (MIT Press, Cambridge, MA).

Leff, H.S. and A.F. Rex, Eds. (1990) *Maxwell's Demon: Entropy, Information, Computing* (Princeton University Press, Princeton, NJ).

Lenski, R.E. and J.E. Mittler (1993). The directed mutation controversy and neo-Darwinism. *Science* **259**, 188.

Leuthäusser, I. (1986). An exact correspondence between Eigen's evolution model and a two-dimensional Ising system. *J. Chem. Phys.* **84**, 1884.

Leuthäusser, I. (1987). Statistical mechanics of Eigen's evolution model. *J. Stat. Phys.* **48**, 343.

Levenstein, V.I. (1966). Binary codes capable of correcting deletions, insertions, and reversals. *Control Theory* **10**, 707.

Luisi, P.L., P. Walde, and T. Oberholzer (1994). Enzymatic RNA synthesis in self-reproducing vesicles: An approach to the construction of a minimal cell. *Ber. Bunsenges. Phys. Chem.* **98**, 1160.

Maes, P. (1994). Modeling adaptive autonomous agents. *Artificial Life* **1**, 135.

MacArthur, R.H. and E.O. Wilson (1967). *The Theory of Island Biogeography* (Princeton University Press, Princeton, NJ).

Mandelbrot, B. (1977). *The Fractal Geometry of Nature* (Freeman, San Francisco).

Maniloff, J. (1997). Nanobacteria: size limits and evidence. *Science* **276**, 1776.

Mattis, D.C. (1976). Solvable spin system with random interactions. *Phys. Lett.* **56A**, 421.

Maxwell, J.C. (1871). *Theory of Heat* (Longmans, London).

Maynard Smith, J. (1970). Natural selection and the concept of a protein space. *Nature* **225**, 563.

Mayr, E. (1970). *Populations, Species, and Evolution* (Harvard University Press, Cambridge, MA).

McCaskill, J.S. and G.J. Bauer (1993). Images of evolution—Origin of spontaneous RNA replication waves. *Proc. Natl. Acad. Sci. USA* **90**, 4191.

McCulloch, W.H. and W. Pitts (1943). A logical calculus of the ideas immanent in nervous activity. *Bull. Math. Biophys.* **5**, 115.

McKay, D.S., E.K. Gibson Jr., K.L. Thomas-Keprta, H. Vali, C.S. Romanek, S.J. Clemett, X.D.F. Chillier, C.R. Maechling, and R.N. Zare (1996). Search for past life on Mars: Possible relic biogenic activity in Martian meteorite ALH84001. *Science* **274**, 924.

Mézard, M., G. Parisi, and M.A. Virasoro (1987). *Spin Glass Theory and Beyond* (World Scientific, Singapore).

Miller, S. (1953). A production of amino acids under possible primitive earth conditions. *Science* **117**, 528.

Mills, D.R., R.L. Peterson, and S. Spiegelman (1967). An extracellular Darwininan experiment with a self-duplicating nucleic acid molecule. *Proc. Natl. Acad. Sci. USA* **58**, 217.

Morowitz, H.J. (1992). *Beginnings of Cellular Life* (Yale University Press, New Haven, CT), p. 60.

Newman, M.E.J., S.M. Fraser, K. Sheppen, and W.A. Tozier. (1997). Comment on "Self-organized criticality in living systems" by C. Adami. *Phys. Lett.* **A 228**, 201.

Oparin. A.I. (1938). *The Origin of Life* (Macmillan, New York).

Pargellis, A.N. (1996a). The spontaneous generation of digital life. *Physica* **D 91**, 86.

Pargellis, A.N. (1996b). The evolution of self-replicating computer organisms. *Physica* **D 98**, 111.

Pesavento, U. (1995). An implementation of von Neumann's self-reproducing machine. *Artificial Life* **2**, 337.

Rasmussen, S., C. Knudsen, R. Feldberg, and M. Hindsholm (1990). The Coreworld: Emergence and evolution of cooperative structures in a computational chemistry. *Physica* **D 42**, 111.

Rasmussen, S. and C.L. Barrett (1995). Elements of a theory of simulation. In *Advances in Artificial Life*. Proc. of 3d Europ. Conf. on Artificial Life, Granada, Spain, June 4–6, 1995, Lecture Notes in Artificial Intelligence Vol. 929. (Springer-Verlag, New York), p. 515.

Raup, D.M. (1986). Biological extinction in earth history. *Science* **231**, 1528.

Raup, D.M. and J.J. Sepkoski (1986). Periodic extinction of families and genera. *Science* **231**, 833.

Ray, T. (1992). An approach to the synthesis of life. In *Artificial Life II*, Ref. [Langton et al., 1992a], p. 371.

Ray, T. (1995). A proposal to create a network-wide biodiversity reserve for digital organisms. ATR Technical Report TR-H-133 (unpublished).

Reggia, J.A., S.L. Armentrout, H.-H. Chou, and Y. Peng (1993). Simple systems that exhibit self-directed replication. *Science* **259**, 1282.

Schmoltzi, K. and H.G. Schuster (1995). Introducing a real time scale into the Bak–Sneppen model. *Phys. Rev.* **E 52**, 5273.

Schrödinger, E. (1945). *What Is Life?* (Cambridge University Press).

Shannon, C.E. and W. Weaver (1949). *The Mathematical Theory of Communication* (University of Illinois Press, Urbana).

Sims, K. (1994). Evolving 3D morphology and behavior by competition. *Artificial Life* **1**, 353.

Sorkin, G.B. (1988). Combinatorial optimization, simulated annealing, and fractals. IBM Research Report RC13674 (No. 61253).

Sornette, D., A. Johansen, and I. Dornic (1995). Mapping self-organized criticality to criticality. *J. de Phys. I* **5**, 325.

Spafford, E. (1994). Computer viruses as Artificial Life. *Artificial Life* **1**, 249.

Stauffer, D. (1979). Scaling theory of percolation clusters. *Phys. Rep.* **54**, 1.

Szilard, L. (1929). Über die Entropieverminderung in einem thermodynamischen System bei Eingriffen intelligenter Wesen. *Z. Phys.* **53**, 840; translated in J.A. Wheeler and W.H. Zurek, Eds. (1983). *Quantum Theory and Measurement* (Princeton University Press, Princeton, NJ), p. 539.

Tarazona, P. (1992). Error thresholds for molecular quasi-species as phase transitions: From simple landscapes to spin-glass models. *Phys. Rev.* **A 45**, 6038.

Terzopoulos, D., X. Tu, and R. Grzesczuk (1994). Artificial fishes: Autonomous locomotion, perception, behavior, and learning in a simulated world. *Artificial Life* **1**, 327.

Theraulaz, G. and E.W. Bonabeau (1995). Coordination in distributed building. *Science* **269**, 687.

Turing, A. (1936). On computable numbers, with an application to the *Entscheidungsproblem*. *Proc. Lond. Math. Soc. Ser. 2* **43**, 544; **42**, 230.

von Neumann, J. (1951). The general and logical theory of automata. In *Cerebral Mechanisms in Behavior—The Hixon Symposium* (John Wiley, New York).

Webb, B. (1996). A cricket robot. *Sci. Am.* **275** (December), 94.

Weinberger, E.D. and P.F. Stadler (1993). Why *some* fitness landscapes are fractal. *J. Theor. Biol.* **163**, 255.

West, B.J. (1995). Fractal statistics in biology. *Physica* **D 86**, 12.

Willis, J.C. (1922). *Age and Area* (Cambridge University Press, Cambridge, UK).

Wolfram, S. (1983). Statistical mechanics of cellular automata. *Rev. Mod. Phys.* **55**, 601.

Wolfram, S. (1984). Universality and complexity in cellular automata. *Physica* **D 10**, 1.

Wright, M.C. and G.F. Joyce (1997). Continuous in vitro evolution of catalytic function. *Science* **276**, 614.

Wright, S. (1932). The roles of mutation, inbreeding, crossbreeding and selection in evolution. In *Proc. of the 6th Intern. Congress on Genetics Vol.1*, p. 356.

Yin, J. and J.S. McCaskill (1992). Replication of viruses in a growing plaque—A reaction-diffusion model. *Biophys. J.* **61**, 1540.

Zurek, W.H. (1990). Algorithmic information content, Church-Turing thesis, physical entropy, and Maxwell's demon. In *Complexity, Entropy and the Physics of Information*, SFI Studies in the science of complexity, vol. VIII, W.H. Zurek, Ed. (Addison-Wesley, Redwood City, CA), p. 73.

Index

NOTE: Principal entries (or pages where terms are defined) are in boldface type.

INTRODUCTION TO
ARTIFICIAL LIFE

REGISTRATION CARD

Since this field is fast-moving, we expect updates and changes to occur that might necessitate sending you the most current pertinent information by paper, electronic media, or both, regarding *Introduction to Artificial Life*. Therefore, in order to not miss out on receiving your important update information, please fill out this card and return it to us promptly. Thank you.

Name: ______________________________

Title: ______________________________

Company: ______________________________

Address: ______________________________

City: ____________________ State: ______ Zip: ____________

Country: ____________________ Phone: ____________

E-mail: ______________________________

Areas of Interest/Technical Expertise: ______________________________

Comments on this Publication: ______________________________

____

____

❑ Please check this box to indicate that we may use your comments in our promotion and advertising for this publication.

Purchased from: ______________________________
Date of Purchase: ______________________________

❑ Please add me to your mailing list to receive updated information on *Introduction to Artificial Life* and other TELOS publications.

❑ I have a ☐ IBM compatible ☐ Macintosh ☐ UNIX ☐ other

Designate specific model ______________________________

TELOS® THE ELECTRONIC LIBRARY OF SCIENCE

Return your postage-paid registration card today!

PLEASE TAPE HERE

FOLD HERE

NO POSTAGE
NECESSARY
IF MAILED
IN THE
UNITED STATES

BUSINESS REPLY MAIL

FIRST CLASS MAIL PERMIT NO. 1314 SANTA CLARA, CA

POSTAGE WILL BE PAID BY ADDRESSEE

TELOS® THE ELECTRONIC LIBRARY OF SCIENCE

3600 PRUNERIDGE AVE STE 200
SANTA CLARA CA 95051-9835